Phosphorene

Physical Properties, Synthesis, and Fabrication

edited by

Yongqing Cai | Gang Zhang | Yong-Wei Zhang

JENNY STANFORD
PUBLISHING

Published by

Jenny Stanford Publishing Pte. Ltd.
Level 34, Centennial Tower
3 Temasek Avenue
Singapore 039190

Email: editorial@jennystanford.com
Web: www.jennystanford.com

British Library Cataloguing-in-Publication Data
A catalogue record for this book is available from the British Library.

Phosphorene: Physical Properties, Synthesis, and Fabrication

ISBN 978-981-4774-64-2 (Hardcover)
ISBN 978-0-203-71061-6 (eBook)

Contents

Preface

A lot of efforts have been devoted to the growth, integration, and application of layered two-dimensional (2D) structures beyond graphene in the last ten years. Improvement in the technique for obtaining the high-quality flake has led to a variety of good-quality sheets and multilayer heterostructures, which are a prerequisite for device fabrication. After several years of intensive research on graphene, interests have gradually shifted toward other 2D materials, notably black phosphorus and molybdenum disulfide. In particular, phosphorene, a 2D form of black phosphorus, has attracted increasing attention due to its high carrier mobility (intrinsic p-type) up to 1000 $cm^2 V^{-1} s^{-1}$ at room temperature and a sizable bandgap which is appropriate for a high ON–OFF ratio in transistors. As a cousin of graphene, phosphorene has a similar honeycomb structure, which is strongly puckered with ridge and accordion structures. The unique structure of phosphorene leads to intriguing electronic and optical properties.

This volume, written by theoretical and experimental experts in the field, is the first book devoted to a systematic discussion of the properties (including electronic, optical, and thermal behaviors), synthesis, and integration of phosphorene. We hope that researchers from various academic backgrounds will benefit from this volume. All the chapters in the book are self-contained, and each concerns a specific issue of the updated progress in the field of phosphorene. In what follows, we limit ourselves to a few general remarks.

The growth, exfoliation, and synthesis of phosphorene are given in Chapter 1. The synthesis of black phosphorus via a physical method (applying high pressure on white and red phosphorus) and chemical catalysis under low-pressure conditions are introduced in this chapter. The bottom-up and top-down methods for the growth the phosphorene flakes are also discussed, followed by mechanical and liquid exfoliations. We also discuss the preparation of nanostructured phosphorene like quantum dots from bulk black phosphorus.

Chapter 2 discusses the methods of characterization, including the approaches for determining the number of layers by using Raman and photoluminescence spectroscopies. We introduce the characterization of few-layer phosphorene, including its layer number identification with phase-shifting interferometry (PSI) method, temperature- and angle-dependent Raman spectroscopy to detect their fine structure and crystalline orientation, layer- and power-dependent PL measurements to determine the band structure, and exciton and trion dynamics of thin-layer phosphorene.

Chapter 3 deals with some basic properties, including the structural, phononic, and defective characteristics of phosphorene. This chapter presents a good summary of the comparison of phosphorene with some other phosphorus polymorphs. The phononic vibrational modes of monolayer phosphorene and the rigid-layer modes of multilayer phosphorene are fully discussed, together with some experimental setups of Raman measurement. We also discuss the mechanical properties of phosphorene, and its structural transformation under high pressure. Lastly, the feature of point defect atomic vacancies of phosphorene is given.

Chapter 4 describes the recent advances in the study of thermal properties of monolayer and few-layer phosphorene. We first discuss some of the interesting features in the thermal conduction of monolayer and multilayer phosphorene and their underlying mechanisms. Next, we present recent understandings in developing phosphorene phononic crystal. Finally, we examine recent experimental works on the thermal conductivity of few-layer phosphorene.

Chapter 5 discusses the electronic and optical properties of phosphorene. The evolution of the bandgaps, and band alignment with the number of layers, electron–phonon coupling, and strain engineering of the electronic structures, renormalized electronic properties due to defects and dopant are presented. In addition, the effects of structural rippling, environmental small molecules, and passivating graphene/BN layers on the bandgap and carrier density of the sheets are also discussed. Lastly, the intriguing excitonic and linear dichroism of phosphorene are introduced.

Chapter 6 deals with the electronic transport properties (mainly transistors) of phosphorene for device applications.

Chapter 7 discusses the promising applications of black phosphorus in energy storage. It also reviews oxygen productions, together with the structural degradation problems of phosphorene and potential strategies on the protection of phosphorene.

Yongqing Cai
April 2019

Chapter 1

Growth, Exfoliation, and Synthesis

Guijian Guan
Institute of High Performance Computing, 1 Fusionopolis Way, #16-16 Connexis North, Singapore 138632, Singapore
guangj@ihpc.a-star.edu.sg

Phosphorene, the single- or few-layer form of black phosphorus, was recently discovered as a two-dimensional (2D) layered material holding great promise for applications in electronics and optoelectronics. Reliable production of atomically thin, layered phosphorene with uniform size and property is essential to translate their superior properties into high-performance device applications. In this chapter, we will first introduce the fabrication of bulk black phosphorus, and then mainly summarize current methods for exfoliating phosphorene from bulk black phosphorus, specifically mechanical exfoliation and solution-phase exfoliation. The preparation of black phosphorus quantum dots and other structured layered black phosphorus is also covered to extend the forms of phosphorene. After overviewing the challenges in exfoliation of phosphorene, the possible opportunities and invaluable perspectives are discussed on production of stable and functionalized phosphorene.

Phosphorene: Physical Properties, Synthesis, and Fabrication
Edited by Yongqing Cai, Gang Zhang, and Yong-Wei Zhang

ISBN 978-981-4774-64-2 (Hardcover), 978-0-203-71061-6 (eBook)
www.jennystanford.com

1.1 Different Allotropes of Elemental Phosphorus and Introduction of Black Phosphorus

As an element, phosphorus exists as several allotropes that exhibit strikingly different properties. It is well known that the two most common allotropes are white phosphorus and red phosphorus.[1] As shown in Fig. 1.1, white phosphorus consists of tetrahedral P_4 molecules, in which each atom is bound to the other three atoms by a single bond. White phosphorus is the least stable, the most reactive, the most volatile, the least dense, and the most toxic allotrope. Under light and heat, it will gradually change to red phosphorus. Red phosphorus is polymeric in structure. It can be viewed as a derivative of P_4 wherein one P–P bond is broken, and one additional bond is formed with the neighbouring tetrahedron resulting in a chain-like structure. Except for exposing white phosphorus to sunlight, red phosphorus may also be formed by heating white phosphorus to 280°C. In addition to white and red phosphorus, phosphorus also features other two crystalline allotropes including violet and black phosphorus.[2] Violet phosphorus can be produced by week-long annealing of red phosphorus above 550°C, and black phosphorus can be synthesized from white phosphorus under high pressure and high temperature.[3] Experimentally, white phosphorus was placed in a high pressure cylinder under kerosene. High pressure up to 0.6 GPa was applied at room temperature. The cylinder was then heated up to 200°C and pressure up to 1.2 GPa. It took 5–30 min for the transition from white phosphorus to black phosphorus under this circumstance. After cooling the cylinder and relieving the pressure, a small volume of black phosphorus was found.[4] The four main allotropes of elemental phosphorus and their traditional transformation/fabrication processes are summarized in Fig. 1.1.[1]

As the most stable phosphorus allotrope at room temperature, black phosphorus has three crystalline phases (orthorhombic, rhombohedral, and simple cubic) and one amorphous form.[5] Among them, semiconducting puckered orthorhombic black phosphorus received much more interest owing to its unique layered structure and a layer-dependent bandgap from 0.3 to 2.0 eV.[6] In the following

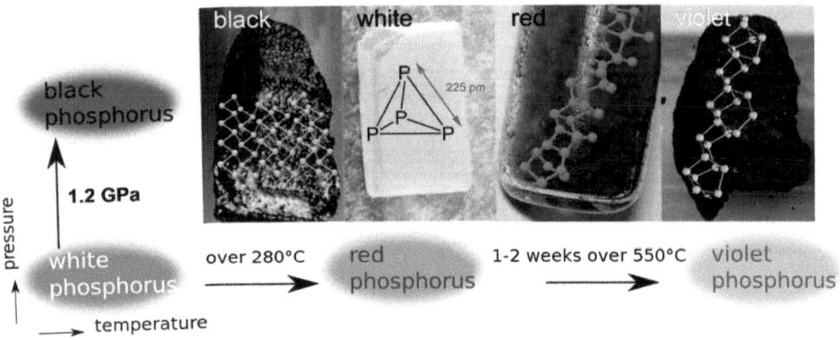

Figure 1.1 Four main allotropes for elemental phosphorus: color, morphologies, atomic structures and traditional transformation/fabrication processes. Adapted from Wiki online.[1]

text, all of black phosphorus are the orthorhombic phase, which belongs to the $D^{18}{}_{2h}$ point group. The bulk black phosphorus is a layered structure with eight atoms per unit cell. The layers are held together by van der Waals forces and strong covalent bonds formed between phosphorus atoms within the layers. The weak interlayer forces in the bulk counterpart suggest that few-layer or even monolayer phosphorus (also called as phosphorene) can be obtained by the exfoliation method.[7] To achieve this exfoliation, an important prerequisite is the availability of highly-quality bulk black phosphorus in large quantity.

1.2 Fabrication of Bulk Black Phosphorus

Since the synthesis of black phosphorus under 1.2 GPa by Bridgman in 1914,[3] black phosphorus was often synthesized by the high-pressure and high-temperature method from white and red phosphorus, which shows higher crystallinity and purity.[8,9] For example, Li et al. reported the growth of bulk black phosphorus crystals under a constant pressure of 1 000 GPa by heating red phosphorus to 1000°C and slowly cooling to 600°C at a cooling rate of 100°C per hour.[8] Zhang et al. synthesized bulk black phosphorus from red phosphorus at pressure of 4 GPa and temperature of 800°C for 10 min.[9] Meanwhile, black phosphorus was also produced by Keyes under the hydrostatic pressure of 1.3 GPa at 200°C.[10] The ingot achieved from this method was polycrystalline and the grain size was roughly around 0.1 mm. In 1981, large single crystals of black phosphorus were first successfully obtained from red phosphorus under 3.8 GPa with temperature higher than 270°C using a wedge-type cubic high-pressure apparatus.[11] Recently, Rissi et al. reported that amorphous red phosphorus could be transformed into crystalline black phosphorus at 7.5 ± 0.5 GPa even at room temperature. By melting black phosphorus at a temperature of 900°C and under a pressure of 1 GPa, black phosphorus single crystals larger than $5 \times 5 \times 10$ mm^3 can be achieved.[12] In Xie's work, they systematically revealed the electrochemical activity of pure black phosphorus synthesized under different pressures and temperatures, and found black phosphorus obtained from white phosphorus exhibited the highest first discharge and charge capacities at 4 GPa and 400°C, and that obtained from red phosphorus exhibited the highest first discharge and charge capacities at 4.5 GPa and 800°C.[13]

Alternative techniques without using high pressure have also been developed, such as the technique involving mercury as a catalyst, developed by Krebs et al.,[14] the bismuth-flux–based method by Brown et al.,[15] and the method based on a chemical transport reaction by Lange et al.[16] that can use a relatively simple setup while avoiding toxic catalysts or "dirty" flux methods. In the bismuth-flux–based method, Brown et al. revealed that black phosphorus crystallized in orthorhombic form and had an infinite puckered layer structure.[15] First, white phosphorus was purified with 15% HNO_3 and then dissolved in liquid bismuth. The mixture was heated up to 400°C and kept for 20 h and slowly cooled to room temperature. By dissolving liquid bismuth with 30% HNO_3, needle or rod-like black phosphorus can be obtained.

The availability of such large black phosphorus crystals under high pressure improved the knowledge of the physical properties drastically but prevented the preparation of large sample amounts for industrial applications. It was not until very recently that black phosphorus was reported to be obtained at lower pressures by using a mineralizer as reaction promoter at non-toxic conditions.[2,16] In the initial experiments,[16] black phosphorus was prepared by the reaction of Au, Sn, red phosphorus, and SnI_4 in evacuated silica ampoules with 0.1 Pa (Fig. 1.2A). The starting materials were heated to 600°C and gradually cooled to room temperature. This reaction usually took several hours (up to 5–10 days) to complete, in which a kinetically controlled mechanism must be favored instead of a thermodynamically controlled formation. In final product, the transformation of red into black phosphorus can be realized via the SnI_4 mineralization, an excess of red phosphorus and Au_3SnP_7 as an important side-product. As further work, a more effective method was developed to used AuSn and SnI_4 as additive agents.[2] Experimentally, after heating to 400°C within 1 h the temperature was held at that temperature for 2 h. In the following the temperature was raised to 600°C (1 h) and kept constant at this temperature for 23 h. In a next step the temperature was reduced to 500°C applying a cooling rate of 40°C/h and kept at that temperature before the oven was switched off. The ampoule was cooled down to room temperature within 4 h. This procedure results in the formation of black phosphorus with crystal sizes larger than 1 cm (Fig. 1.2B). This process is further simplified by Köpf using Sn/SnI_4 as the only mineralization additive.[17] They reported that single crystals of orthorhombic black phosphorus can be grown by a short way transport reaction from red phosphorus and Sn/SnI_4 as mineralization additive. Crystal growth was initiated after cooling the starting materials down from elevated temperatures at 500°C. Sizes of several millimeters can be realized with high crystal quality and purity, because black phosphorus is formed directly via gas phase without the occurrence of any other intermediate phase.

Currently, the low-pressure procedure has widely been used to fabricate bulk black phosphorus. For example, Latiff et al. synthesized black phosphorus crystals by using an AuSn and SnI_4 as a vapour transport medium in a sealed ampoule.[18] Experimentally, 500 mg of AuSn alloy was prepared by melting stoichiometric amounts of Au and Sn under high vacuum directly into a quartz ampoule. SnI_4 was prepared by direct synthesis using tin and iodine in chloroform under reflux, and subsequently purified by recrystallization using chloroform. 720 mg of red phosphorus and 15 mg of SnI_4 were added to the ampoule and were subsequently

Figure 1.2 The preparation of black phosphorus in silica ampules from red phosphorus under low-pressure conditions at 600°C. (A) Small quantities of gold, tin, and tin(IV) iodide (SnI_4) were used as additive agents and the reaction time was 5–10 days. ① Silica ampoules; ② SnI_4 on top of Au_3SnP_7 bulk; ③ BP crystals; ④ representative batch of black phosphorus grown on top of a Au_3SnP_7/AuSn bulk. Reprinted with permission from Lange et al.[16] Copyright (2007) American Chemical Society. (B) AuSn and SnI_4 were used as additive agents and the reaction time was 23 h. **1**, **2** and **3** represent the bulk residue, violet phosphorus and the main product black phosphorus. Reprinted with permission from Nilges et al.[2] Copyright (2008) Elsevier.

sealed by an oxygen/hydrogen torch. The ampoule was placed in a muffle furnace and heated to 400°C for 1 h. After 2 h at 400°C, the ampoule was heated to 600°C for 24 h. The furnace was then cooled to room temperature overnight. The formed crystals of black phosphorus in the form of plates with size up to 5 × 2 mm were removed from the ampoule and washed with CS_2 to remove white phosphorus formed as a byproduct. The low-pressure transport route was similarly used to prepare bulk black phosphorus by Wang et al.[19] 500 mg of red phosphorus, 20 mg of tin, and 10 mg of SnI_4 were sealed in an evacuated Pyrex tube. The tube was heated at 650°C for 5 h with a heating ramp rate of about 1.35°C per min and then the temperature was reduced to 500°C with a cooling rate of 0.33°C per min, followed with a natural cooling process. The black product was then collected and washed with hot toluene and acetone for several times to remove the residual mineralizer. In Woomer's work,[20] red phosphorus (0.5 g), tin (10 mg), and tin iodide (15 mg) were loaded into a partially sealed quartz tube in a glovebox. The tube was 14 cm long, with a 1 cm inner diameter and a 1.4 cm outer diameter. The tube and its contents were evacuated to ~0.4 Pa and sealed with an oxyhydrogen torch while maintaining vacuum within the ampule. They ensured that the ends of the quartz tubes were >0.4 cm and without entrapped gas bubbles. The reaction was carried out in a Lindberg Blue three-zone furnace using the temperature profile described by Nilges.[16] After proceeding the high temperature, they maintained a 60°C temperature differential across the 15 cm tube by inserting insulation between the two zones. The tube was opened in their glovebox, and the black phosphorus was washed sequentially in anhydrous acetone, *N*-methyl-2-pyrrolidone, and isopropyl alcohol and subsequently dried prior to use.

It should also be noted that the high quality of black phosphorus crystals is now commercially available from *Smart Elements*.[21] In lots of research works, the layered bulk black phosphorus were directly purchased from this company, which have a purity of 99.998% and are sealed in ampoule under argon.

1.3 Growth and Exfoliation Methods for Phosphorene

As a new and important 2D material, the growth and fabrication of phosphorene have been extensively exploited, which are essential for using the superior properties of black phosphorus (BP) in practical applications. Referring to the synthesis of other 2D materials such as graphene and transition metal dichalcogenides (TMDs), phosphorene can be fabricated via two routes: bottom-up methods and top-down methods. First, we will comment these fabrication methods and discuss their applicability for obtaining phosphorene. As typical bottom-up methods, the development of wafer-scale synthesis methods via chemical vapor deposition (CVD) on metal substrates and epitaxial growth on insulating substrates has enabled large-scale device fabrication based on graphene and TMDs.[7] However, direct chemical growth strategies for phosphorene are still lacking at present, probably due to its chemically active surface that is fragile when exposed to air, as well as the absence of a suitable substrate for its CVD growth. Its development and implementation will require concerted efforts from the materials and chemistry communities.

In the top-down method, the mechanical cleavage is a traditional approach for exfoliation of layered bulk crystals to obtain 2D flakes. The isolation of single-layer graphene by mechanical exfoliation has unleashed a new research field devoted to the study of the properties of 2D materials.[22] One important precondition for mechanical exfoliation is that the interlayer interaction in the bulk counterpart is dominated by weak van der Waals forces, making possible the cleavage of the materials using just an adhesive tape. The structural characteristic of BP, where its layered structure is held together by weak interlayer forces with a significant character of van der Waals, suggests that few-layer or even monolayer phosphorene can be obtained by the exfoliation method. The energy required to exfoliate layered crystals can be quantified by the surface energy, which is the energy to remove a monolayer from the crystal divided by twice the monolayer surface area. Although no measured data on BP is currently available, the theoretically predicted binding energy is similar to those values determined for hexagonal carbon nitride (h-BN), graphene and TMDs.[7]

Experimental results also indicate that atomically thin flakes of phosphorene can be peeled from their parent bulk BP crystal by micromechanical cleavage using an adhesive tape. The obtained pristine ultrathin 2D nanomaterials have excellent crystal quality with few defects and sizes of up to several tens of micrometers, which are ideal for the fundamental study of their intrinsic properties and the demonstration of electronic devices. However, several limitations of this method in its current form restrict their practical application. First, the production yield is quite low, and it cannot meet the requirements of real applications for which large amounts of materials are required. Second, a substrate is always necessary to support the produced nanosheets. Third, it is hard to control the thickness, size, and shape of the obtained 2D nanomaterials. Last, this method is only applicable to materials for which large, layered crystals are available. Therefore, to fully utilize this method, the production rate and yield must be improved to meet the large quantities required for practical applications.[23] For this, some improved fabrication methods have been suggested. One example is the introduction of plasma-assisted process, as was also used for graphene and MoS_2.[24,25] This method combined with mechanical exfoliation provides an improved way for controlling thickness of phosphorene, but the requirement for laser raster scanning makes it challenging for scale-up applications. Another alternative approach employs a silicone-based transfer layer to optimize the yield of atomically thin BP flakes,[7] can improve the samples to some degree, but this is still only suitable for small-scale production for fundamental research.

To obtain larger quantities of exfoliated phosphorene nanosheets, liquid-phase preparation techniques are very promising. This approach has been successfully applied for fabrication of h-BN, graphene and TMDs, which have shown good performances in devices.[7] Thus, it is also expected that solution-based phosphorene production will have similarly good prospects for making flexible electronics, wherein the interlayer intercalation of ionic species in BP expands the interlayer distance to allow the layers to be

Table 1.1 Summary of fabrication methods for phosphorene in comparison with those for graphene and transition metal dichalcogenides (TMDs)[7]

	Bottom-up methods		Top-down methods			
	CVD growth	**Hydrothermal synthesis**	**Cleavage with tap**	**Plasm-assisted fabrication**	**Liquid-phase exfoliation**	**Lithiation**
Graphene	✓	✓	✓	✓	✓	✓
TMDs	✓	✓	✓	✓	✓	✓
Phosphorene	×	×	✓	✓	✓	×

exfoliated in liquid. Typically, a chunk of black phosphorus crystal is immersed into different solvents to allow the ionic species to intercalate. The sample is then sonicated to break down the interlayer van der Waals bonding to form exfoliated BP nanoflakes. The exfoliated nanoflakes show competitive electrical properties compared to mechanically exfoliated BP flakes. This opens up new possibilities for the atomically thin BP layers to be formed into thin films and composites on large scales with a wide range of applications such as flexible electronics and optoelectronics.

Another promising alternative method that is faster and more controllable is lithiation, where an electrochemical cell with a lithium foil anode and a layered-material-containing cathode are used. The resulting Li-intercalated material is exfoliated by sonication in water like liquid phase preparation, yielding monolayers. This method has been demonstrated for BN, graphene and TMDs,[7] but its use for phosphorene fabrication has not yet been reported. One possible problem of this method for phosphorene fabrication is that the element of lithium may strongly interact with phosphorene and that it is not easy to move. In addition, phosphorene was found to degrade in the presence of moisture (see Section 1.7 for more information), which can also restrict the use of lithium method to exfoliate phosphorene in aqueous solution.

Table 1.1 summarizes the recent progress of phosphorene fabrication in comparison with the more mature and well-studied graphene and TMDs. It is clear that the top-down method, which has been extensively used for graphene and TMDs, is successful in phosphorene fabrication except lithiation. This means that the small-scale fabrication of monolayer or few-layer phosphorene is relatively mature, which can be used for device demonstrations, such as field-effect transistors (FETs) and optoelectronics. In contrast, the bottom-up approach such as chemical synthesis is almost totally blank for phosphorene. The large unexplored area for chemical synthesis of phosphorene inhibits large-area and uniform phosphorene growth and its practical applications. Advanced chemical synthesis of phosphorene may draw inspiration from the techniques used for graphene and TMDs, such as hydrothermal synthesis or CVD growth.

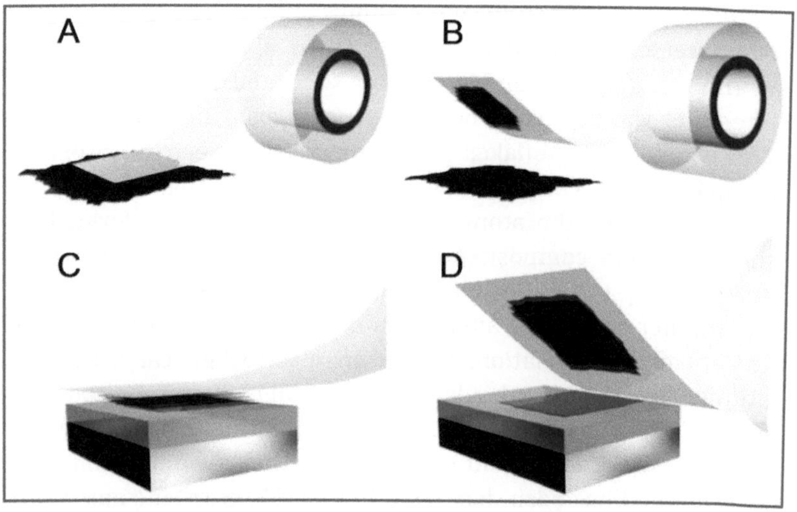

Figure 1.3 Micromechanical exfoliation of 2D crystals via cleavage with tap. (A) Adhesive tape is pressed against a 2D crystal so that the top few layers are attached to the tape (B). (C) The tape with crystals of layered material is pressed against a surface of choice. (D) Upon peeling off, the bottom layer is left on the substrate. Reprinted with permission from Novoselov and Neto.[26] Copyright (2012) IOP Publishing.

1.4 Mechanical Exfoliation of Phosphorene from Bulk BP Crystals

For exfoliation of layered bulk BP crystals to form phosphorene, the mechanical cleavage method was developed, after referring to the mechanical exfoliation of graphene and TMDs. In a typical process (Fig. 1.3), the bulk crystal is first attached onto a piece of Scotch tape, and another piece of Scotch tape is adhered onto the other crystal surface. After that, one of the Scotch tape pieces is detached from the crystal. This process can be repeated several times in order to obtain flakes that are thin enough and can be transferred onto a target substrate (e.g., SiO_2/Si). Then, the Scotch tape with the thin flake is attached onto a target substrate under gentle pressure. Finally, the Scotch tape is peeled off from the substrate and single- or few-layer nanosheets can be easily obtained on the substrate.[23,26] Mu et al. simulated the sliding processes of bilayer phosphorene based on first-principles calculations and found that the interlayer Coulomb interactions dictate the optimal sliding pathway, leading to the minimal energy barrier as low as ~60 meV, which gives a comparable surface energy of ~59 mJ/m^2 in experiment.[27] This means that black phosphorus can be exfoliated by the sliding approach using an adhesive tape. This study builds up a fundamental understanding of how bulk BP is exfoliated to few-layer forms, providing a good guide to experimental research.

In the work by Ye's group,[28] Scotch-tape-based microcleavage of the layered bulk BP crystals was used for fabrication of all 2D devices containing phosphorene, followed by transfer onto the SiO_2/Si substrate. After few-layer crystals of phosphorene were transferred onto the substrate, all samples were sequentially cleaned by acetone, methanol, and isopropyl alcohol to remove any scotch tape residue. This procedure has been followed by a 180°C postbake process to remove solvent residue to obtain highly-purity phosphorene. After few-layer phosphorene was mechanically exfoliated onto a SiO_2 substrate, they also synthesized monolayer MoS_2 on the substrate with phosphorene by CVD method.[29] Due to van der Waals interactions, 2D BP nanosheet-monolayer MoS_2 heterojunctions can be formed at the overlapped regions. Further, electron-beam lithography was used to define the contact patterns. Finally, 20/60 nm of Ni/Au were deposited as metal contacts for developing the gate-tunable p–n diode based on a p-type BP/n-type monolayer MoS_2 heterojunction. In another work,[30] they studied the transistor behavior on mechanically exfoliated few-layer phosphorene with the top-gate. By standard scotch-tape techniques, multi-layer of phosphorene with thickness around 4–8 nm were transferred to heavily doped silicon substrate which was capped with 90 nm SiO_2. They further deposited Al_2O_3 by atomic layer deposition and studied the effects of dielectric capping. They observed that the polarity of the transistors alternated from p-type to ambipolar with Al_2O_3 grown on the top, attributing to the changes of the effective Schottky barrier heights for both electrons and holes at the metal contact edges. In Gehring's work,[31] BP flakes with lateral size up to several μm and thickness down to 10 nm were successfully exfoliated from bulk BP crystals. The next thermal evaporation of a 100 nm thick SiO_x film at the contact regions was to avoid the formation of a trivial

Schottky barrier between the electrical contacts to phosphorene and the underlying GaAs substrate. Recently, not on Si/SiO_2 substrate, BP flakes were also released onto transparent polymer films after exfoliation and then for optical examination under a microscope.[32]

In another important paper on Nat Nanotechnology,[8] Li et al. fabricated FETs based on few-layer BP crystals with thickness down to a few nanometers. A scotch tape-based mechanical exfoliation method was used to peel those thin flakes from bulk crystal onto degenerately doped silicon wafer covered with a layer of thermally grown silicon dioxide. Metal contacts were then deposited on BP thin flakes by sequential electron-beam evaporation of chromium and gold (typically 5 nm and 60 nm, respectively) through a stencil mask aligned with the sample. A standard electron-beam lithography process and other contact metals such as titanium/gold were also used to fabricate few-layer phosphorene FETs with a backgate electrode. Finally, reliable transistor performance was achieved at room temperature in samples thinner than 7.5 nm and the charge-carrier mobility was found to be thickness-dependent, with the highest values obtained for a thickness of ~10 nm. Similarly, Wu et al. also used micromechanical exfoliation method to obtain few-layer phosphorene on p-doped silicon wafers with 300 nm SiO_2.[33] The wafers have been previously treated with O_2 plasma to clean and activate the surface. Immediately after locating the few-layer flake of BP, they spin-coated electron beam resist poly-(methyl methacrylate) (PMMA) to protect the flake because of its fast degradation in ambient (see Section 1.7 for more information). Similar procedure also was used by Koenig et al. to develop gate dielectric for field effect devices.[34] The exfoliated flakes were quickly covered with PMMA once being identified under optical microscopy in order to minimize the effect of crystal degradation. The average time to cover a flake with PMMA after exfoliation was less than 30 min. To avoid the degradation of BP, exfoliated few-layer BP on 300 nm SiO_2/Si substrate or glass substrate was also coated by parylene (~100 nm) immediately, to replace PMMA.[35]

To better control the exfoliation of bulk BP on different substrates, some optimized strategies derived from the scotch-tape exfoliation technique have been developed. Typically, Buscema et al. micromechanically exfoliated phosphorene from a synthetic bulk BP onto a viscoelastic stamp (GelFilm by GelPak) with blue Nitto tape (Nitto Denko Co., SPV 224P) for FETs.[36] Thin BP flakes are then identified by optical transmission and transferred onto a SiO_2 (285 nm)/Si substrate by pressing the viscoelastic stamp surface against the SiO_2/Si substrate and then slowly releasing it. In Castellanos-Gomez's work,[37] they found that conventional mechanical exfoliation with adhesive tape yielded a low density of few-layer BP flakes and leaved traces of adhesive on the surface. Employing an intermediate viscoelastic surface to exfoliate the flakes substantially increased the yield and reduced the contamination of the fabricated flakes. Based on this supposition, they developed the polydimethylsiloxane (PDMS) stamp-based method: a piece of bulk BP crystal is cleaved several times using blue Nitto tape. The tape containing the thin BP crystallites is then slightly pressed against a PDMS-based substrate and peeled off rapidly. The thin flakes on the surface of the PDMS substrate can be transferred to other substrates by simply putting the PDMS substrate in gentle contact with the new acceptor substrate and peeling it off slowly (it takes about 5–10 min to peel off completely the stamp from the surface). The viscoelastic stamp-based transfer process does not employ any wet chemistry step (all-dry transfer), therefore, is very advantageous to freely suspend these materials as there are no capillary forces involved.[38] Moreover, the whole fabrication process is quick, efficient, clean, and it can be performed with high yield.

Due to the more controllable feature on different substrates, the PDMS-based procedure has been widely used to mechanically exfoliate phosphorene from bulk BP crystals for the experimental characterization or facile applications. For example, Li et al. micromechanically exfoliated bulk BP crystal to produce BP thin films and then transferred them to optical fiber ends.[39] BP films were produced directly onto a viscoelastic PDMS stamp. The selected BP film on the PDMS stamp was then placed on a fiber end with the help of optical microscope and micromanipulator. Due to viscoelastic properties of PDMS, the BP film adhered to the fiber end when the PDMS stamp was gently lifted off. In Tayari's work,[40] they fabricated BP-naked quantum wells in a back-gated field effect transistor geometry with thicknesses ranging from

6 ± 1 nm to 47 ± 1 nm, which was performed using a PDMS stamp technique within a nitrogen glove box environment to minimize exposure to water and oxygen. In detail, a multi-step process was used, wherein BP was first exfoliated directly to a strip of adhesive tape. The BP was then repeatedly exfoliated 15 times with the strip of adhesive tape. A flat PDMS stamp was then used to exfoliate BP from the adhesive tape five times, and finally the flat PDMS stamp was used to deliver BP to target sample substrates. The substrates were pre-patterned with metal alignment marks and were annealed at 150°C for 15 min to desorb water before BP exfoliation. Once the Ti/Au (5 nm/ 80 nm) contacts on BP flakes fabricated, the BP-based FETs were encapsulated in a glove box environment by spin-coating 300 nm of copolymer (methyl methacrylate) and 200 nm of polymer (polymethyl methacrylate) followed by an annealing step at 170°C for 15 min. Recently, Favron et al. achieved the mechanical exfoliation of mono-, bi- and multilayer phosphorene from bulk BP by the modified PDMS-based method in a glove box under a clean nitrogen flow.[41] First, the BP crystal was exfoliated onto a flat PDMS stamp and then transferred onto a curved PDMS stamp. The second PDMS stamp covered with flakes was then rolled onto either a 305 nm or a 291 nm SiO_2/Si substrate. After transfer, the thinnest flakes were identified inside the glove box for ensuring that the gas components leading to degradation are significantly reduced.

For achieving the thickness control of exfoliated phosphorene, the combination of mechanical exfoliation with other techniques has also been developed. One effective approach is that the introduction of plasma treatment after mechanical exfoliation provides an improved way for controlling thickness of phosphorene. As shown in Fig. 1.4, thickness-controlled BP films can be synthesized through the modulated Ar^+ plasma treatment of exfoliated BP flakes by Lee's group.[42] Ar gas was used to maintain the pressure at 30 mTorr, and 350 W power delivered at 13.56 MHz was applied to a four-turn spiral coil to discharge the high-density plasma. As shown in Fig. 1.4A, after patterning of source/drain electrodes, the exposed BP films were plasma-treated and passivated with PMMA. Not only does the plasma treatment control the thickness of the BP film, it also removes the chemical degradation of the exposed oxidized BP surface, which results in enhanced FET performance. To find an

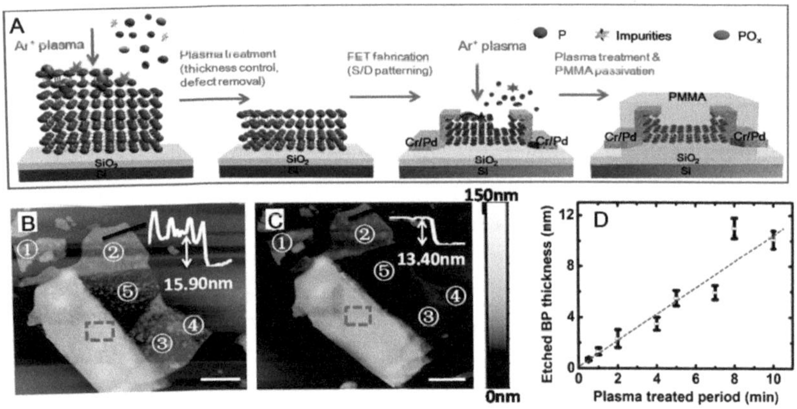

Figure 1.4 Thickness control of BP flakes by using Ar^+ plasma. (A) Schematic diagram of the effects of the plasma treatment of a BP flake: thickness control, surface defect removal, and device fabrication process. (B,C) AFM mapping images (scale bars: 2 μm) of an exfoliated BP flake after 24 h of storage (B) and after plasma treatment for 4 min (C). (D) The etched BP thickness as a function of the duration of the plasma treatment. Reprinted with permission from Jia et al.[42] Copyright (2015) American Chemical Society.

optimal plasma treatment condition with which the BP thickness can be controlled by varying the plasma treatment duration without inducing damages on the morphology and crystal structure of the BP films, they further examined various plasma power conditions at fixed pressure of 30 mTorr for 5 min. As a result, in the case of the plasma power below 350 W, no significant optical microscopy (OM) contrast changes were observed after plasma treatments. After plasma treatment at 400 W, the OM contrast changed too much and the surface became rough. Moderate OM contrast change was observed after the plasma treatments at 350 W. Considering uniform etching result on BP films and damages induced on the BP surface, the plasma treatment at 350 W was finally chosen. Figure 1.4B and C show atomic force microscopy (AFM) images of BP flakes: after 24 h of storage in low vacuum (10^{-1} Torr) and after plasma treatment for 4 min. The thicknesses of the as-exfoliated BP films varied in the range 4–50 nm. Figure 1.4D shows the uniform decrease in the BP layer thickness on plasma treatment. After plasma treatment for

4 min, the BP thicknesses of the as-exfoliated sample decreased by ~6 nm. And, the thickness gradually decreased with increase the plasma treatment period, leading to the controllable BP thickness in a monotonic manner. Meanwhile, the plasma treatment of the same BP sample was found to completely remove the bubbles from the BP top surface, producing a clean surface with a reduced surface roughness of less than 1 nm.

In Lu's work, they demonstrated that the focused Ar^+ plasma (commercial 13.56 MHz RF source with a power of 30 W and a pressure of 30 Pa) can be used to thin few layer phosphorene down to monolayer thickness by thermal ablation with micrometer-scale resolution,[43] which can be identified by optical contrast spectra and AFM. Experimentally, thin phosphorene films were firstly exfoliated onto SiO_2/Si substrate by mechanical exfoliation. With hundreds of few-layer sheets checked, the thinnest forms were bilayer flakes. The absence of monolayer phosphorene might be due to the strong interlayer coupling in multilayer flakes or the instability of single layer. However, the monolayer BP nanosheets were observed via using plasma treatment on exfoliated phosphorene. Besides the control of layer thickness, plasma can also be used to form special pattern for device application. For example, in Tao's work,[44] after fabricating few-layer BP flakes on a silicon substrate with a 200 nm thick thermal oxide layer, the flakes were trimmed into orthogonal strips along the principal lattice axes by Ar^+ plasma (recipe: Ar flow ~80 sccm, pressure ~20 Pa, RF power ~80 W).

1.5 Liquid-Phase Exfoliation of Phosphorene

To obtain a large quantity of phosphorene, liquid-phase preparation techniques are very promising, which has been extensively used as a simple and effective technique to prepare 2D nanosheets from the layered bulk crystals, such as h-BN, graphene and TMDs.[45] It is thus expected that liquid-phase preparation of phosphorene will also have the good prospects to make flexible electronics for good performances in devices. Theoretically, the interlayer intercalation of ionic species in BP expands the interlayer distance to allow the layers to be exfoliated in liquid. Meanwhile, the solvent with proper surface energy can provide stable dispersions of layered materials. For carrying out this process, a chunk of BP crystal is immersed into different solvents (such as alcohols, chloroorganic solvents, ketones, cyclic or aliphatic pyrrolidones, *N*-alkyl-substituted amides, and organosulfur compounds) to allow the ionic species to intercalate. The sample is then sonicated to break down the interlayer van der Waals bonding and form 2D BP nanosheets. This strategy opens up new possibilities for the atomically thin BP layers to be formed into thin films and composites on large scales with a wide range of applications from ultrafast saturable absorbers to gas sensors to fillers for composite reinforcement.

Sresht et al. reported the design rules for liquid-phase exfoliation of phosphorene in different solvents via molecular dynamics simulations.[46] In their work, the exfoliation of phosphorene, in the solvents dimethyl sulfoxide (DMSO), dimethylformamide (DMF), *N*-methyl-2-pyrrolidone (NMP), isopropyl alcohol (IPA), and *N*-cyclohexyl-2-pyrrolidone was investigated using three molecular-scale "computer experiments". The solvent-phosphorene interactions were modeled using an atomistic force field, based on ab initio calculations and lattice dynamics, that accurately reproduces experimental mechanical properties. The solvent molecule ordering at phosphorene/solvent interfaces was probed and discovered that planar molecules such as NMP preferentially orient parallel to the interface (Fig. 1.5). The energy required to peel a single phosphorene monolayer from a stack of BP was subsequently measured and the role of "wedges" of solvent molecules was also analyzed for intercalating between phosphorene sheets in initiating exfoliation. As a result, the exfoliation efficacy was enhanced when either molecular planarity "sharpens" this molecular wedge or strong phosphorene-solvent adhesion stabilizes the newly exposed phosphorene surfaces. Furthermore, they examined the colloidal stability of exfoliated flakes by simulating their aggregation and showed that dispersion is favored when the cohesive energy between the molecules in the solvent monolayer confined between the phosphorene is high and is hindered when the adhesion between these molecules and phosphorene is strong; and the molecular planarity in solvents enhances the cohesive energy. These molecular insights into the significant role of solvent molecular geometry and

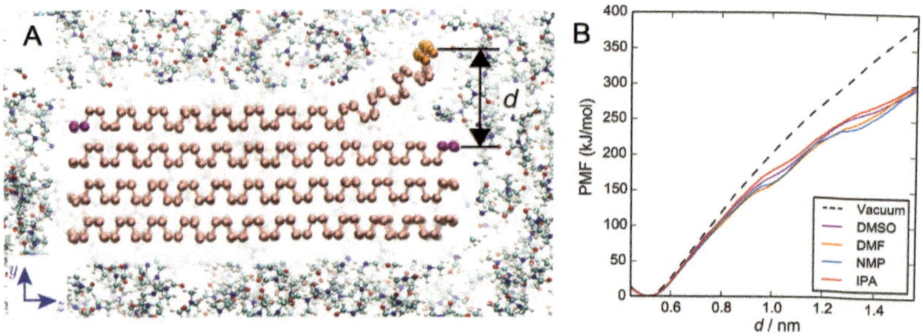

Figure 1.5 MD simulations of phosphorene monolayers peeled apart in vacuum, DMSO, DMF, NMP, and IPA. (A) Snapshot of the simulation box when phosphorene is exfoliated while immersed in NMP. The rows of atoms indicated in purple are restrained to their initial positions using a harmonic potential, while another harmonic is used to pull the rows of atoms indicated in yellow in the $+y$ direction. The distance between the yellow atoms and their below phosphorene layer defines the reaction coordinate d for the peeling process. The infiltration of NMP molecules into the interlayer gap is immediately discernible. (B) Potential of mean force (PMF) exerted on the phosphorene monolayer being peeled away from bulk BP as a function of d for the different exfoliation environments. These PMF curves represent the energy needed to peel the upper phosphorene layer away to a given distance d. Reprinted with permission from Sresht et al.[46] Copyright (2015) American Chemical Society.

ordering should complement prevalent solubility-parameter-based approaches in establishing design rules for effective nanomaterial exfoliation media. As an important conclusion, if one ignores the solvent's molecular shape, the optimal liquid-phase exfoliation medium is one where the cohesion between solvent molecules is as strong as the adhesion between the solvent molecules and phosphorene, and the planarity of solvent is also important in explaining the superior performance of NMP and DMF. This more microscopic view should guide the discovery and optimization of solution environments for low agitation nanomaterial exfoliation.

In the experimental fields, Yasaei et al. surveyed several solvents from different chemical families such as alcohols, chloro-organic solvents, ketones, cyclic or aliphatic pyrrolidones, NMP, and organosulfur compounds, covering a wide range of surface tensions (21.7–42.78 dyne/cm) and polar interaction parameters (2.98–9.3 $MPa^{1/2}$), and examined their performance for BP exfoliation.[47] In detail, bulk BP of ~0.2 mg was immersed in 10 mL of solutions in the form of either a single chunk, or a ground powder. The prepared samples were sonicated in a Sonics Vibra-Cell sonicator (130 W) for an appropriate time. Afterward, the solution was centrifuged in an Eppendorf 5424 Centrifugation machine (250 W) for 30 min in 2000 rpm and the top 50% of the solution was collected and filtered using a Sigma–Aldrich vacuum filtration assembly on a polytetrafluoroethylene (PTFE) membrane filter of 0.1 μm pore size. The filtered film was then thoroughly washed many times by ethanol, water, and IPA to remove the solvent residue. The stacked flakes were then dispersed in IPA for further usage. For the case of ground bulk of 0.2 mg, 6 hours of sonication at 130 W, was the appropriate time to obtain high concentrations of atomically thin BP flakes. For the single chunk, 15 hours of sonication in the same experimental conditions was required to obtain similar results. As the most important result, aprotic and polar solvents such as DMF and DMSO are appropriate solvents for the synthesis of atomically thin BP nanoflakes and can produce uniform and stable dispersions after the sonication. In their another work,[48] an BP film-sensor was developed using BP nanosheets produced in large quantities by the liquid exfoliation method via sonication of the ground bulk BP (powder) in DMF or DMSO solvents. Liquid exfoliation started with grinding and immersing 15 mg of bulk BP into 20 mL of appropriate aprotic and polar solvents (DMF or DMSO). Next, the samples were sonicated for 12 h and centrifuged for 30 min at 500 rpm, and then the top 90% of the solution was collected. The lateral size distributions of the flakes produced in DMF and DMSO are in the ranges of 100–250 and 200–1200 nm, respectively. The statistical height distributions of the flakes produced in DMF and DMSO are centered around 5–10 and 15–20 nm, respectively. Although the use of DMF produces thinner flakes, DMSO might be preferred due to its lower toxicity. The prepared solutions were then filtered (with vacuum) on a hydrophilic PTFE membrane filter with 0.1 μm pore size and thoroughly washed by ethanol and IPA to remove the solvent residue. For cross-section imaging, the wet film was immersed in IPA, sonicated to separate the flakes from the film, and then refiltered on a mixed cellulose membrane filter. The films were then cooled in liquid nitrogen to enhance fragility and were broken to access an intact cross-section view. For the sensing experiments and characterizations, the films were dried in vacuum and cutted into desired pieces. They explored the sensing characteristics of BP films and observed an ultrasensitive and selective response toward humid air with a trace-level detection capability and a very minor drift over time.

On the basis of the liquid exfoliation method, Lu et al. prepared a series of dispersions with BP nanoplatelets in three different sets of organic solvents NMP, IPA and ethyl alcohol (EA),[49] which can exfoliate and disperse the as-fabricated multi-layer BP flakes. The entire preparation procedure included four steps (using IPA as model): firstly, the BP powder was prepared from the BP bulk crystal by grinding. Then the powder was dispersed into IPA for 1 mg/mL and ultra-sonicated for 2 h. In the following, the dispersions were settled for more than 24 h in order to remove the large size sedimentations with a centrifugation at a speed of 1500 rpm for 20 min. The top one third of the dispersions was collected and put in 1 mm cuvette as the sample for experiment testing. As observed, more than 51% of the nanoplatelets out of the 52 objects have an averaged thickness between 15 nm and 20 nm. The broadband and enhanced saturable absorption of these multi-layer BP was experimentally verified by wide-band Z-scan measurement technique, and multi-layer BP was anticipated to be developed as another new type of two-dimensional saturable absorber with operation bandwidth ranging from the visible (400 nm) towards mid-IR (at least 1930 nm). In Lin's work,[50] solution exfoliated few-layer BP was reported to serve as an effective electron transport layer (ETL) in organic photovoltaics (OPVs) for the first time, which is different with that most of 2D materials are served as hole transport layers in organic devices. In their experiments, various solvents such as ethanol, acetone, NMP, and IPA were used for the ultrasonication process to exfoliate BP. In detail, about 20 mg of BP crystals was added into acetone (4 mL), IPA (4 mL), NMP (4 mL) and ethanol (8 mL) solutions. The samples were ultrasonicated using an ultrasonic bath of 400 W at a temperature of 28°C for 48 h. The obtained BP solution was centrifuged with rates of 1000–4000 rpm for 60 min in order to remove larger particles. The exfoliated BP in different solvents exhibit slightly different colors which are due to the variation in concentration of BP in the solvents, which can be adjusted by tuning the centrifugation process. They finally found that the thickness of the BP flakes prepared in IPA and ethanol led to the highest concentration of BP flakes and the thickness ranges were from 1 nm to 21 nm, the most frequent thickness was between 2 and 7 nm.

Woomer et al. systematically surveyed the conditions for liquid exfoliation of BP to achieve the first large-scale production of monolayer, bilayer, and few-layer phosphorene.[20] In their initial experiment, BP solution was sonicated in electronic grade IPA for 16 h. During sonication, the BP was suspended in solution and its color changed from black to reddish brown, indicating a profound change in the electronic structure of the material. Over several weeks, there was limited reaggregation and no further change in color, suggesting that these suspensions comprised small phosphorus particulates. Those SEM images confirmed the presence of thin BP flakes with lateral dimensions between 50 nm and 50 μm. After this, they surveyed 18 solvents for their ability to exfoliate BP. BP crystals were first ground into granular pieces using a mortar and pestle in a nitrogen glovebox. Then, 10 mg of BP pieces was added to 20 mL of each solvent and sonicated for 13 h under anhydrous and air-free conditions. In this experiment, the scintillation vials including BP solutions were tightly capped and wrapped with parafilm to prevent air exposure before placing into a Branson 5800 bath sonicator. The bath sonicator was outfitted with a test tube rack to allow for controlled placement of vials. Vials were systematically moved through several locations during the course of sonication to minimize vial-to-vial variations in phosphorus dispersion. The samples were subjected to 8 to 10 cycles of sonication at high power, each lasting 99 min. Bath water was changed after each cycle to maintain a temperature between 22 and 30°C (during sonication, bath temperature increased dramatically). During the sonication process, the BP crystals dispersed into the solution and the suspension acquired a brown appearance. After sonication, the vials were returned to the glovebox. With the goal of isolating phosphorene with well defined thicknesses and optical properties, centrifugation was used to fractionate the phosphorus suspensions, which was performed in a Sorvall RC-5B

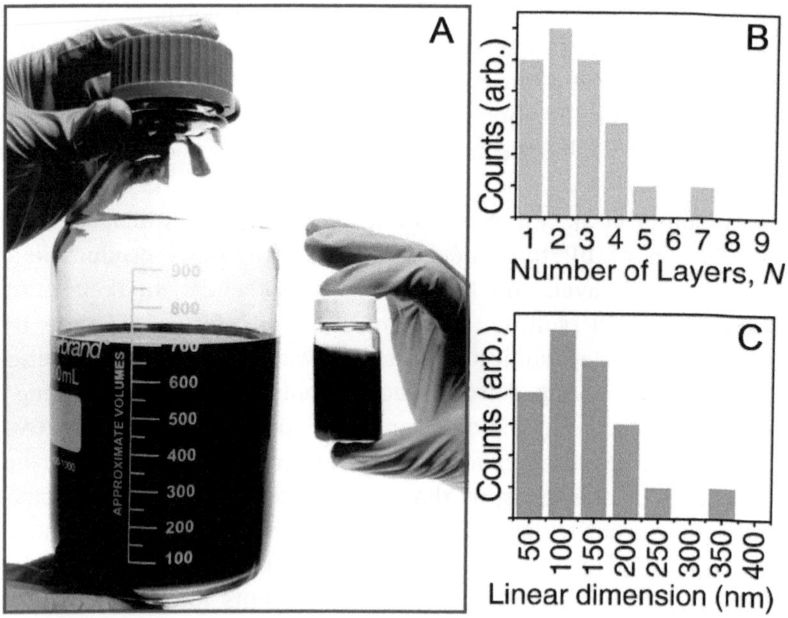

Figure 1.6 Scaled-up production of phosphorene by using a combination of shear mixing and sonication. (A) Photograph of exfoliated solutions, which were obtained via using 6 g of BP and 800 mL of NMP, followed by centrifugation at 20 200g to isolate a highly concentrated suspension containing thin pieces (right). (B,C) The distribution in layer number (B) and size (C) for exfoliated phosphorene. Reprinted with permission from Woomer et al.[20] Copyright (2015) American Chemical Society.

superspeed refrigerated centrifuge with a rotor radius of 10.7 cm. Dispersions of sonicated BP were transferred into Nalgene Oak Ridge FEP 10 or 50 mL centrifuge tubes. They centrifuged the mixture at a rotational centrifugal force (rcf) as low as 120 g and then centrifuged the supernatant at a slightly greater rcf, reaching values of up to 48 000 g. The sediment from the second centrifugation was collected and redispersed in pure solvent. The exfoliated phosphorene can further be purified via dialysis to remove small (<2.5 nm) phosphorus fragments. Finally, they analyzed the thicknesses and lateral sizes of the suspended 2D BP flakes (Fig. 1.6 shows the phosphorene solution exfoliated in NMP) and found that this centrifugation approach could systematically isolate flakes with varying size and thickness distributions. When high centrifugation speeds were used, for example, they collected macroscopic quantities of flakes with size distributions centered near one-layer and two-layer thicknesses, in which monolayers comprised up to 45% of the sample.

The obtained suspensions were also characterized with inductively coupled plasma mass spectroscopy and UV-vis transmission spectroscopy to measure a dispersed concentration. As a result, they found that the best solvent was benzonitrile, which achieved a mean concentration of 0.11 ± 0.02 mg/mL. The monolayer and few-layer flakes produced by the approach are crystalline and unoxidized. With large quantities of 2D BP now available, the first quantitative measurements of the material's absorption edge was performed as a function of flake thickness, which was nearly identical to the material's bandgap under their experimental conditions.

For obtaining high-purity phosphorene from the liquid phase exfoliation, Kang et al. designed and prepared a customtip ultrasonication setup that minimized exposure to ambient air during processing.[51] As shown in Fig. 1.7A,B, those pristine, electronic-grade BP dispersions were produced with anhydrous organic solvents in a sealed-tip ultrasonication system, which circumvents BP degradation that would otherwise occur via solvated O_2 or H_2O. In this experiment, all BP crystal storage containers were first annealed in the Ar glovebox (>100°C) to remove adventitious H_2O. For solvent exfoliation of BP, a custom-tip sonicator setup was prepared by perforating the plastic lid of a 50 mL conical tube with a 0.125 in. sonicator tip. The interface between the tip and the lid was sealed with PDMS several times, preventing O_2 and H_2O penetration into the tube. Additionally, Parafilm and Teflon tapes were used to further suppress potential O_2 or H_2O pathways between the lid and container. Therefore, only BP crystals, anhydrous solvent, and Ar gas were present in the conical tube during sonication (Fig. 1.7). The BP crystal and NMP solvent were placed in this sealed conical tube with the sonicator tip in an Ar glovebox. The container was then connected to the sonicator (Fisher Scientific model 500 sonic dismembrator) in ambient conditions, after which the BP crystal was exfoliated via ultrasonication. As prepared BP dispersions were centrifuged with different speeds to remove unexfoliated BP crystals

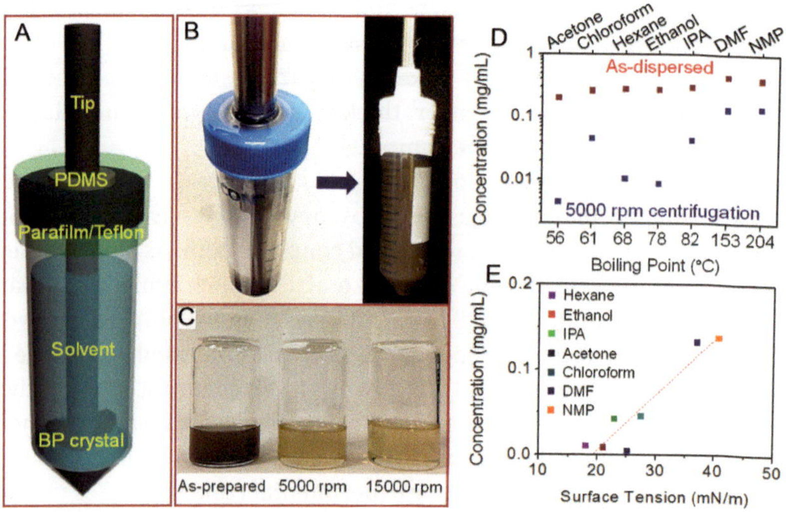

Figure 1.7 Solvent exfoliation of BP in various solvents via tip ultrasonication. (A) Schematic and (B) photograph of the customtip ultrasonication setup that minimized exposure to ambient air during processing. (C) Photograph of a BP dispersion in NMP after ultrasonication, 5000 rpm centrifugation, and 15,000 rpm centrifugation (left to right). (D) BP concentration plot for various solvents with different boiling points before and after 5000 rpm centrifugation and (E) with different surface tensions after 5000 rpm centrifugation. Reprinted with permission from Kang et al.[51] Copyright (2015) American Chemical Society.

and isolate well-exfoliated 2D BP nanosheets using an Eppendorf tabletop centrifuge, ultimately resulting in the phosphorene solution color evolving from brown to yellow (Fig. 1.7C). In this work, different solvents such as acetone, chloroform, hexane, ethanol, IPA, DMF, and NMP were used to estimate the optimal solvent for BP exfoliation, and NMP was found to provide stable, highly concentrated (~0.4 mg/mL) BP dispersions. Meanwhile, residual NMP from the liquid-phase processing suppressed the rate of BP oxidation in ambient conditions.

Recently, liquid phase exfoliation of phosphorene in NMP has been attracted more and more research interests, which was confirmed to be greatly successful for exfoliate and disperse graphene, TMDs, et al.[45] For example, Qin et al. successfully fabricated multilayer BP by liquid phase exfoliation in NMP and coated them on a gold-coated mirror as reflection-type saturable absorber mirror.[52] Experimentally, they first mixed bulk BP (30 mg) with NMP solution (30 mL) together and sonicated at 40 kHz frequency and 300 W power for 10 h. Then, the supernatant liquor was obtained after centrifuging at 1500 rpm for 10 min. The exfoliated multilayer BP flakes had a thickness distribution from 5–20 nm, and more than 51% of the flakes had a thickness between 15 nm and 20 nm. Brent et al. also reported the liquid exfoliation of BP in NMP to form few-layer phosphorene nanosheets.[53] Bulk BP was exfoliated in NMP (5 mg/mL BP in NMP) using bath ultrasonication (820 W across four horns operating at 37 kHz frequency and 30% power) for 24 h. The temperature of the bath was maintained below 30°C through using a water cooling coil. Turbid dispersions were finally obtained from the exfoliation step, which were further purified by centrifugation to remove larger solids. Luo et al. demonstrated that few-layer BPs fabricated in NMP can be developed as a new and practical saturable absorber by depositing few-layer BPs with microfiber.[54] In their work, 30 mg of bulk BP was first added to NMP solution (30 ml). Then, the mixture was bath sonicated at 40 kHz frequency and 300 W for 10 h to carry on the liquid exfoliation of bulk BP. After the sonication step, the as-prepared few-layer BPs was centrifuged at 1500 rpm for 10 min and then the supernatant was taken for the testing samples.

In Guo's work,[55] a basic NMP liquid exfoliation method was described to produce phosphorene with excellent water stability, controllable size and layer number, as well as a high yield. As shown in Fig. 1.8A, bulk BP was added to a saturated NaOH NMP solution, which was put in a sonicator for 4 h to conduct liquid exfoliation. After sonication, the phosphorene in NMP was separated and transferred to water by centrifugation. In detail, 15 mg of bulk BP was added to the saturated NaOH/NMP solution (30 mL). The mixture was put in a sonicator operated at 40 kHz frequency and 80% power for 4 h to conduct the liquid exfoliation of the bulk BP. After exfoliation, the solution was centrifuged at 3000 rpm for 10 min to remove any nonexfoliated bulk BP. The supernatant was then centrifuged at 12,000 rpm for another 20 min to separate the relatively thick phosphorene (5–12 layers) from

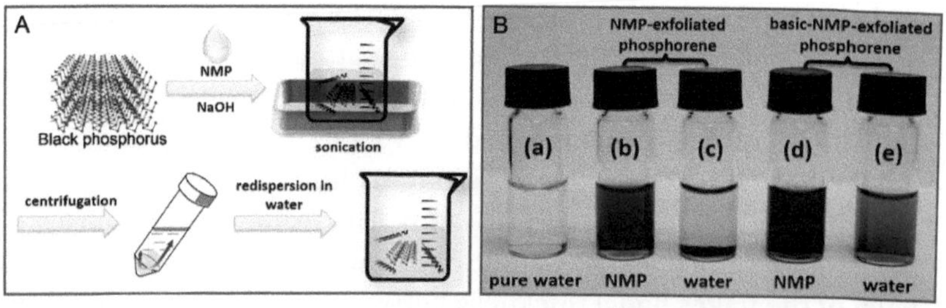

Figure 1.8 Basic-NMP exfoliation of phosphorene from BP crystals. (A) Schematic illustration of the fabrication process. (B) Photographs of phosphorene dispersed in NMP and water: (a) pure water, (b,c) NMP-exfoliated phosphorene in NMP (b) and water (c); (d,e) basic-NMP-exfoliated phosphorene in NMP (d) and water (e). Reprinted with permission from Guo et al.[55] Copyright (2015) John Wiley and Sons.

the NMP. The supernatant was further centrifuged at 18,000 rpm for 20 min to separate the thinner phosphorene (1–7 layers) from the supernatant. The precipitations obtained by the two separation processes were redispersed in water and the solutions were washed by deionized water. As shown in Fig. 1.8B, the obtained phosphorene via the basic NMP liquid exfoliation had higher water stability, as compared with NMP-based liquid exfoliation. In addition, the basic NMP-based phosphorene composed of one to four layers also exhibited layer-dependent Raman scattering characteristics thus providing a fast and efficient means for the in situ determination of the thickness (layer number) of phosphorene.

In addition of NMP, another organic amide solvent *N*-cyclohexyl-2-pyrrolidone (CHP) was also used to produce high-quality, few-layer BP nanosheets. As a typical work, Hanlon et al. demonstrated that few-layer BP nanosheets with controllable size and observable photoluminescence can be produced in large quantities by liquid phase exfoliation in solvent CHP under ambient conditions.[56] Experimentally, BP was lightly ground with pestle and mortar and immersed in CHP (concentration 2 g/L). The dispersion was sonicated for 5 h at 60% amplitude with a horn-probe sonic tip (VibraCell CVX, 750W) under cooling, yielding a stock dispersion. Aliquots of the stock dispersion were centrifuged at 1000 rpm (106 g) for time periods varying from 5 to 240 min in a Hettich Mikro

220R centrifuge equipped with a fixed-angle rotor 1016. The supernatant was decanted and subjected to absorbance spectroscopy. The supernatant with centrifugation conditions 1000 rpm for 180 min was denoted as std-BP. The std-BP was subsequently separated into small and large stable nanosheets. For this purpose, aliquots of std-BP dispersion were subjected to an additional centrifugation of 5000 rpm (2660 g) for 120 min. The supernatant (containing small flakes) was decanted and characterized as S-BP, while the sediment (containing large flakes) was redispersed in fresh CHP and characterized as L-BP. Finally, various brown dispersions with phosphorene nanosheets were yielded in CHP, wherein the nanosheets had different size and thickness. The nanosheets were surprisingly stable in CHP, probably due to the solvation shell protecting the nanosheets from reacting with water or oxygen. Moreover, the liquid-exfoliated BP nanosheets were demonstrated to be potentially useful in a range of applications from ultrafast saturable absorbers to gas sensors.

Although these organic solvents demonstrated above can efficiently exfoliate the bulk BP, their adsorption on the surface of as-exfoliated nanosheets is too intimate to be removed. For obtaining the clean phosphorene, Wang et al. prepared the ultrathin BP nanosheets by direct exfoliation of corresponding bulk sample in water via a simple water exfoliation strategy.[19] Experimentally, 50 mg of the bulk BP was dispersed in 100 mL of distilled water that bubbled with argon to eliminate the dissolved oxygen molecules for avoiding the oxidation, and the mixture solution was then sonicated in ice water for 8 h and the ice water was used to keep the system in a relatively low temperature. Afterward, the resultant brown suspension was centrifuged at 1500 rpm for 10 min to remove the residual unexfoliated particles and the supernatant was collected for further use. The ultrathin BP nanosheets with clean surface were finally obtained in water with concentration of the nanosheets estimated to be ~0.02 mg/mL. The exfoliated BP nanosheets were demonstrated to be effective photosensitizers for the generation of singlet oxygen with a high quantum yield of about 0.91, rendering their attractive applications in catalysis and photodynamic therapy. Meanwhile, the water dispersible BP nanosheets showed notable cancer therapy ability through in vitro and in vivo studies.

Recently, an electrochemically exfoliation of BP nanosheets in ionic solution was also reported for prospective field emitters,[57] in which the high-quality BP crystal material was used as the working electrode and source. Experimentally, the bulk BP crystals were clamped to a conducting stainless steel clip and were immersed in ionic solution. The platinum wire was used as the counter-electrode. The distance between the counter electrode and the working electrode was maintained at 0.5 cm, and both the electrodes were held parallel to each other. The ionic solution was an aqueous solution of Na_2SO_4 (0.5 m) prepared by mixing sodium sulfate (3.15 g) in distilled water (50 mL). The electrochemical processes was carried out by applying a positive bias voltage of about +7 V and a current of about 0.2 A to the working electrode. The rates of reaction were estimated by the amount of gas bubbles produced on the platinum electrode (H_2 gas) and transferred to the working electrode (O_2 gas). When reaction started, the solution was initially colorless, but it turned to faint yellow after 25 min. After the reaction for a period of 50 min, the exfoliated BP nanosheets were collected in a beaker, which was further kept under stable conditions for 2 h to enable the development of a precipitate. The as-synthesized precipitate was then washed several times with distilled water with the help of ultrasonication for 15 min and centrifuged for 20 min and followed by washing with ethanol. The obtained electrochemically exfoliated BP sample showed sheet-like morphology with lateral dimensions between 5 and 10 μm. The typical AFM image and corresponding height profile (approximately 1.4 nm) indicated the bi-layered nature of the as-synthesized BP nanosheets.

1.6 Preparation of Other Nanostructured BP Materials from Bulk BP

Mechanical and liquid exfoliation methods have been adopted to prepare BP nanosheets with different numbers of layers and sizes, and the products have fascinating applications in electronics and photoelectric devices.[58] At the same time, some modified methods have also been developed to synthesize other nanostructured BP

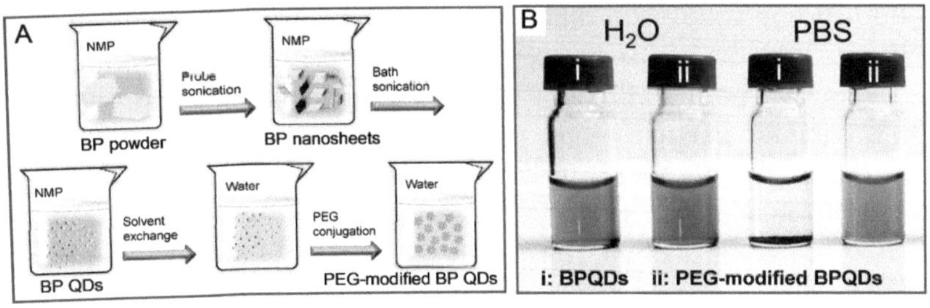

Figure 1.9 Synthesis and characterization of the BP QDs. (A) Schematic synthesis and surface modification. (B) Photographs of BP QDs (i) and PEG-modified BP QDs (ii) dispersed in water or PBS solution. Reprinted with permission from Sun et al.[58] Copyright (2015) John Wiley and Sons.

materials for a wider application. As the typical example, the production of BP quantum dots (QDs) provides a suitable synthesis strategy for the application of BP in biomedicine.

It has recently been reported that the combination of grinding and sonication processes conveniently led to the exfoliation of bulk BP crystals into high-yield BP QDs. In Zhang's work,[9] the BP QDs were synthesized by grinding followed by ice-bath sonication of micrometer-sized BP crystals. Except for the sonication and centrifugation processes, other experiments were conducted in a glovebox. Experimentally, 5 mg of BP powder was added into 1 mL of NMP in a mortar and then ground for 20 min. The mixture was transferred to a glass vial containing 3 mL of NMP. After it was sealed carefully, the vial was taken out from glove box and sonicated in an ice-bath for 3 h at the power of 200 W. The resultant dispersion was centrifuged for 20 min at speed of 7 000 rpm. The supernatant containing BP QDs was decanted gently. The obtained BP QDs have a lateral size of 4.9 ± 1.6 nm and thickness of 1.9 ± 0.9 nm (ca. 4 ± 2 layers). In another work by Sun et al.,[58] BP QDs were synthesized using a liquid exfoliation method that combined probe sonication and bath sonication, which have a lateral size of approximately 2.6 nm and a thickness of about 1.5 nm. As illustrated in Fig. 1.9A, the liquid exfoliation involves ultrasound probe sonication followed by ice-bath sonication of bulk BP powder in NMP. The ultrasmall BP QDs obtained by centrifugation were dispersed in water, and PEG was conjugated to enhance their stability in physiological medium.

Both probe sonication and bath sonication are commonly used in the exfoliation of 2D layered materials including BP. Although it seems that bath sonication is more efficient in breaking bulk BP into smaller particles than probe sonication, the use of only one of them can only result in irregular BP nanosheets. Herein, probe sonication and bath sonication were combined to exfoliate bulk BP crystals into ultrasmall BP QDs. By compared with the combination of grinding and sonication, probe sonication is more reproducible than grinding and produce the more uniform QDs with smaller size. These ultrasmall BP QDs exhibited an excellent NIR photothermal performance with a large extinction coefficient of 14.8 Lg^{-1} cm^{-1} at 808 nm and a photothermal conversion efficiency of 28.4%, as well as good photostability. Moreover, the BP QDs showed enhanced stability in physiological medium PBS after PEG conjugation (Fig. 1.9B), and there was no observable toxicity to different types of cells.

Different with BP QDs, Yang et al. reported another unexplored nanoscale form of BP, namely amorphous BP (a-BP) ultrathin films with the thickness ranging from 2 to 10 nm.[5] Such wafer-scale a-BP ultrathin films can be made at the temperature as low as 150°C by conventional pulsed laser deposition (PLD), which is in contrast with extreme conditions of high temperature and high pressure for the preparation of bulk BP. Experimentally, BP ultrathin films were grown on graphene/copper or SiO_2/Si substrates by PLD technique. The distance between the target of bulk BP crystal and substrates is 4 cm. Prior to PLD deposition, the chamber was evacuated to a base pressure of around 1.5×10^{-7} Torr. To synthesize a-BP thin films, the substrate temperature was controlled at 150°C. The BP target was ablated by a KrF pulsed laser ($\lambda = 248$ nm) with 5 Hz repetition rate. Both BP target and substrates were rotated during the deposition process for achieving uniform film growth. The as-deposited ultrathin films were then cooled down to room temperature naturally in high vacuum chamber. The a-BP ultrathin films showed thickness dependence of direct bandgap, namely, the bandgap decreases from 0.80 to 0.21–0.26 eV as the thickness of the film becomes thicker from 2 to 8 nm. Furthermore, the a-BP ultrathin films were also used to develop BP-based FETs for demonstrating the application of a-BP ultrathin films in the devices.

1.7 Degradation of Phosphorene under Ambient Condition

Although bulk crystals of BP are stable under ambient conditions for at least a few months, BP in its single- and few-layer forms is found to be unstable in the presence of the moisture and oxygen in air. The formation of oxidized phosphorus species lead to the degradation of these BP material.[6] The oxidation kinetics is consistent with a phenomenological model involving electron transfer and quantum confinement as key parameters. As reported, samples of 10 nm thickness without a proper protection can degrade in days whereas single-layer and few-layer samples may degrade within hours. For observing the degradation process, Favron et al. exfoliated mono-, bi- and multilayer phosphorene using a procedure derived from the scotch-tape exfoliation technique and investigated their oxidation chemistry using in situ Raman and transmission electron spectroscopies,[41] which is not easy to achieve due to a fast degradation of the thinnest layers on exposure to ambient conditions. Through AFM images, small bumps were seen shortly after exfoliation in the ambient laboratory conditions. The apparent heights of these bumps can often exceed the flake thickness (about five monolayers ($n \approx 5$) using $\sim$0.53 nm per layer), and the bumps were uniformly distributed and seem to grow on top rather than from the edges, indicating that a nucleation step was required before growth. After a few days, the degradation leads to large droplets located in the vicinity of the flake, suggesting a complete transformation into phosphoric acid. These experiments revealed a thickness-dependent photoactivated oxidation by oxygen dissolved in adsorbed water. Overall, the severe degradation clearly illustrates the hurdles in processing the BP nanosheets.

Recent researches on exfoliation of BP materials frequently observed the degradation of as-fabricated phosphorene. In Liu's work,[59] the atomic-scale microscopic and spectroscopic study were performed to characterize the thermal degradation of mechanically exfoliated 2D BP. From in situ scanning/transmission electron microscopy, decomposition of 2D BP was observed to occur at $\sim$400°C in vacuum, in contrast to the 550°C bulk BP sublimation

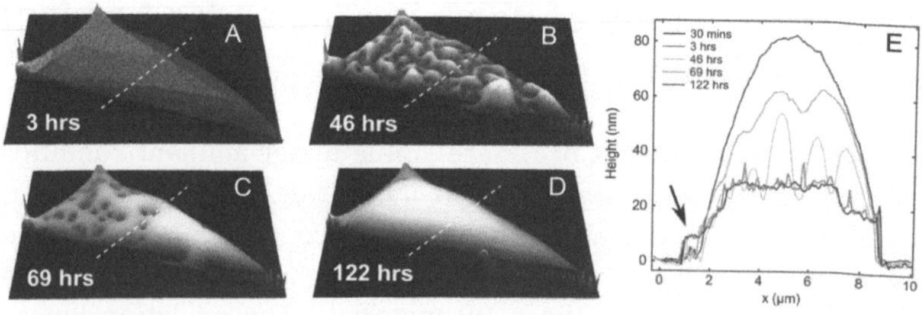

Figure 1.10 Environmental instability of few-layer BP flake. (A–D) Selected AFM scans of a BP flake in air taken at (A) 3 h, (B) 46 h, (C) 69 h, and (D) 122 h after exfoliation. (E) Line profiles taken across the BP flake from the scans in (A–D) as well as a line profile for the AFM scan taken directly after exfoliation (30 mins). Reprinted with permission from Island et al.[60] Copyright (2015) IOP Publishing.

temperature. This decomposition initiated via eye-shaped cracks along the [001] direction and then continued until only a thin, amorphous red phosphorus like skeleton remained. Meanwhile, in situ electron energy loss spectroscopy, energy-dispersive X-ray spectroscopy, and energy-loss near-edge structure changes further provided quantitative insight into this chemical transformation process.

The environmental instability of mechanically exfoliated few-layer BP was systematically studied by Island et al.[60] As shown in Fig. 1.10, from continuous measurements of flake topography over several days, they observed an increase of over 200% in volume due to the condensation of moisture from air. The BP's strong affinity for water greatly modified the performance of fabricated BP FETs measured in ambient conditions. Upon exposure to air, they differentiated between two timescales for changes in BP FET transfer characteristics: a short timescale (minutes) in which a shift in the threshold voltage occured due to physisorbed oxygen and nitrogen, and a long timescale (hours) in which strong p-type doping occured from water absorption. Continuous measurements of BP FETs in air revealed eventual degradation and break-down of the channel material after several days due to the layer-by-layer etching process, which was resulted from the long-term exposure of BP flakes to ambient conditions. In another report, Castellanos-

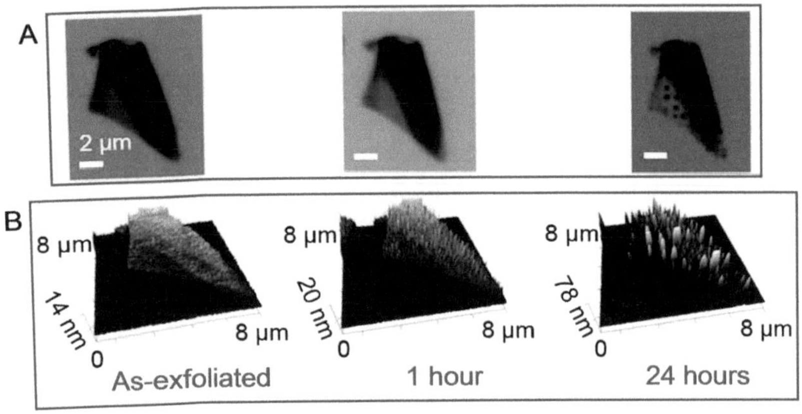

Figure 1.11 Observation on instability of BP flakes in air. (A) Optical microscope images and (B) AFM images illustrating the physical degradation of an exfoliated sample in ambient. Reprinted with permission from Kim et al.[61] Copyright (2015) Nature Publishing Group.

Gomez et al. also studied the environmental stability of BP flakes and found that the flakes were very hydrophilic and that long-term exposure to air moisture etched BP away.[37] This could be related to hydrogen (H) and hydroxyl (OH) adsorption which form strong bonds with P atoms and act as "chemical scissors" to break the P–P backbone bonds. Meanwhile, the oxygen defects (e.g., dangling oxygen) were considered as another reason for surface degradation because oxygen defects are low energy metastable forms which introduce deep donor and/or acceptor levels in the bandgap and the dangling oxygen atom can be a starting point for forming more complex defects. Herein, density functional theory calculations were also performed to give us insight into the water-induced changes of the structural and electronic properties of BP. In Kim's study,[61] short term degradation of BP upon exposure to air was also observed. As shown in Fig. 1.11, uncapped BP samples under the normal lighted in-door ambient conditions (temperature 24–27°C, relative humidity 40–45%) experienced material degradation in a matter of hours, which can be seen in the optical and AFM images of a representative flake with initially smooth terraces of different thicknesses. Additional AFM data collected within 24 h indicated moisture accumulation on the surface, which resulted in substantial local volume expansion and a cleated surface with sharp spikes. Corresponding optical images showed increasing optical transparency during the process, indicative of continuous thinning of the flake. As the surrounding Al_2O_3 surface area showed no discernible moisture adsorption, a conclusion can be given that the degradation was localized to the BP flake.

As described above, despite the successful preparation of BP flakes by mechanical exfoliation, it is inevitable that O_2 and H_2O upon ambient exposure irreversibly reacts with BP to form phosphoric acid or oxidized phosphorous compounds and this rapid surface oxidation takes place over a short period of time. In Wood's work, water droplets and oxidation were detected on the surface of BP after several hours of ambient exposure, and the surface was more severely oxidized if exposed for longer than 1 week, causing etching of some thin BP flakes.[62] This surface degradation not only increases the surface roughness but also results in the formation of chemisorbed species, which has several adverse effects including physical changes such as volume expansion and cleated surface, chemical changes towards a liquid phase degrading mobility as well as large Schottky barrier heights due to the metal-induced gap states. Thus, the BP-based device performances were influenced significantly by surface degradation. For example, Hersam et al. reported that exposure of BP FET in an ambient atmosphere induced an increase in threshold voltage and decrease in on/off ratio and mobility due to the p-type doping from ambient adsorbates, in which the charge transfer to trap sites or the surface oxidation layer may be the reason for observing large hysteresis in the V_{bg}–I_{ds} curve.[63] Overall, its instability of phosphorene in air is of paramount concern for practical applications of exfoliated phosphorene. It is thus critical to develop an efficient isolation or passivation layer for the BP surface to avoid surface degradation of BP nanosheets and maintain intrinsic electronic properties for a longer time.

The strong hydrophilic and photo-oxidative properties of BP make scientists to develop the corresponding strategies to address the concerns regarding its chemical and physical stabilities during the device fabrication. The oxidation rate was predicted to depend exponentially on the square of the energy gap of the layer. Thus, developing protection methods will be effective to slow down

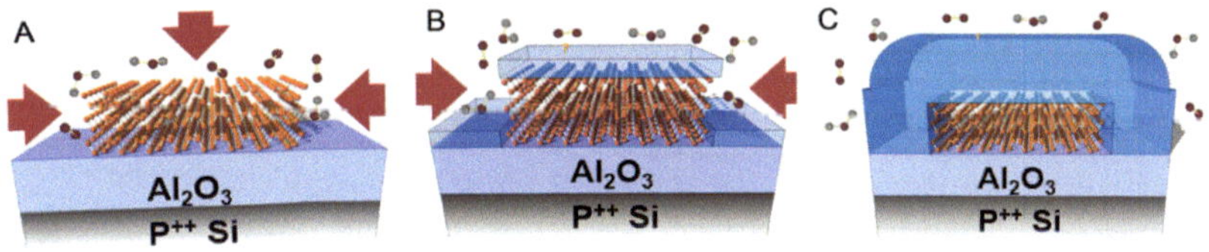

Figure 1.12 The capping of Al_2O_3 layers on BP flakes for their air-stability. (A–C) Schematics of BP samples on Al_2O_3/Si substrates against oxygen and moisture in the ambient: (A) without, (B) with a thin Al_2O_3 capping layer, and (C) with a thick Al_2O_3 capping layer. Red arrows indicate the possible pathways of oxygen and moisture to react with the sample. Reprinted with permission from Kim et al.[61] Copyright (2015) Nature Publishing Group.

and eliminate the degradation process. Several recent experiments demonstrated the use of oxidized aluminum as a passivation layer to isolate the BP surface from the ambient, which works effectively in reducing the degradation of a relatively thick sample (~5 nm). As depicted in Fig. 1.12, in order to improve the air-stability of BP films, a thin (~2–3 nm) layer of aluminum was deposited on the exfoliated flakes and oxidized into Al_2O_3 to serve as a barrier to moisture adsorption.[61] Furthermore, the thick capping of Al_2O_3 afforded BP devices and transistors with indefinite air-stability (Fig. 1.12C), overcoming a critical material challenge for applied research and development. The degradation process of BP transistors with thin and thick Al_2O_3 capping layers was further systemically investigated over a period of one week using an AFM-based microwave impedance microscope (MIM). The observed change in local resistance indicated that the degradation mostly influences the electronic rather than structural properties. After being encapsulated by Al_2O_3 overlayers, the BP flakes were stable for at least several weeks in an ambient environment.[6] Currently, most of typical BP-based transistors are fabricated by using Al_2O_3 as a dielectric layer for the efficient isolation or passivation layer of the BP surface to maintain intrinsic electronic properties for a longer time.[64] The work by Luo et al. specifically mentioned that a few-layer BP metal oxide semiconductor field-effect transistor (MOSFET) with Al_2O_3 passivation can be ambient stable as long as 100 h.[65] However, a negative shift of I_d–V_g curves and an increasingly severe hysteresis were also observed, suggesting that Al_2O_3 may not be able to provide long-term protection. They further developed a double-layer superstrate by coating fluoropolymer film, which showed strong hydrophobicity and results in the most air-stable device.

Other techniques, such as PMMA coating, have also been proposed similarly to suppress the fast degradation of few-layer BP in ambient.[6] For example, in Wu's work,[33] micromechanical exfoliation method was used to obtain few-layer BP on p-doped silicon wafers with 300 nm SiO_2. Immediately after locating the few-layer flake of BP based on the optical contrast and Raman measurements, they spin-coated electron beam resist PMMA to protect the flake. In Koening's work,[34] in order to minimize the effect of crystal degradation, the exfoliated flakes which were used for device preparation were quickly covered with PMMA once being identified under optical microscopy. The average time to cover a flake with PMMA after exfoliation was less than 30 min. Overall, phosphorene has good intrinsic thermal stability and the material is also stable at high temperature if isolated from water and oxygen.

It should also be noted that Kang et al. recently reported that residual NMP from the liquid-phase processing can suppress the rate of BP oxidation in ambient conditions.[54] As the solvation shell can act as a barrier to prevent oxidative species reaching the nanosheet surface, the oxidation could be effectively minimized by this exfoliation method. Thus, NMP provides stable and highly concentrated BP dispersions, as well as the residual NMP leads to the reduction in the ambient degradation kinetics. In addition, it was also shown that pristine surfaces of BP could be safely prepared by sample cleavage in dry nitrogen gas followed by transfer into vacuum.[40]

1.8 Conclusion and Perspectives

As a new member of the layered 2D materials family, the production of phosphorene has been attracted more and more interest for using their excellent performance in devices such as FETs. The layer-stacked configuration of bulk BP allows relatively easy fabrication of few-layer phosphorene via mechanical exfoliation. Nonetheless, the phosphorus has a trouble of degradation due to oxidation. The exfoliated phosphorene has been observed to inevitably show the fast oxidation and laser-induced degradation in ambient conditions. In order to manage this undesired phenomenon, two effective ways were adopted including passivation and solvent exfoliation. The first approach is to deposit Al_2O_3 encapsulation layers acting as barriers against moisture adsorption by atomic layer deposition to slow down surface degradation. In addition to Al_2O_3, an encapsulating copolymer capping (such as PMMA) has also been reported to effectively suppress oxidation and the obtained device showed excellent performances. Another way to avoid surface degradation reported recently was by direct liquid exfoliation in anhydrous and oxygen-free organic solvents. For example, it was found that residual NMP from the liquid-phase processing can suppress the rate of BP oxidation in ambient conditions.

Interestingly, coating of other 2D materials on phosphorene was recently reported to suppress the oxidation or degradation of phosphorene. For example, Özyilmaz et al. first fabricated BP passivated in an inert gas environment without chemical precursors or exposure to water, and then covered graphene or h-BN on BP flakes.[66] As a result, no significant roughness change was detected after covering graphene or h-BN, and no reaction between phosphorene and the passivating crystal was confirmed by Raman spectra. First-principles calculations were further performed to study the structural and electronic properties of hybrid graphene and phosphorene nanocomposites, and they showed that weak van der Waals interactions dominate between graphene and phosphorene with their intrinsic electronic properties preserved. In Hu's work,[67] it was reported that the capping of phosphorene by the h-BN layers will be an excellent solution to protect the intrinsic properties of the phosphorene. Moreover, the h-BN contributed to enhancing the bandgap of the phosphorene layer. Comparing the bandgap of the pristine phosphorene layer, the bandgap of the phosphorene/BN (1 layer) system was enhanced by 0.15 eV, and further enhanced by 0.31 eV in the BN/phosphorene/BN trilayer structure. In addition, annealing was found to be an effective approach to managing the surface degradation as it could successfully eliminate the adsorbates on BP surfaces which work as trapping centres for electrons.[63]

Another important concern on exfoliation of bulk BP is how to fabricate or isolate large-area and uniform single-layer phosphorene. This goal has not still been achieved until now, although the combination of mechanical exfoliation and plasma-assisted method has the potential to obtain phosphorene with partial single layer structure. One of possible problem is their instability, because single layers can rapidly react with the air to degrade. In addition, there still lack alternative strategies to fabricate uniform phosphorene with single-layer structure. The more effective and controllable lithiation method was widely used to exfoliate 2D materials such as graphene and TMDs in single-layer, however, it is not suitable for exfoliating phosphorene due to the special chemical activity and structure property of BP. Phaedon Avouris, a chemical physicist at IBM's Thomas J. Watson Research Center in Yorktown Heights, New York, suspected that phosphorene's success in electronics will finally depend on whether researchers can find efficient ways to extract single layers and deposit them on substrates.[68] For this purpose, some of recent advances on preparation of 2D materials may provide some valuable guidance and give some research directions. For example, density functional theory simulation investigated the interaction of phosphorene with oxygen to form 2D phosphorene oxide,[69] indicating that it is also possible to obtain single-layer phosphorene by reduction of phosphorene oxide in proper solvent, similar with graphene oxide. Our recent work demonstrated that single-layer graphene and TMDs can be obtained via layer-by-layer exfoliation using the strong binding of protein on nanosheets,[70] which is also hopeful to be used in exfoliation of bulk BP to form single-layer functionalized phosphorene. Overall, the future efforts toward this aim will continuously be required for developing some novel fabrication method and ideal substrate to obtain the stable single-layer phosphorene.

References

1. Allotropes of phosphorus. Wikipedia (https://en.wikipedia.org/wiki/Allotropes_of_phosphorus).
2. Nilges T, Kersting M, Pfeifer T, (2008). A fast low-pressure transport route to large black phosphorus single crystals. *J. Solid State Chem.*, **181**:1707–1711.
3. Bridgman PM, (1914). Two new modifications of phosphorus. *J. Am. Chem. Soc.*, **36**:1344–1363.
4. Liu H, Du Y, Deng Y, Ye PD, (2015). Semiconducting black phosphorus: synthesis, transport properties and electronic applications. *Chem. Soc. Rev.*, **44**:2732–2743.
5. Yang Z, Hao J, Yuan S, Lin S, Yau HM, Dai J, Lau SP, (2015). Field-effect transistors based on amorphous black phosphorus ultrathin films by pulsed laser deposition. *Adv. Mater.* **27**:3748–3754.
6. Ling X, Wang H, Huang S, Xia F, Dresselhaus MS, (2015). The renaissance of black phosphorus. *Proc. Natl. Acad. Sci. U.S.A.*, **112**:4523–4530.
7. Kou L, Chen C, Smith SC, (2015). Phosphorene: fabrication, properties, and applications. *J. Phys. Chem. Lett.*, **6**:2794–2805.
8. Li L, Yu Y, Ye GJ, Ge Q, Ou X, Wu H, Feng D, Chen XH, Zhang Y, (2014). Black phosphorus field-effect transistors. *Nat. Nanotechnol.*, **9**:372–377.
9. Zhang X, Xie H, Liu Z, Tan C, Luo Z, Li H, Lin J, Sun L, Chen W, Xu Z, Xie L, Huang W, Zhang H, (2015). Black phosphorus quantum dots. *Angew. Chem. Int. Ed.*, **54**:3653–3657.
10. Keyes RW, (1953). The electrical properties of black phosphorus. *Phys. Rev.* **92**:580–584.
11. Maruyama Y, Suzuki S, Kobayashi K, Tanuma S (1981). Synthesis and some properties of black phosphorus single crystals. *Physica B*, **105**:99–102.
12. Rissi EN, Soignard E, McKiernan KA, Benmore CJ, Yarger JL, (2012). Pressure-induced crystallization of amorphous red phosphorus. *Solid State Commun.*, **152**:390–394.
13. Sun LQ, Li MJ, Sun K, Yu SH, Wang RS, Xie HM, (2012). Electrochemical activity of black phosphorus as an anode material for lithium-ion batteries. *J. Phys. Chem. C*, **116**:14772–14779.

14. Krebs H, Weitz H, Worms KH, (1955). Die katalytische darstellung des schwarzen phosphors [The catalytic preparation of black phosphorus]. *Z. Anorg. Allg. Chem.*, **280**:119–133.
15. Brown A, Rundqvist S, (1965). Refinement of the crystal structure of black phosphorus. *Acta Crystallogr.*, **19**:684–685.
16. Lange S, Schmidt P, Nilges T, (2007). Au_3SnP_7@black phosphorus: an easy access to black phosphorus. *Inorg. Chem.*, **46**:4028–4035.
17. Köpf M, Eckstein N, Pfister D, Grotz C, Krüger I, Greiwe M, Hansen T, Kohlmann H, Nilges T, (2014). Access and in situ growth of phosphorene-precursor black phosphorus. *J. Cryst. Growth*, **405**: 6–10.
18. Latiff NM, Teo WZ, Sofer Z, Fisher AC, Pumera M, (2015). The cytotoxicity of layered black phosphorus. *Chem. Eur. J.*, **21**:13991–13995.
19. Wang H, Yang X, Shao W, Chen S, Xie J, Zhang X, Wang J, Xie Y, (2015). Ultrathin black phosphorus nanosheets for efficient singlet oxygen generation. *J. Am. Chem. Soc.*, **137**:11376–11382.
20. Woomer AH, Farnsworth TW, Hu J, Wells RA, Donley CL, Warren SC, (2015). Phosphorene: synthesis, scale-up, and quantitative optical spectroscopy. *ACS Nano*, **9**:8869–8884.
21. Smart Elements: http://www.smart-elements.com/
22. Novoselov KS, Jiang D, Schedin F, Booth TJ, Khotkevich VV, Morozov SV, Geim AK, (2005). Two-dimensional atomic crystals. *Proc. Natl. Acad. Sci. U.S.A.*, **102**:10451–10453.
23. Zhang H, (2015). Ultrathin two-dimensional nanomaterials. *ACS Nano*, **9**:9451–9469.
24. Yang X, Tang S, Ding G, Xie X, Jiang M, Huang F, (2012). Layerby-layer thinning of graphene by plasma irradiation and postannealing. *Nanotechnology*, **23**:025704.
25. Liu YL, Nan H, Wu X, Pan W, Wang W, Bai J, Zhao W, Sun L, Wang X, Ni Z, (2013). Layer-by-layer thinning of MoS_2 by plasma. *ACS Nano*, **7**:4202–4209.
26. Novoselov KS, Neto AHC, (2012). Two-dimensional crystals-based heterostructures: materials with tailored properties. *Phys. Scr.*, **T146**: 014006.
27. Mu Y, Si MS, (2015). The mechanical exfoliation mechanism of black phosphorus to phosphorene: a first-principles study. *EPL*, **112**:37003.

28. Liu H, Neal AT, Zhu Z, Luo Z, Xu X, Tománek D, Ye PD, (2014). Phosphorene: an unexplored 2D semiconductor with a high hole mobility. *ACS Nano*, **8**:4033–4041.
29. Deng Y, Luo Z, Conrad NJ, Liu H, Gong Y, Najmaei S, Ajayan PM, Lou J, Xu X, Ye PD, (2014). Black phosphorus–monolayer MoS_2 van der Waals heterojunction p–n diode. *ACS Nano*, **8**:8292–8299.
30. Liu H, Neal AT, Si M, Du Y, Ye PD, (2014). The effect of dielectric capping on few-layer phosphorene transistors: tuning the Schottky barrier heights. *IEEE Electron Device Lett.*, **35**:795–797.
31. Gehring P, Urcuyo R, Duong DL, Burghard M, Kern K, (2015). Thin-layer black phosphorus/GaAs heterojunction p-n diodes. *Appl. Phys. Lett.*, **106**:233110.
32. Luo Z, Maassen J, Deng Y, Du Y, Garrelts RP, Lundstrom MS, Ye PD, Xu X, (2015). Anisotropic in-plane thermal conductivity observed in few-layer black phosphorus. *Nat. Commun.*, **6**:8572.
33. Wu J, Koon GKW, Xiang D, Han C, Toh CT, Kulkarni ES, Verzhbitskiy I, Carvalho A, Rodin AS, Koening SP, Eda G, Chen W, Neto AHC, Özyilmaz B, (2015). Colossal ultraviolet photoresponsivity of few-layer black phosphorus. *ACS Nano*, **9**:8070–8077.
34. Koenig SP, Doganov RA, Schmidt H, Neto AHC, Özyilmaz B, (2015). Electric field effect in ultrathin black phosphorus. *Appl. Phys. Lett.*, **104**:103106.
35. Ling X, Liang L, Huang S, Puretzky AA, Geohegan DB, Sumpter BG, Kong J, Meunier V, Dresselhaus MS, (2015). Low-frequency interlayer breathing modes in few-layer black phosphorus. *Nano Lett.*, **15**:4048–4088.
36. Buscema M, Groenendijk DJ, Blanter SI, Steele GA, van der Zant HSJ, Castellanos-Gomez A, (2014). Fast and broadband photoresponse of few-layer black phosphorus field-effect transistors. *Nano Lett.*, **14**:3347–3352.
37. Castellanos-Gomez A, Vicarelli L, Prada E, Island JO, Narasimha-Acharya KL, Blanter SI, Groenendijk DJ, Buscema M, Steele GA, Alvarez JV, Zandbergen HW, Palacios JJ, van der Zant HSJ, (2014). Isolation and characterization of few-layer black phosphorus. *2D Mater.*, **1**:025001.
38. Castellanos-Gomez A, Buscema M, Molenaar R, Singh V, Janssen L, van der Zant HSJ, Steele GA, (2014). Deterministic transfer of two-dimensional materials by all-dry viscoelastic stamping. *2D Mater.*, **1**:011002.

39. Li D, Jussila H, Karvonen L, Ye G, Lipsanen H, Chen X, Sun Z, (2015). Polarization and thickness dependent absorption properties of black phosphorus: new saturable absorber for ultrafast pulse generation. *Sci. Rep.*, **5**:15899.

40. Tayari V, Hemsworth N, Fakih I, Favron A, Gaufrès E, Gervais G, Martel R, Szkopek T, (2015). Two-dimensional magnetotransport in a black phosphorus naked quantum well. *Nat. Commun.*, **6**:7702.

41. Favron A, Gaufrès E, Fossard F, Phaneuf-L'Heureux AL, Tang NYW, Lévesque PL, Loiseau A, Leonelli R, Francoeur S, Martel R, (2015). Photooxidation and quantum confinement effects in exfoliated black phosphorus. *Nat. Mater.*, **14**:826–833.

42. Jia J, Jang SK, Lai S, Xu J, Choi YJ, Park JH, Lee S, (2015). Plasma-treated thickness-controlled two-dimensional black phosphorus and its electronic transport properties. *ACS Nano,* **9**:8729–8736.

43. Lu W, Nan H, Hong J, Chen Y, Zhu C, Liang Z, Ma X, Ni Z, Jin C, Zhang Z, (2014). Plasma-assisted fabrication of monolayer phosphorene and its Raman characterization. *Nano Res.*, **7**:853–859.

44. Tao J, Shen W, Wu S, Liu L, Feng Z, Wang C, Hu C, Yao P, Zhang H, Pang W, Duan X, Liu J, Zhou C, Zhang D, (2015). Mechanical and electrical anisotropy of few-layer black phosphorus. *ACS Nano*, **9**:11362–11370.

45. Coleman JN, Lotya M, O'Neill A, Bergin SD, King PJ, Khan U, Young K, Gaucher A, De S, Smith RJ, et al. (2011). Two-dimensional nanosheets produced by liquid exfoliation of layered materials. *Science*, **331**:568–571.

46. Sresht V, Pádua AAH, Blankschtein D, (2015). Liquid-phase exfoliation of phosphorene: design rules from molecular dynamics simulations. *ACS Nano*, **9**:8255–8268.

47. Yasaei P, Kumar B, Foroozan T, Wang C, Asadi M, Tuschel D, Indacochea JE, Klie RF, Salehi-Khojin A, (2015). High-quality black phosphorus atomic layers by liquid-phase exfoliation. *Adv. Mater.*, **27**:1887–1892.

48. Yasaei P, Behranginia A, Foroozan T, Asadi M, Kim K, Khalili-Araghi F, Salehi-Khojin A, (2015). Stable and selective uumidity sensing using stacked black phosphorus flakes. *ACS Nano*, **9**:9898–9905.

49. Lu SB, Miao LL, Guo ZN, Qi X, Zhao CJ, Zhang H, Wen SC, Tang DY, Fan DY, (2015). Broadband nonlinear optical response in multilayer black phosphorus: an emerging infrared and mid-infrared optical material. *Opt. Express*, **23**:11183–11194.

50. Lin S, Liu S, Yang S, Li Y, Ng TW, Xu Z, Bao Q, Hao J, Lee C-S, Surya C, Yan F, Lau SP, (2016). Solution-Processable Ultrathin black phosphorus as an

effective electron transport layer in organic photovoltaics. *Adv. Funct. Mater.*, **26**:864–871.

51. Kang J, Wood JD, Wells SA, Lee J-H, Liu X, Chen K-S, Hersam MC, (2015). Solvent exfoliation of electronic-grade, two-dimensional black phosphorus. *ACS Nano*, **9**:3596–3604.
52. Qin Z, Xie G, Zhang H, Zhao C, Yuan P, Wen S, Qian L, (2015). Black phosphorus as saturable absorber for the Q-switched Er:ZBLAN fiber laser at 2.8 μm. *Opt. Express*, **23**:24713–24718.
53. Brent JR, Savjani N, Lewis EA, Haigh SJ, Lewis DJ, O'Brien P, (2014). Production of few-layer phosphorene by liquid exfoliation of black phosphorus. *Chem. Commun.*, **50**:13338–13341.
54. Luo ZC, Liu M, Guo ZN, Jiang XF, Luo AP, Zhao CJ, Yu XF, Xu WC, Zhang H, (2015). Microfiber-based few-layer black phosphorus saturable absorber for ultra-fast fiber laser. *Opt. Express*, **23**:20030–20039.
55. Guo Z, Zhang H, Lu S, Wang Z, Tang S, Shao J, Sun Z, Xie H, Wang H, Yu XF, Chu PK, (2015). From black phosphorus to phosphorene: basic solvent exfoliation, evolution of Raman scattering, and applications to ultrafast photonics. *Adv. Funct. Mater.*, **25**:6996–7002.
56. Hanlon D, Backes C, Doherty E, Cucinotta CS, Berner NC, Boland C, Lee K, Harvey A, Lynch P, Gholamvand Z, Zhang S, Wang K, Moynihan G, Pokle A, Ramasse QM, McEvoy N, Blau WJ, Wang J, Abellan G, Hauke F, Hirsch A, Sanvito S, O'Regan DD, Duesberg GS, Nicolosi V, Coleman JN, (2015). Liquid exfoliation of solvent-stabilized few-layer black phosphorus for applications beyond electronics. *Nat. Commun.*, **6**:8563.
57. Erande MB, Suryawanshi SR, More MA, Late DJ, (2015). Electrochemically exfoliated black phosphorus nanosheets–prospective field emitters. *Eur. J. Inorg. Chem.*, **2015**:3102–3107.
58. Sun Z, Xie H, Tang S, Yu XF, Guo Z, Shao J, Zhang H, Huang H, Wang H, Chu PK, (2015). Ultrasmall black phosphorus quantum dots: synthesis and use as photothermal agents. *Angew. Chem. Int. Ed.*, **54**:11526–11530.
59. Liu X, Wood JD, Chen KS, Cho E, Hersam MC, (2015). In situ thermal decomposition of exfoliated two-dimensional black phosphorus. *J. Phys. Chem. Lett.*, **6**:773–778.
60. Island JO, Steele GA, van der Zant HSJ, Castellanos-Gomez A, (2015). Environmental instability of few-layer black phosphorus. *2D Mater.*, **2**:011002.
61. Kim J-S, Liu Y, Zhu W, Kim S, Wu D, Tao L, Dodabalapur A, Lai K, Akinwande D, (2015). Toward air-stable multilayer phosphorene thin-films and transistors. *Sci. Rep.*, **5**:8989.

62. Wood JD, Wells SA, Jariwala D, Chen K, Cho E, Sangwan VK, Liu X, Lauhon LJ, Marks TJ, Hersam MC, (2014). Effective passivation of exfoliated black phosphorus transistors against ambient degradation. *Nano Lett.*, **14**:6964–6970.

63. Du H, Lin X, Xu Z, Chu D, (2015). Recent developments in black phosphorus transistors. *J. Mater. Chem. C*, **3**:8760–8775.

64. Zhu H, Qin X, Azcatl A, Addou R, McDonnell S, Ye PD, Wallace RM, (2015). Surface and interfacial study of half cycle atomic layer deposited Al_2O_3 on black phosphorus. *Microelectron. Eng.*, **147**:1–4.

65. Luo X, Rahbarihagh Y, Hwang JCM, Liu H, Du YC, Ye PD, (2014). Temporal and thermal stability of Al_2O_3-passivated phosphorene MOSFETs. *IEEE Electron Device Lett.*, **35**:1314–1316.

66. Hu W, Wang T, Yang J, (2015). Tunable Schottky contacts in hybrid graphene-phosphorene nanocomposites. *J. Mater. Chem. C*, **3**:4756–4761.

67. Hu T, Hong J, (2015). Anisotropic effective mass, optical property, and enhanced band gap in BN/phosphorene/BN heterostructures. *ACS Appl. Mater. Interfaces*, **7**:23489–23495.

68. Reich ES, (2014). Phosphorene excites materials scientists (News). *Nature*, **506**:19.

69. Wang G, Pandey R, Karna SP, (2015). Phosphorene oxide: stability and electronic properties of a novel two-dimensional material. *Nanoscale*, **7**:524–531.

70. Guan G, Zhang S, Liu B, Cai Y, Low M, Teng CP, Phang IY, Cheng Y, Duei KL, Srinivasan BM, Zheng Y, Zhang YW, Han MY, (2015). Protein induces layer-by-layer exfoliation of transition metal dichalcogenides. *J. Am. Chem. Soc.*, **137**:6152–6155.

Chapter 2

Phosphorene Characterization

Jiong Yang and Yuerui Lu
Research School of Engineering, College of Engineering and Computer Science, the Australian National University, Canberra, ACT 2601, Australia
jiong.yang@anu.edu.au, yuerui.lu@anu.edu.au

2.1 Introduction

Phosphorene is a recently developed two-dimensional (2D) material that has attracted tremendous attention owing to its unique anisotropic manner,[1–7] layer-dependent direct bandgaps,[8–10] and quasi-one-dimensional (1D) excitonic nature,[11–13] which are all in drastic contrast with the properties of other 2D materials, such as graphene[14] and transition metal dichalcogenide (TMD) semiconductors.[15–17] Monolayer phosphorene has been of particular interest in exploring technological applications and investigating fundamental phenomena, such as 2D quantum confinement and many-body interactions.[11,18] However, this unique 2D material is unstable in ambient conditions and degrade quickly.[9,19–24] Particularly, monolayer phosphorene is expected to be much less stable than few-layer phosphorene,[19] and hence its identification and characterization are extremely challenging. So far,

Phosphorene: Physical Properties, Synthesis, and Fabrication
Edited by Yongqing Cai, Gang Zhang, and Yong-Wei Zhang

ISBN 978-981-4774-64-2 (Hardcover), 978-0-203-71061-6 (eBook)
www.jennystanford.com

to identify the layer number of thin-layer phosphorene, atomic force microscopy (AFM),[2,5,6,9,13,19,22,24–27] phase-shifting interferometry (PSI),[7,9,10,12,21] Raman spectroscopy,[5,19–21,27–30] photoluminescence (PL) spectroscopy[9–11,21,31,32] and optical contrast method[6,27,29,31,33] have been used by researchers. And by combining results from above-mentioned techniques, the layer number of thin-layer phosphorene can be precisely determined. The other intriguing property of phosphorene is its layer-dependent direct bandgap in the infrared range[9–11,21,31,32] With this unique property, phosphorene can bridge the zero-bandgap graphene[14,34] and comparatively large-bandgap TMD semiconductors,[35–37] leading it to potential applications in infrared range, such as infrared photodetectors and absorber.[28,30,38–40] Besides, unlike TMD semiconductors, which show an indirect-to-indirect bandgap transition only when the layer number is thinned down to monolayer,[35–37] phosphorene is a direct-bandgap semiconductor from monolayer to its bulk form,[5,9,31] resulting in high internal quantum efficiency and making it a promising candidate for future optoelectronic applications. The anisotropic nature of phosphorene, another unique property of phosphorene originated from its puckering structure, has also been characterized by using Raman spectroscopy.[2,6,13,25,33,40–42] Furthermore, X-ray diffraction (XRD)[43] X-ray photoelectron spectroscopy (XPS),[23,28,43–45] scanning tunneling microscopy (STM),[46] and transmission electron microscopy (TEM)[20,23,25,26,42,45,47] are also applied to characterize the structure and component of fresh phosphorene and to detect its degradation in ambient atmosphere. In this chapter, we will focus on the layer number characterization of few-layer phosphorene with PSI method, and temperature/angle-dependent Raman spectroscopy and layer-dependent PL measurements of few-layer phosphorene.

2.2 Layer Number Characterization

Both AFM and Raman spectroscopy have been used to reliably determine the sample thickness of TMD semiconductors with monolayer precision.[48] However, these two methods are not reliable for the identification of very-few-layer phosphorene (one or two layers). The scanning rate of AFM is slow compared to the fast degradation of very-few-layer phosphorene in ambient conditions and AFM can easily generate an error of one or even two layers, owing to the large surface roughness in very-few-layer phosphorene samples due to the bubbles formed on the sample surface during the fast degradation. AFM can also introduce potential contaminants that might affect further characterizations on the same sample. Unlike in TMD semiconductors, where Raman mode frequency has a monotonic dependence on the layer number, phosphorene has a non-monotonic dependence owing to the complicated Davydov-related effects.[20] Moreover, the relatively high-power laser used in Raman spectroscopy can significantly damage the phosphorene samples.

To overcome the aforementioned challenges, phase-shifting interferometry, a rapid, non-invasive and highly accurate approach, has been proposed and implemented to determine the layer number. Specifically, the optical path length (OPL) of the light reflected from the phosphorene that was mechanically exfoliated onto a SiO_2/Si substrate (275 nm thermal oxide) is measured (Fig. 2.1). The OPL is determined from the relation

$$\mathrm{OPL_{BP}} = -(\lambda/2\pi) * (\varphi_{\mathrm{BP}} - \varphi_{\mathrm{SiO2}}), \qquad (2.1)$$

where λ is the wavelength of the light source and is equal to 535 nm, and φ_{BP} and φ_{SiO2} are the phase shifts of the light reflected from the phosphorene flake and the SiO_2/Si substrate (Fig. 2.1c inset), respectively. The direct relationship between the OPL and the layer number is firmly established by a first-principle calculation and experimental calibration, as shown in Fig. 2.1d. Even though the thickness of monolayer phosphorene is less than 1 nm, its OPL is significantly larger than 20 nm owing to the multiple interfacial light reflections. That is, the virtual thickness of a phosphorene flake is amplified by more than 20 times in the optical interferometry, making the flakes easily identifiable. In the experiment, PSI is used to measure the OPL by analyzing the digitized interference pattern. In contrast to the highly focused and relatively high-power laser used in Raman system, PSI uses almost non-focused and very low-density light from a light-emitting diode (LED) source to achieve fast

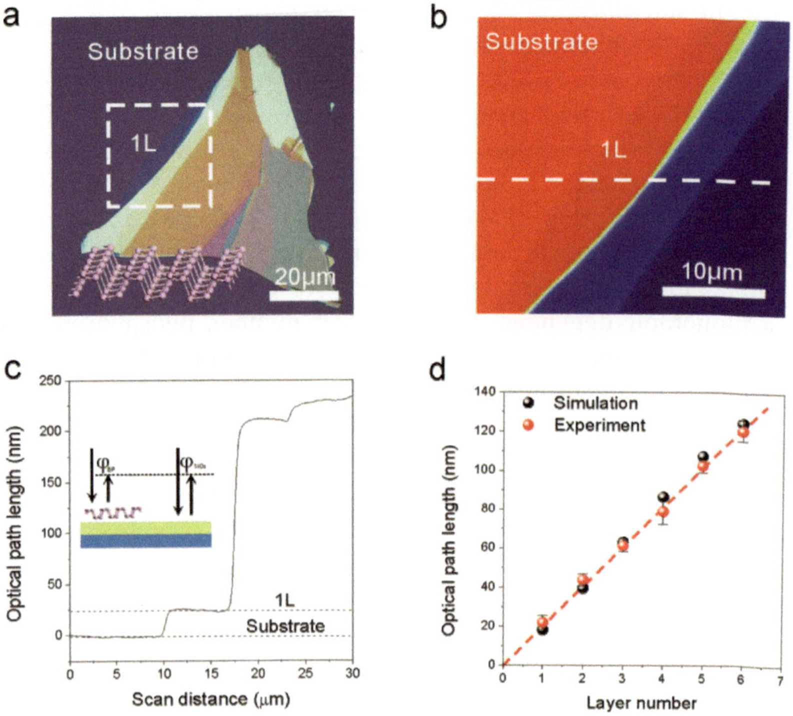

Figure 2.1 Monolayer phosphorene characteristics. (a) Optical microscope image of monolayer phosphorene (labeled "1L"). Inset is the schematic of single-layer phosphorene molecular structure. (b) PSI image of the dash line box area indicated in (a). (c) PSI measured OPL values along the dash line indicated in (b). Inset is the schematic plot showing the PSI measured phase shifts of the reflected light from the phosphorene flake (φ_{BP}) and the SiO_2/Si substrate (φ_{SiO2}). (d) OPL values from simulation and experiment PSI measurements for phosphorene samples from 1L to 6L. For each layer number of phosphorene, at least five different samples were characterized for the statistical measurements. The red dash line is the linear trend for statistical data measured with the PSI system. Reprinted with permission from Yang et al.[10] Copyright (2015) by CIOMP.

imaging, which inflicts no damage to the phosphorene samples. The step change of the OPL is ~20 nm for each additional phosphorene layer, as indicated by the red dots in Fig. 2.1d. Considering that the accuracy of the instrument is ~0.1 nm, a step change of 20 nm yields extremely robust identification of the layer number. Statistical OPL values for phosphorene from mono- to six-layer (1L to 6L) were collected and at least five different samples were measured with the PSI system for each layer number. The measured OPL values agree very well with theoretical calculations (Fig. 2.1d).

2.3 Raman Spectroscopy

Raman spectroscopy is a powerful tool to characterize the structural and electronic properties of few-layer phosphorene. In particular, temperature-dependent Raman spectroscopy can help to further understand the fine structure and properties of the material, such as atomic bonds, thermal expansion, and thermal conductivity.[9] And angle-dependent Raman measurements can quickly determine the crystalline orientation of a phosphorene flake due to its strong anisotropic nature.

2.3.1 Temperature-Dependent Raman Spectroscopy

Temperature-dependent Raman study for few-layer phosphorene is important to further understand the fine structure and properties of the material, such as atomic bonds, thermal expansion, and thermal conductivity. The temperature dependence of the Raman spectra, measured at temperature ranging from 20°C to −160°C in a 5L phosphorene, is shown in Fig. 2.2. The decreasing temperature leads to the blue shift of the Raman phonon modes, A_g^1, B_{2g}, and A_g^2. The measured temperature dependence of the Raman mode frequency shift in few-layer phosphorene can be characterized by the linear equation

$$\omega = \omega_0 + \chi T, \tag{2.2}$$

where ω_0 is the mode frequency at zero K and χ is the first-order temperature coefficient. The measured χ values for modes A_g^1, B_{2g}, and A_g^2 in a 5L phosphorene are -0.023 cm^{-1}/°C, -0.018 cm^{-1}/°C, and -0.023 cm^{-1}/°C, respectively (Fig. 2.2b). The change of the Raman shift with temperature is determined by the anharmonic terms in the lattice potential energy, which is related to the anharmonic potential constants, the phonon occupation number,

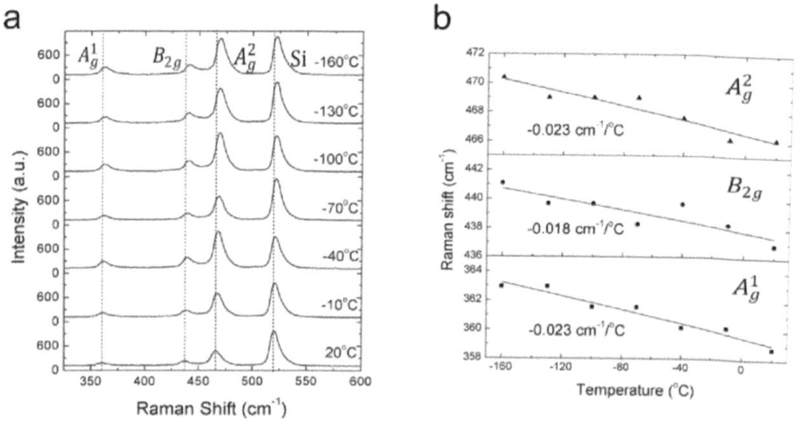

Figure 2.2 Temperature-dependent Raman spectra of 5L phosphorene. (a) Raman spectra of 5L phosphorene at temperature ranging from 20°C to −160°C. (b) Temperature dependence of A_g^1, B_{2g}, and A_g^2 Raman peak positions. Reprinted with permission from Zhang et al.[9] Copyright (2014) by American Chemical Society.

and the thermal expansion of the crystal.[49] These measured χ values in 5L phosphorene are all larger in absolute value than those from both bi-layer graphene (-0.0154 cm^{-1}/°C for G peak)[50] and few-layer MoS_2 (-0.0123 cm^{-1}/°C and -0.0132 cm^{-1}/°C for A_{1g} and E_{2g}^1 modes, respectively).[51] This indicates that the phonon frequencies in few-layer phosphorene are more sensitive to temperature modulation than those in graphene and MoS_2, which could be due to the superior mechanical flexibility of phosphorene originated from its unique puckered crystal structure.[52]

2.3.2 Angle-Dependent Raman Spectroscopy

Compared with other 2D materials, phosphorene shows strong anisotropic properties, allowing for various unique applications in opto-electronics. Normally, TEM is used to analyze the crystalline orientation of crystals. But the high-energy electrons in TEM is likely to introduce lots of defects in phosphorene, which can expedite the reaction of phosphorene with moisture and other possible reactants from the environment.[19] Thus, it is important to develop optical means to characterize the crystalline properties. It has been recently shown that second-harmonic generation is used to determine the crystalline orientation of MoS_2.[53] Both polarized reflection spectroscopy[1] and polarized IR spectroscopy[2] are capable of providing simple and nondestructive optical way to determine the crystalline orientation of phosphorene. Angle-dependent Raman spectroscopy is an even simpler optical means to determine the crystalline orientation of phosphorene. This process is critically important as it has minimal damage to phosphorene and significantly shortened the time of characterization, allowing for more accurate analysis of intrinsic phosphorene properties.

Linearly polarized Raman measurements were performed to quickly determine the crystalline orientation of a phosphorene flake (15L) (Fig. 2.3). The normally incident laser was in the z direction. The polarization angle θ is relative to the zero degree reference, which can be arbitrarily selected at the beginning. Raman intensities of B_{2g} and A_g^2 were observed to be significantly dependent on the polarization angle (Fig. 2.3a). The angle-dependent intensities of B_{2g} and A_g^2 modes both show an angle period of 180°and are out-of-phase (Fig. 2.3b). The Raman intensity of A_g^1 is less sensitive to the polarization angle (Fig. 2.3b). The measurement results can be perfectly explained by the vibration directions of these three Raman modes in crystalline black phosphorous. Figure 2.3c shows a schematic of the vibration direction of the phosphorus atoms in the different Raman modes.[54] In A_g^2, B_{2g}, and A_g^1 vibrational modes, the phosphorus atoms oscillate along the x (zigzag), y (armchair), and z (out-of-plane) directions, respectively. When the laser polarization is parallel to x direction (y direction), the intensity of the A_g^2 mode will reach the maximum (minimum) value. Meanwhile, the intensity of the B_{2g} mode will reach the minimum (maximum) value. On the basis of this polarization-dependent Raman measurement, the crystalline orientation of this 15L phosphorene flake was quickly determined, as indicated in Fig. 2.3d. This technique provides a fast and precise determination of the crystalline orientation, without the need of complicated and high-resolution imaging systems, like STM or TEM.

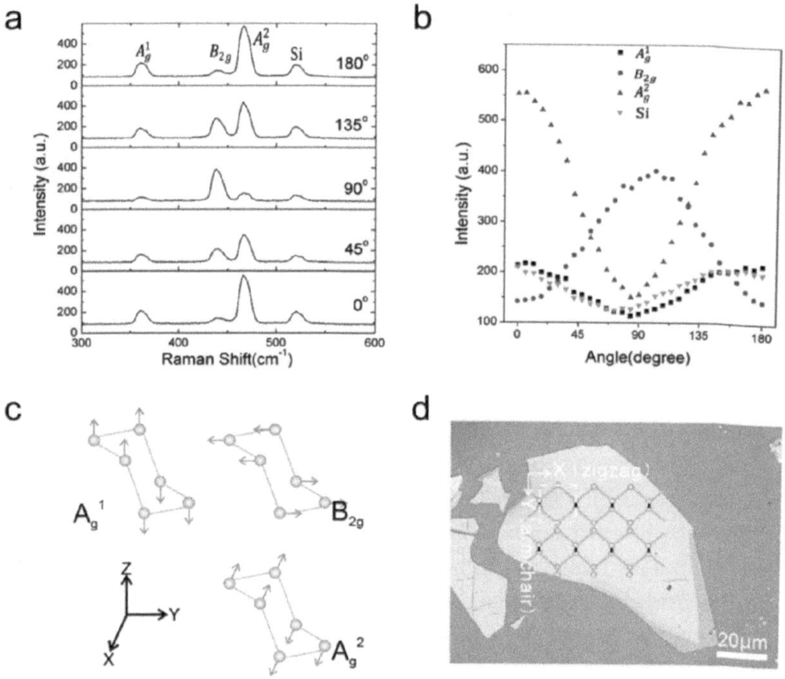

Figure 2.3 Phosphorene crystalline orientation determination by polarization Raman spectra. (a) Raman spectra of 15L phosphorene under different polarization angles. (b) Polarization dependence of A_g^1, A_g^2, B_{2g} modes in 15L phosphorene and the Raman peaks in silicon. (c) Schematic plot showing the vibration directions of A_g^1, A_g^2, B_{2g} Raman modes. (d) Crystalline orientation of the 15L phosphorene flake, determined by angle-dependent Raman measurement. Reprinted with permission from Zhang et al.[9] Copyright (2014) by American Chemical Society.

2.4 Photoluminescence Measurements

2.4.1 Layer-Dependent PL

Because of the strongly layer-dependent peak energies and the direct bandgap nature of phosphorene, layer number identification can be further confirmed by measuring their corresponding peak energies of the PL emission (Fig. 2.4). Figure 2.4a shows the normalized PL spectra of the mono- to five-layer phosphorene samples.

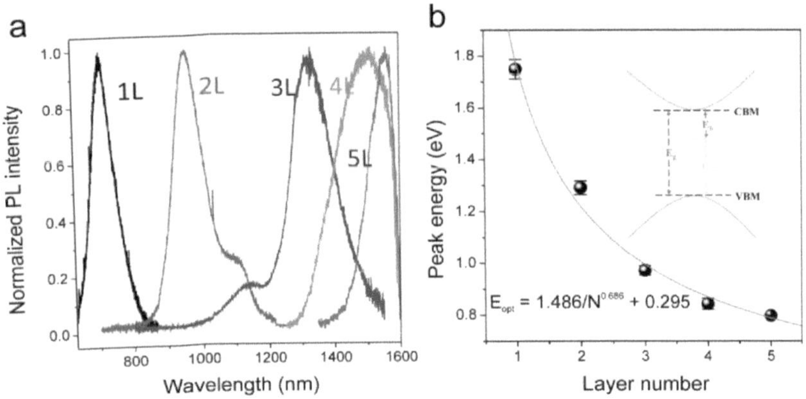

Figure 2.4 Photoluminescence (PL) characteristics of mono- and few-layer phosphorene. (a) PL spectra of the mono- to five-layer phosphorene samples. Each PL spectra is normalized to its peak intensity and system background. (b) Evolution of PL peak energy with layer number of phosphorene from experimental PL spectra, showing a rapid increase in peak energy as the layer number decreases. The solid grey line is the fitting curve $E_{\text{opt}} = 1.486/N^{0.686} + 0.295$, where E_{opt} is the optical gap in unit of eV and N is the layer number. Inset: Schematic energy diagram showing the electronic bandgap (E_g), and the exciton binding energy (E_b). Reprinted with permission from Yang et al.[10] Copyright (2015) by CIOMP.

The emission peak of the PL spectrum for monolayer phosphorene is at 711 nm, corresponding to a peak energy of 1.75 eV. This PL peak energy value was measured at $-10°$C, and it is expected not to vary too much at room temperature. Combining the results of previous work on few-layer phosphorene (2L to 5L)[9] with the results obtained from recent samples (1L to 5L),[10] it can be clearly observed that the peak energy of PL emission shows unambiguous layer dependence (Fig. 2.4b). For each layer number, at least three samples were characterized; the measured peak energies for 1L to 5L phosphorene are 1.75 ± 0.04, 1.29 ± 0.03, 0.97 ± 0.02, 0.84 ± 0.02, and 0.80 ± 0.02 eV, respectively. The PL emission energy of the phosphorene samples with higher layer number (>5) is beyond the measurement wavelength range (up to 1600 nm) of a normal PL system.

The peak energy of PL emission, also termed as optical gap (E_{opt}), is the difference between the electronic bandgap (E_a) and the exciton binding energy (E_{b}) (Fig. 2.4b inset). Owing to the strong quantum confinement effect, free-standing monolayer phosphorene is expected to have a large exciton binding energy of ~0.8 eV,[11,18] whereas this value is expected to be only ~0.3 eV for monolayer phosphorene on a SiO_2/Si substrate because of the increased screening from the substrate. By using the measured electronic energy gap of ~2.05 eV by Pan et al.[46] using scanning tunneling spectroscopy and the measured optical gap of 1.75 eV in monolayer phosphorene, the exciton binding energy of monolayer phosphorene on SiO_2/Si substrate is determined to be ~0.3 eV, which agrees very well with the prediction.[18] The optical gaps in phosphorene increase rapidly with decreasing layer number because of the strong quantum confinement effect and the van der Waals interactions between the neighboring sheets in few-layer phosphorene.[11,55] A power law form was used to fit the experimental data and obtained the fitting curve of

$$E_{\mathrm{opt}} = 1.486/\mathrm{N}^{0.686} + 0.295 \tag{2.3}$$

where E_{opt} is the optical gap in unit of eV and N is the layer number (Fig. 2.4b). The layer-dependent optical gaps, as indicated in Fig. 2.4b, agree very well with the theoretical predictions.[8,11] For bulk phosphorene sample with large N value, its optical gap approaches the limit value of ~0.295 eV, which matches very well with the measured energy gap (~0.3 eV) of bulk phosphorene.[1,56] Previously, Ye et al.[5] observed bright exciton PL emission at ~1.45 eV from a monolayer phosphorene that is coated with a protection layer of PMMA. This protection layer might introduce some defect states to the sample, which could change the PL emission energy.

2.4.2 Power-Dependent PL

The dynamics of excitons and trions have been of considerable interest for fundamental studies of many-body interactions, such as carrier multiplication and Wigner crystallization.[57] Monolayer phosphorene, whose excitons are predicted to be confined in a quasi-one-dimensional (1D) space, provides an ideal platform for investigating remarkable exciton and trion dynamics in a reduced dimension. Exciton and trion dynamics has been successfully probed in monolayer phosphorene by controlling the photocarrier injection in power-dependent PL measurements (Fig. 2.5). The measured PL spectra exhibit two clear peaks with central wavelengths at ~705 nm (labeled "A") and ~760 nm (labeled "X"), whose intensities and peak positions are highly dependent on the excitation power (Fig. 2.5a). After Lorentzian fitting of the measured PL spectra, the spectral components of these two peaks were extracted, indicated by red and blue curves in Fig. 2.5a. The intensity of higher-energy peak A increases almost linearly with laser power (Fig. 2.5b), while that of lower-energy peak X gradually saturates at a relatively high photocarrier injection. When the excitation power changed from 0.19 to 1.15 μW (Fig. 2.5c), the optical gap of A emission (E_A) monotonically decreased by ~15 meV, while that of X emission (E_X) increased by ~35 meV. The higher-energy peak A is attributed to exciton emission.

The origin of peak X is particularly interesting. The experimental observations do not support that peak X comes from the emission of localized excitons. First, the energy difference between the PL emissions from the localized and free excitons is not sensitive to the relatively low density of photocarrier injection.[58–60] But in the experiment, this energy difference decreases significantly from ~150 meV to ~100 meV when the power of excitation laser increases from 0.19 to 1.15 μW. Furthermore, localized excitons are expected to have a longer lifetime than free excitons because of the localization effect of squeezing an exciton in the zero-dimension-like state.[58,61] However, time-resolved PL measurements revealed that the carrier lifetime from X state is less than that from A state. On the other hand, the origin of X state can be attributed to trions. The trion peak intensity saturates at high photocarrier injection (Fig. 2.5b) because the free carriers coming from initial doping are almost depleted during the formation of trions. The trion density estimated from the optical injection is comparable to the estimated initial doping level of the monolayer phosphorene sample.

The difference between the peak energies of exciton and trion PL emission is predicted to be[16,62]

$$E_A - E_X = E_{Tb} + \Delta E \tag{2.4}$$

where E_{Tb} is the binding energy of trions and ΔE is the average energy needed to add one carrier into the free-carrier system. The energy difference $E_A - E_X$ represents the minimum energy for the removal of one electron (hole) from a negative (positive) trion. To convert a trion to an exciton, one of the two electrons (holes) in the negative (positive) trion will be unbound first ($+E_{Tb}$) and then added into the system ($+\Delta E$). During PL excitation, the same amount of electrons and holes will be injected into the conduction band and valence band, respectively. At low excitation power (power close to zero), the system is in an equilibrium state and ΔE is approximately equal to the Fermi energy E_F that is determined at the initial doping level. At higher PL excitation power with more carrier injection, the initial doping becomes less important because of the enhanced screening effect from the injected photocarriers.

In most conventional PL measurements with very high excitation power,[12,17,63] ΔE can be neglected and the trion binding energy can be approximately taken as the energy difference $E_A - E_X$. However, at a very low power range, the unique effect of ΔE cannot be observed. During the experiment, the start power excitation was set at 0.19 μW, which has an injected photocarrier concentration comparable to the initial doping level. With the increasing of the excitation power, the energy difference $E_A - E_X$ decreased, indicating a reduction of ΔE. At high excitation power (1.15 μW), the initial doping becomes less important and ΔE can be approximately neglected; the energy difference $E_A - E_X$ of ~100 meV at high excitation power (1.15 μW) can be taken to be the upper limit of the trion binding energy in monolayer phosphorene. This value agrees well with the estimated trion binding energy in monolayer phosphorene of $E_{Tb} \approx 0.3E_b$[55] and the calculated exciton binding energy in monolayer phosphorene on SiO_2/Si substrate is $E_b \approx 0.3$ eV.[11] In monolayer phosphorene, both positive and negative trions are expected to have similar binding energies, since the effective masses of electrons and holes are almost equal.[8,11]

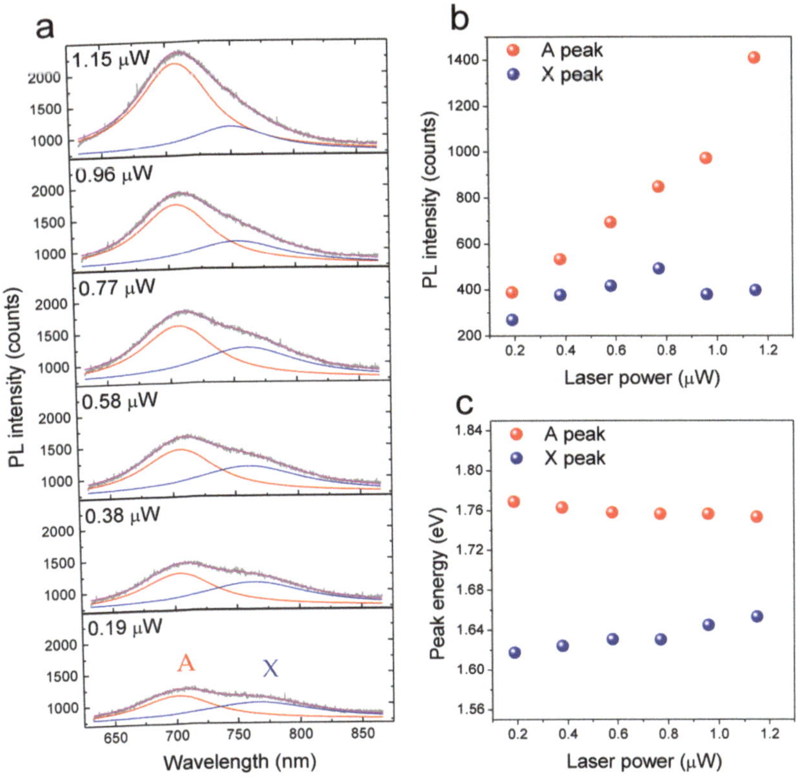

Figure 2.5 Exciton and trion dynamics in monolayer phosphorene. (a) Measured PL spectra (solid grey lines) under various excitation laser power. PL spectra are fit to Lorentzians (solid red lines are the exciton components, solid blue lines are the trion components, and solid pink lines are the cumulative fitting results). (b) PL intensity of exciton (A) and trion (X) as a function of laser power. (c) PL peak energy of exciton (A) and trion (X) as a function of laser power. Reprinted with permission from Yang et al.[10] Copyright (2015) by CIOMP.

2.5 Conclusion

In this chapter, we have discussed the characterization of few-layer phosphorene, including its layer number identification with PSI method, temperature- and angle-dependent Raman spectroscopy on few-layer phosphorene to detect their fine structure and crystalline orientation, and the layer- and power-dependent PL measurements to determine the band structure, and exciton and trion dynamics of thin-layer phosphorene. In particular, PSI method is a fast, non-invasive and accurate way to determine the layer number of few-layer phosphorene. It can overcome the challenges from AFM measurements and Raman spectroscopy and determine layer number of phosphorene quickly and precisely. By analyzing the different vibration modes in Raman spectra, the fine structure and other properties like crystalline orientation can be further understood. Temperature-dependent Raman spectroscopy on thin-layer phosphorene indicates that the phonon frequencies in few-layer phosphorene are more sensitive to temperature modulation than those in graphene and MoS_2, which could be due to the superior mechanical flexibility of phosphorene originated from its unique puckered crystal structure. Angle-dependent Raman measurements give another quick and non-destructive way to determine the crystalline orientation of thin-layer phosphorene. PL measurements can help understand the band structure of phosphorene, and exciton and trion dynamics can be detected by optical or electrical tuning in PL measurements. In this chapter, few-layer phosphorene was characterized by PL measurements and it layer-dependent PL peaks clearly indicate the layer-dependent bandgaps. Besides, exciton and trion dynamics was detected by means of PL measurements with optical tuning and the trion binding energy is measured to be $\sim$100 meV, larger than that of TMD materials, possibly due to the quasi-1D nature of exciton and trion in phosphorene.

References

1. Li L, Yu Y, Ye GJ, Ge Q, Ou X, et al. (2014). Black phosphorus field-effect transistors. *Nat. Nanotechnol.*, **9**:372–377.
2. Xia F, Wang H, Jia Y, (2014). Rediscovering black phosphorus as an anisotropic layered material for optoelectronics and electronics. *Nat. Commun.*, **5**:4458.
3. Fei R, Yang L, (2014). Strain-engineering the anisotropic electrical conductance of few-layer black phosphorus. *Nano Lett.*, **14**:2884–2889.

4. Hong T, Chamlagain B, Lin W, Chuang H-J, Pan M, et al. (2014). Polarized photocurrent response in black phosphorus field-effect transistors. *Nanoscale*, **6**:8978–8983.
5. Liu H, Neal AT, Zhu Z, Luo Z, Xu X, et al. (2014). Phosphorene: An unexplored 2D semiconductor with a high hole mobility. *ACS Nano*, **8**:4033–4041.
6. Mao N, Tang J, Xie L, Wu J, Han B, et al. (2016). Optical anisotropy of black phosphorus in the visible regime. *J. Am. Chem. Soc.*, **138**:300–305.
7. Xu R, Yang J, Zhu Y, Yan H, Pei J, et al. (2016). Layer-dependent surface potential of phosphorene and anisotropic/layer-dependent charge transfer in phosphorene-gold hybrid systems. *Nanoscale*, **8**:129–135.
8. Qiao J, Kong X, Hu Z-X, Yang F, Ji W, (2014). High-mobility transport anisotropy and linear dichroism in few-layer black phosphorus. *Nat. Commun.*, **5**:4475.
9. Zhang S, Yang J, Xu R, Wang F, Li W, et al. (2014). Extraordinary photoluminescence and strong temperature/angle-dependent Raman responses in few-layer phosphorene. *ACS Nano*, **8**:9590–9596.
10. Yang J, Xu R, Pei J, Myint YW, Wang F, et al. (2015). Optical tuning of exciton and trion emissions in monolayer phosphorene. *Light Sci. Appl.*, **4**:e312.
11. Tran V, Soklaski R, Liang Y, Yang L, (2014). Layer-controlled band gap and anisotropic excitons in few-layer black phosphorus. *Phys. Rev. B*, **89**:235319.
12. Xu R, Zhang S, Wang F, Yang J, Wang Z, et al. (2016). Extraordinarily bound quasi-one-dimensional trions in two-dimensional phosphorene atomic semiconductors. *ACS Nano*, **10**:2046–2053.
13. Wang X, Jones AM, Seyler KL, Tran V, Jia Y, et al. (2015). Highly anisotropic and robust excitons in monolayer black phosphorus. *Nat. Nanotechnol.*, **10**:517–521.
14. Geim AK, Novoselov KS, (2007). The rise of graphene. *Nat. Mater.*, **6**:183–191.
15. Radisavljevic B, Radenovic A, Brivio J, Giacometti V, Kis A, (2011). Single-layer MoS_2 transistors. *Nat. Nanotechnol.*, **6**:147–150.
16. Mak KF., He K, Lee C, Lee GH, Hone J, et al. (2013). Tightly bound trions in monolayer MoS_2. *Nat. Mater.*, **12**:207–211.

17. Ross JS, Wu S, Yu H, Ghimire NJ, Jones AM, et al. (2013). Electrical control of neutral and charged excitons in a monolayer semiconductor. *Nat. Commun.*, **4**:1474.
18. Rodin AS, Carvalho A, Castro Neto AH, (2014). Excitons in anisotropic two-dimensional semiconducting crystals. *Phys. Rev. B*, **90**:075429.
19. Andres C-G, Leonardo V, Elsa P, Joshua OI, Narasimha-Acharya KL, et al. (2014). Isolation and characterization of few-layer black phosphorus. *2D Mater.*, **1**:025001.
20. Favron A, Gaufres E, Fossard F, Phaneuf-Lheureux A-L, Tang NYW, et al. (2015). Photooxidation and quantum confinement effects in exfoliated black phosphorus. *Nat. Mater.*, **14**:826–832.
21. Pei J, Gai X, Yang J, Wang X, Yu Z, et al. (2016). Producing air-stable monolayers of phosphorene and their defect engineering. *Nat. Commun.*, **7**:10450.
22. Wood, JD, Wells SA, Jariwala D, Chen K-S, Cho E, et al. (2014). Effective passivation of exfoliated black phosphorus transistors against ambient degradation. *Nano Lett.*, **14**:6964–6970.
23. Kang J, Wood JD, Wells SA, Lee J-H, Liu X, et al. (2015). Solvent exfoliation of electronic-grade, two-dimensional black phosphorus. *ACS Nano*, **9**:3596–3604.
24. Doganov RA, O/'Farrell ECT, Koenig SP, Yeo Y, Ziletti A, et al. (2015). Transport properties of pristine few-layer black phosphorus by van der waals passivation in an inert atmosphere. *Nat. Commun.*, **6**:6647.
25. Lu W, Ma X, Fei Z, Zhou J, Zhang Z, et al. (2015). Probing the anisotropic behaviors of black phosphorus by transmission electron microscopy, angular-dependent Raman spectra, and electronic transport measurements. *Appl. Phys. Lett.*, **107**:021906.
26. Late DJ, (2015). Temperature dependent phonon shifts in few-layer black phosphorus. *ACS Appl. Mater. Inter.*, **7**:5857–5862.
27. Luo X, Lu X, Koon GKW, Castro Neto AH, Özyilmaz B, et al. (2015). Large frequency change with thickness in interlayer breathing mode—significant interlayer interactions in few layer black phosphorus. *Nano Lett.*, **15**:3931–3938.
28. Hanlon D, Backes C, Doherty E, Cucinotta CS, Berner NC, et al. (2015). Liquid exfoliation of solvent-stabilized few-layer black phosphorus for applications beyond electronics. *Nat. Commun.*, **6**:8563.
29. Lu W, Nan H, Hong J, Chen Y, Zhu C, et al. (2014). Plasma-assisted fabrication of monolayer phosphorene and its Raman characterization. *Nano Research*, **7**:853–859.

30. Guo Z, Zhang H, Lu S, Wang Z, Tang S, et al. (2015). From black phosphorus to phosphorene: Basic solvent exfoliation, evolution of raman scattering, and applications to ultrafast photonics. *Adv. Funct. Mater.*, **25**:6996–7002.

31. Li L, Kim J, Jin C, Ye G, Qiu DY, et al. (2016). Direct observation of layer-dependent electronic structure in phosphorene. *arXiv preprint arXiv:1601.03103.*

32. Cai Y, Zhang G, Zhang Y-W, (2014). Layer-dependent band alignment and work function of few-layer phosphorene. *Sci. Rep.*, **4**:6677.

33. Ling X, Liang L, Huang S, Puretzky AA, Geohegan DB, et al. (2015). Low-frequency interlayer breathing modes in few-layer black phosphorus. *Nano Lett.*, **15**:4080–4088.

34. Novoselov KS, Geim AK, Morozov SV, Jiang D, Katsnelson MI, et al. (2005). Two-dimensional gas of massless dirac fermions in graphene. *Nature*, **438**:197–200.

35. Mak KF, Lee C, Hone J, Shan J, Heinz TF, (2010). Atomically thin MoS_2: A new direct-gap semiconductor. *Phys. Rev. Lett.*, **105**:136805.

36. Zhang Y, Chang T-R, Zhou B, Cui Y-T, Yan H, et al. (2014). Direct observation of the transition from indirect to direct bandgap in atomically thin epitaxial $MoSe_2$. *Nat. Nanotechnol.*, **9**:111–115.

37. Zhao W, Ghorannevis Z, Chu L, Toh M, Kloc C, et al. (2013). Evolution of electronic structure in atomically thin sheets of WS_2 and WSe_2. *ACS Nano*, **7**:791–797.

38. Youngblood N, Chen C, Koester SJ, Li M, (2015). Waveguide-integrated black phosphorus photodetector with high responsivity and low dark current. *Nat. Photon.*, **9**:247–252.

39. Zhang B, Lou F, Zhao R, He J, Li J, et al. (2015). Exfoliated layers of black phosphorus as saturable absorber for ultrafast solid-state laser. *Opt. Lett.*, **40**:3691–3694.

40. Viti L, Hu J, Coquillat D, Knap W, Tredicucci A, et al. (2015). Black phosphorus terahertz photodetectors. *Adv. Mater.*, **27**:5567–5572.

41. Ribeiro HB, Pimenta MA, de Matos CJS, Moreira RL, Rodin AS, et al. (2015). Unusual angular dependence of the Raman response in black phosphorus. *ACS Nano*, **9**:4270–4276.

42. Yasaei P, Kumar B, Foroozan T, Wang C, Asadi M, et al. (2015). High-quality black phosphorus atomic layers by liquid-phase exfoliation. *Adv. Mater.*, **27**:1887–1892.

43. Uk Lee H, Lee SC, Won J, Son B-C, Choi S, et al. (2015). Stable semiconductor black phosphorus (BP)@titanium dioxide (TiO_2) hybrid photocatalysts. *Sci. Rep.*, **5**:8691.
44. Edmonds MT, Tadich A, Carvalho A, Ziletti A, O'Donnell KM, et al. (2015). Creating a stable oxide at the surface of black phosphorus. *ACS Appl. Mater. Inter.*, **7**:14557–14562.
45. Jia J, Jang SK, Lai S, Xu J, Choi YJ, et al. (2015). Plasma-treated thickness-controlled two-dimensional black phosphorus and its electronic transport properties. *ACS Nano*, **9**:8729–8736.
46. Liang L, Wang J, Lin W, Sumpter BG, Meunier V, et al. (2014). Electronic bandgap and edge reconstruction in phosphorene materials. *Nano Lett.*, **14**:6400–6406.
47. Wu RJ, Topsakal M, Low T, Robbins MC, Haratipour N, et al. (2015). Atomic and electronic structure of exfoliated black phosphorus. *J. Vac. Sci. Technol. A*, **33**:060604.
48. Li H, Zhang Q, Yap CCR, Tay BK, Edwin THT, et al. (2012). From bulk to monolayer MoS_2: Evolution of Raman scattering. *Adv. Funct. Mater.*, **22**:1385–1390.
49. Postmus C, Ferraro JR, Mitra SS, (1968). Pressure dependence of infrared eigenfrequencies of KCl and KBr. *Phys. Rev.*, **174**:983–987.
50. Calizo I, Balandin AA, Bao W, Miao F, Lau CN, (2007). Temperature dependence of the Raman spectra of graphene and graphene multilayers. *Nano Lett.*, **7**:2645–2649.
51. Sahoo S, Gaur APS, Ahmadi M, Guinel MJF, Katiyar RS, (2013). Temperature-dependent Raman studies and thermal conductivity of few-layer MoS_2. *J. Phy. Chem. C*, **117**:9042–9047.
52. Wei Q, Peng X, (2014). Superior mechanical flexibility of phosphorene and few-layer black phosphorus. *Appl. Phys. Lett.*, **104**:251915.
53. Yin X, Ye Z, Chenet DA, Ye Y, O'Brien K, et al. (2014). Edge nonlinear optics on a MoS_2 atomic monolayer. *Science*, **344**:488–490.
54. Sugai S, Shirotani I, (1985). Raman and infrared reflection spectroscopy in black phosphorus. *Solid State Commun.*, **53**:753-755.
55. Thilagam A, (1997). Two-dimensional charged-exciton complexes. *Phys. Rev. B*, **55**:7804–7808.
56. Keyes RW, (1953). The electrical properties of black phosphorus. *Phys. Rev.*, **92**:580–584.
57. Wigner E, (1934). On the interaction of electrons in metals. *Phys. Rev.*, **46**:1002–1011.

58. Miyauchi Y, Iwamura M, Mouri S, Kawazoe T, Ohtsu M, et al. (2013). Brightening of excitons in carbon nanotubes on dimensionality modification. *Nat. Photon.*, **7**:715–719.

59. Iwamura M, Akizuki N, Miyauchi Y, Mouri S, Shaver J, et al. (2014). Nonlinear photoluminescence spectroscopy of carbon nanotubes with localized exciton states. *ACS Nano*, **8**:11254–11260.

60. Xu R, Yang J, Myint YW, Pei J, Yan H, et al. (2016). Exciton brightening in monolayer phosphorene via dimensionality modification. *Adv. Mater.*, **28**:3493–3498.

61. Wang H, Zhang C, Chan W, Manolatou C, Tiwari S, et al. (2016). Radiative lifetimes of excitons and trions in monolayers of the metal dichalcogenide MoS_2. *Phys. Rev. B*, **93**:045407.

62. Huard V, Cox RT, Saminadayar K, Arnoult A, Tatarenko S, (2000). Bound states in optical absorption of semiconductor quantum wells containing a two-dimensional electron gas. *Phys. Rev. Lett.*, **84**:187–190.

63. Kheng K, Cox RT, d' Aubigné MY, Bassani F, Saminadayar K, et al. (1993). Observation of negatively charged excitons X^- in semiconductor quantum wells. *Phys. Rev. Lett.*, **71**:1752–1755.

Chapter 3

Structure, Phonons, and Defects

Yongqing Cai, Gang Zhang, and Yong-Wei Zhang
*Institute of High Performance Computing, The Agency for Science, Technology and Research (A*STAR), Singapore 138632, Singapore*
zhangg@ihpc.a-star.edu.sg, zhangyw@ihpc.a-star.edu.sg

3.1 Structure of Phosphorus Polymorphs

Since its first isolation by Hennig Brand in 1669, the elemental phosphorus is found to show more allotropic modifications than any other element.[1,2] At ambient condition, well-known stable phosphorus phases include white, red, and black allotropes (Table 3.1). At nonambient condition, more allotropes like crystalline violet, red, brown, and gray can be synthesized. In particular, a scarlet phase can be formed under x-ray radiation of white phosphorus. With a valence configuration of $3s^2 3p_x 3p_y 3p_z$, phosphorus allows twofold, threefold, and fourfold coordinated neighbors, which accounts for its large number of polyphosphides. In the following sections, the introduction of the common phases of phosphorus is given.

Phosphorene: Physical Properties, Synthesis, and Fabrication
Edited by Yongqing Cai, Gang Zhang, and Yong-Wei Zhang

ISBN 978-981-4774-64-2 (Hardcover), 978-0-203-71061-6 (eBook)
www.jennystanford.com

Table 3.1 Comparison of the properties of different phosphorus allotropes

	Number of atoms in unit cell	Bandgap (eV)	Density ($g\,cm^{-3}$)	Stability
Red phosphorus		2.5–3.0	2.37	very stable
White phosphorus	232	~3.9		very reactive
Black phosphorus	8	0.3–1.5	2.69	degradation in air
Hittorf's phosphorus	84	~1.5	2.37	stable

3.1.1 Black Phosphorus

Black phosphorus is the most thermodynamically stable and densest allotropes under ambient temperature and pressure. It can be synthesized by heating several other forms of phosphorus under high pressure above 10 kbar. Brown and Rundqvist first synthesized crystalline black phosphorus by a growth medium of liquid bismuth.[4]

Similar to graphite and hexagonal boron nitride, bulk black phosphorus has a space group of Cmca and a layered form with neighboring layers are connected with each other through van der Waal's interaction (Fig. 3.1a). Each phosphorene layer consists of a puckered honeycomb structure with six-membered rings in chair conformation, and pairs of rings are organized in a cis-decalin

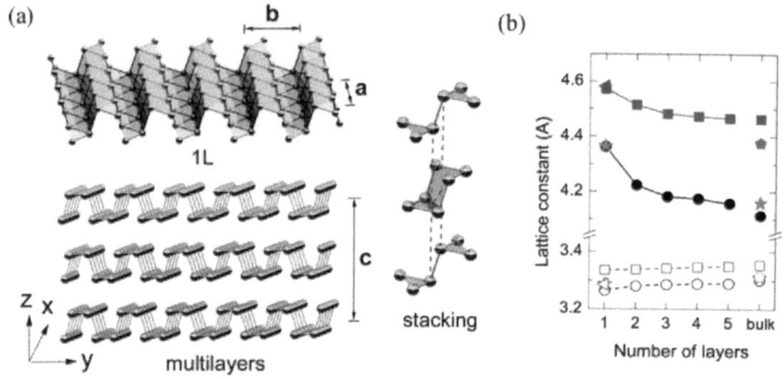

Figure 3.1 (a) Atomic structure of black phosphorus. (b) Thickness dependent lattice constant of few-layer phosphorene. Reprinted with permission from Cai et al.[3] Copyright (2015) John Wiley and Sons.

fashion. Thus armchair and zigzag edges, orthogonal to each other, can be identified in the phosphorene sheet. As shown in Fig. 3.1a, with a shift of half-lattice constant along the armchair direction, the stacking sequence of the layers in black phosphorus adopts AC stacking with the vertical phosphorus dimmer in one layer being located above the center of the puckered hexagon of the neighboring layer. This is different from the AB and AA′ stackings in graphite and h-BN, respectively, where at least one atom in a layer is directly below an atom of the neighboring layer.

All the P atoms in monolayer phosphorene are threefold coordinated through hybridization of 3p orbitals, leaving one lone electron pair, and the three bonds are 2.224, 2.224, and 2.244 Å, respectively (theoretical values). The fourth nearest P atoms are distributed across neighboring layers and bond length of 3.592 and 3.801 Å. First-principles calculations show that there is a strong thickness-dependent lattice constant, as shown in Fig. 3.1b, reflecting a strong interlayer coupling. The LDA calculations[3] show that the lattice constant along the zigzag direction keeps almost rigid, whereas the lattice constant along the armchair direction reduces dramatically from 4.36 Å of 1L to 4.16 Å of 5L. The angular dependent evolution of the lattice constant with thickness, which is different from graphene and MoS_2, indicates an anisotropic effect of the interlayer coupling.

Figure 3.2a–d shows the atomic scale STM image of black phosphorus at 4.3 K at sample bias $U = -0.315$ and -1.3 V.[5] There is a clear difference between the phosphorus atoms in the upper and lower ridges, reflected by the contrasts with different brightness and size labeled as P1 (brighter) and P2 (darker), respectively. More distinct contrast between the P1 and P2 atoms at -0.315 V than -1.3 V was observed. The line profiles show that the difference in the height of the two atoms can reach up to ~100 pm. Both these observations can be rooted in the different electronic states in the P1 and P2 atoms due to a puckered structure, which is absent in graphene.

The structural evolution of mechanically exfoliated phosphorene under vacuum condition with heating at 400°C for 5, 8, and 12 min is shown in Fig. 3.2e–g. According to the in situ scanning/transmission electron microscopy, the thermal degradation of phosphorene

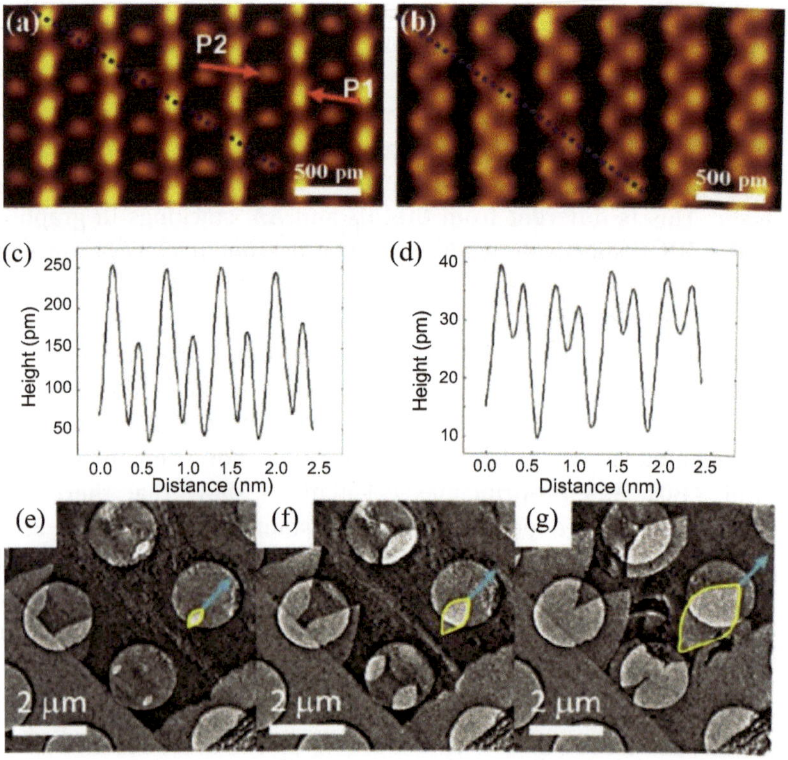

Figure 3.2 STM image of black phosphorus with sample bias at −0.315 V (a) and −1.3 V (b) taken at 4.3 K. (c, d) Line profiles along the dashed lines in (a) and (b). (e–g) Eye-shaped void (yellow) forms in phosphorene after heating for 5 min (e), 8 min (f), and 12 min (g) at 400°C in vacuum and propagates along the blue arrow direction. Reprinted with permission from Zhang et al.[5] and Liu et al.[6] Copyright (2015) American Chemical Society.

initiates at around 400°C, slightly smaller than the critical temperature of 550°C. This can be attributed to the nanoscale thickness and high surface area of exfoliated phosphorene layers. The atomic-scale microscopic study shows that the decomposition starts from forming eye-shaped cracks along the [001] direction, producing an amorphous red phosphorus–like skeleton in the end. Due to the presence of the skeleton structure without liquid-like characteristics, the mechanism of structural degradation is believed to not proceed via melting but sublimation, which predominantly occurs at edges and defective sites. Similar to the graphene case, during sublimation the cracks in phosphorene grow in size and coalesce frequently. However, all the cracks in phosphorene maintain a regular shape which is different from graphene.

Owing to their tunable bandgap and small effective mass, phosphorene and phosphorene-based materials are highly promising for transistor applications. Field-effect transistors (FETs) made from phosphorene show high mobility up to 1000 $cm^2V^{-1}s^{-1}$ at room temperature and a satisfying ON/OFF ratio ($10^3 - 10^5$). Its unique structure implies potential applications in flexible nanoelectronics. In addition, with an ambipolar conducting behavior and controllable bandgap and work function, phosphorene is promising in various photovoltaic applications. For instance, its p-type conduction allows it to act as a hole-transporting layer alternative to the conventional poly(3,4-ethylenedioxythiophene):polystyrene sulfonate (PEDOT:PSS) in organic photovoltaic devices. As its work function is properly aligned with that p-NiO, it can also work together with p-NiO to improve the carriers' transport efficiency in organic photovoltaic devices. Besides the above applications in electronic devices, phosphorene is also useful for applications in photocatalysis, chemical sensors, supercapacitors, and saturable absorbers. Its puckered structure of black phosphorus and absence of dangling bonds on surface allow the possibility of growing ordered molecular films or nanostructures above its surface. Its asymmetric transport of electrons and phonons makes it an especially promising candidate for thermoelectric applications. However, the bottleneck in phosphorene's application is related to its poor stability in ambient condictions. Various strategies have been developed to protect phosphorene and improve its stability. This issue will be thoroughly discussed in the following chapters.

3.1.2 White Phosphorus

Bulk white phosphorus, grown by the P_4 vapor condensation, has a cubic form containing 58 tetrahedral P_4 molecules in the unit cell. All the phosphorus atoms are covalently bonded within each P_4 molecule, while the P_4 units are held together bound by van der Waals interaction. Thus the electronic properties of white phosphorus can be similar to the electronic structures of a single P_4 molecule, with the gap of its HOMO–LUMO around 3.9 eV. White phosphorus is very reactive and can be strongly oxidized and burns in the presence of oxygen despite having a very high vapor pressure. This can be attributed to the very weak P–P bonds (200 kJ mol^{-1}) in the P_4 unit, the length of the P–P bond being 2.22 Å, and a low-barrier to oxidation. As the P–O bonds are strong, the oxidization of white phosphorus releases a lot of energy. In addition, thermal photodegradation and x-ray of P_4 to black or red phosphorus are thermodynamically favored. When heating up to 1070 K, P_4 dissociates to P_2 fragment, which is 49 kcal/mol higher in energy than neutral P_4 and not observed at ambient condition. The room temperature phase of white phosphorus is the α structure, which reversibly transforms into the β phase with a P–P bond length of 2.209 Å at 196 K. With a cream and waxy color, white phosphorus is extensively used for incendiary, signaling, and screening in munitions. It is also used in pesticides and fireworks and in the production of phosphoric acid, fertilizers, and cleaning compounds. Various chemical reactions can occur with intercalation of external chemical species into the white phosphorus. For instance, a host–guest $(P_4)_2C_{60}$ structure was obtained via integration with fullerenes.[7]

3.1.3 Red Phosphorus

Structural characterization of red phosphorus is difficult compared to black and white phosphorus due to its tendency to form polymeric amorphous structures. Red phosphorus can be synthesized via thermal, ultraviolet, or γ-ray degradation of white phosphorus. Several structures of red phosphorus were proposed, including amorphous structure and fibrous red phosphorus, which is a polymeric structure of linked pentagonal rings. Thurn and Krebs suggest that red phosphorus is composed of layered pentagonal tubes which are aligned parallel to each other.[8,9] As illustrated in Fig. 3.3, red phosphorus has a closed structural relationship with violet phosphorus. Boecker and Haeser theoretically suggested that there exist various polymeric structures of red phosphorus with a similar energetic stability.[10] Such predictions were partially

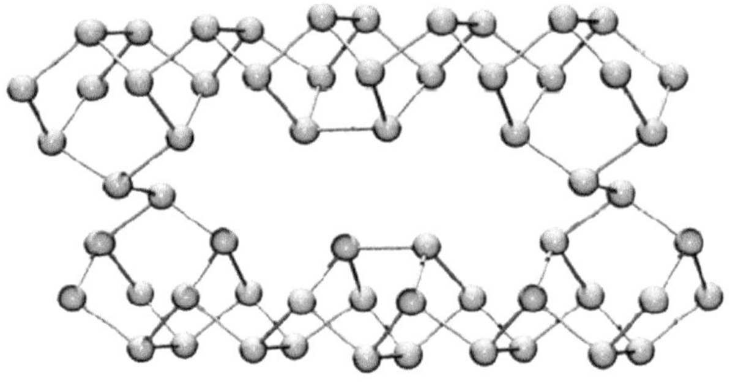

Figure 3.3 Atomic structure of fibrous red phosphorus.

supported experimentally with the commonly observed phosphorus subunit structures in CuI hosts. For instance, various CuI–phosphorus composites such as $(CuI)_3P_{12}$, $(CuI)_8P_{12}$, and $(CuI)_2P_{14}$ structures were synthesized, where phosphorus nanorods can be obtained through removing the CuI by a KCN solution.[11] Amorphous red phosphorus is particularly suitable for the anode materials of sodium batteries.[12]

3.1.4 Hittorf's Phosphorus

Hittorf's phosphorus (violet phosphorus) was first discovered by Hittorf in 1865.[13] It crystallizes from a phosphorus-lead metal. Also, Hittorf's violet phosphorus can be obtained by heating up the red phosphorus at 530°C. Bulk Hittorf's phosphorus has a monoclinic lattice with 84 atoms per unit cell.[9] It has a structure similar to fibrous red phosphorus consisting of layered pentagonal tubes arranged in perpendicular to each other (Fig. 3.4). There are four phosphorus atoms which connect the two neighboring perpendicular tubes. All the P atoms are threefold coordinated with the bond length around 2.219 Å and an average bond angle of 100.9°. Both red and Hittorf's phosphorus contain tubes with pentagonal cross sections and a sequence of three units along the tube direction: eight (P_8) and nine (P_9) phosphorus atoms and dumbbells of two phosphorus atoms (P_2). The tubes are similar to zigzag chains forming a puckered ladder. Concerning the growth mechanism, experiments show that in a lead melt the Hittorf's

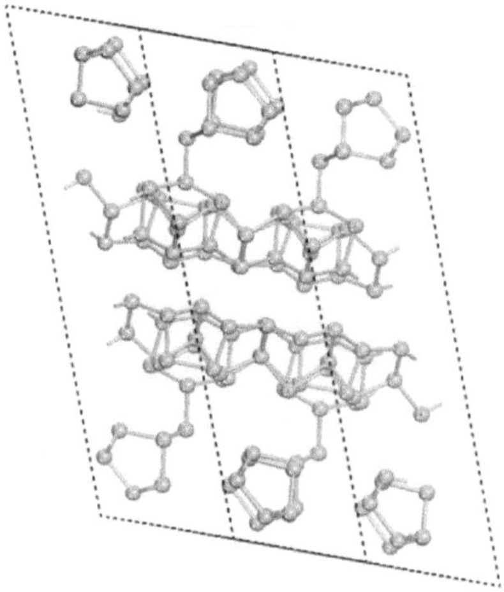

Figure 3.4 Atomic models of Hittorf's phosphorus.

phosphorus shows a homogeneous nucleation while the phosphorus fibers tend to show a heterogeneous nucleation.

It is hard to visibly distinguish Hittorf's phosphorus from fibrous red phosphorus, since both structures show red to deep red color and a platelike shape. However, by applying mechanical stress, the fibrous-type phosphorus splits into thin filaments whereas Hittorf's phosphorus splits into smaller platelets. Transmission electron microscope images show Hittorf's phosphorus with perpendicular strands and fibrous phosphorus with parallel strands.[8] The fast Fourier transform of the Hittorf's phosphorus phase shows the orthogonal rows of pseudo-Bragg reflections (plane hk0) and a nearly equal reciprocal lattice constant of $\mathbf{a}^* = \mathbf{b}^* = (920\ \text{pm})^{-1}$. In contrast, the relationship of the reciprocal lattice of fibrous phosphorus follows $|\mathbf{a}^* + 2\mathbf{b}^*| = (440\ \text{pm})^{-1}$ and $\mathbf{c}^* = (580\ \text{pm})^{-1}$.

3.2 0D and 1D Phosphorus

3.2.1 Phosphorus Cluster

The presence of rich structural allotropes for bulk phosphorene suggests that there are a great variety nanosized phosphorus structures. The exact structures of low-dimensional phosphorus are very difficult to characterize. For phosphorus clusters, although the structures of dimer and tetramer are well known, the exact structures of the larger-sized clusters are still not fully understood. This complexity also lies in the fact that different charge states of the clusters normally lead to different structures for a certain sized phosphorus cluster. P_n^+ clusters with n up to 24 were investigated mass spectrometrically.[14] The vapor phase of P, As, Sb, and Bi mainly contains tetramers and multiples of four atoms for the clusters (n up to 300), as revealed by the mass spectra and photoionization experiments.[15] Theoretical calculations showed that many of the larger phosphorus clusters are energetically stable with respect to P_4.[16] The most stable structures of P_5, P_6, and P_7 correspond to the "roof" or "butterfly" structure of P_4 with the addition of one, two, and three more phosphorus atoms.[17]

Early first-principles calculations mainly focused on even-numbered neutral clusters.[16–20] The presence of a large number of even-numbered phosphorus clusters suggests that the coordination number of phosphorus is generally 3. For anionic clusters, there exist phosphorus double bonds due to the presence of additional electrons. For cationic clusters, besides the threefold P atoms, there exists one fourfold phosphorus atom with a valence of 5. Calculations of cationic phosphorus clusters showed that there is higher stability for odd-numbered clusters.[21] These cationic phosphorus clusters are prone to the polymerized of cuneane P_8 to form a long tubular shape with one extra phosphorus atom at the end. Large or medium-sized charged phosphorus clusters may dissociate into small sized clusters. Through the collision-induced dissociation mass spectrometry,[18] it was found that the primary dissociating event occurs through the loss of P_4 unit for cationic clusters P_{2m+1} ($6 \leq m \leq 11$). For larger clusters, there coexist different mechanisms with P_{8k+1} ($k = 3$) via the loss of P_4 unit, P_{8k+1} ($k = 4,5$) via the loss of P_8 unit, and P_{8k+1} ($k = 6,7,8$) through the loss of P_{10} unit.

There exists another group of zero-dimensional phosphorus motifs that are the ring-shaped structures[19] up to hundreds of phosphorus atoms. These ring structures are developed based on chain structures which are grouped into two families of the chain structures with P_{18+10n} and P_{18+12n}. The latter which contains a set of 2^{n-1} isomers of almost identical energy is found to be more stable the former. Through the deletion of the end P_8 groups and ring closure, the ring P_{10n} structures with D_{nh} symmetry and P_{24n} with D_{nd} symmetry are formed. The ring structures with n between 12 and 20 are more stable than their corresponding chain structures. Moreover, the P_{18+12n} chain structures are more stable than the P_{10n} ring structures. Therefore, for large-sized phosphorus structures, the most stable structures correspond to P_{24n} with D_{nd} symmetry such as P_{120}, P_{144}, and P_{168} structures.

Experimentally, singly charged phosphorus cluster anions with n up to 9 were synthesized by a pulsed arc cluster ion source.[22] All the clusters are three-dimensional from $n = 4$. Structural transitions between anionic and neutral structures were observed. Due to the different electrons in the system, the anionic structures are prone to be more open than the neutral counterparts. For instance, neutral P_4 cluster has a high symmetrical tetrahedral structure. However, the P_4^- undergoes a structural distortion to have a lower symmetrical structure due to Jahn–Teller mechanism. On the contrary, the Jahn–Teller triggers a higher symmetrical D_{3h} (D_{5h}) structure for P_3^- (P_5^-) clusters than neutral P_3(P_5) clusters. It seems that the increased population of one electron in anionic structure can facilitate the stabilization of structures that are unstable in the neutral cases.

The neutral and cationic phosphorus clusters (up to P_{40}) have been synthesized in the gas phase by laser ablation of red phosphorus.[23] The neutral clusters were found to be rich of even-numbered clusters like $n = 8$, 10, 14, and 40, implying a threefold coordination number of each P atom. There is a local abundance minimum at $n = 22$. In contrast, the cationic clusters were highly populated with odd-numbered clusters. For large-sized clusters ($n > 21$), there exists a four-periodicity for magic-sized clusters with $n = 25$, 37, 41, 49, 53, 57, 65, 77, 81, and 89. It was predicted that dodecahedral fullerene-like P_{20} should be particularly stable. According to the mass spectrum, small cationic clusters ($n = 3 - 7$) are more populated than larger-sized ones. The P^+ and P_2^+ seem to be not stable as they were not detected. The observation of a nearly equal cluster velocities for a wide range of mass suggests that large-sized clusters are formed via gas-phase condensation in the laser plasma instead of direct ejection from the target. This efficient generation of large phosphorus clusters may be due to the presence of some basic blocks like P_7 and P_8, which are the main components of red phosphorus. It should be noted that these phosphorus clusters can be easily oxidized as oxidized phosphorus clusters were commonly recorded.

3.2.2 Phosphorus Fullerene

Since sixfold puckered rings and fivefold rings exist in the various phosphorus allotropes, constructing phosphorus-based fullerenes similar to the carbon fullerene structures in the form of molecular cages made up of hexagons and pentagons will be highly interesting. Stability of these phosphorous fullerene structures were simulated through Hartree–Fock (HF) calculation.[16] By screening various phosphorus clusters with size up to P_{28}, it was found that these structures are unstable. The smallest possible fullerene structure, dodecahedral P_{20} (3.2.3a) with an icosahedral I_h symmetry and 12 pentagons, automatically dissociates into P_4 molecules, which is the smallest polyhedral cage (tetrahedron) and a well-known structural motif for white phosphorous. By using density functional tight-binding (DFTB) calculations, Seifert et al.[24] found that phosphorus fullerenes are unstable, with the larger fullerenes more likely to decompose into P_4 molecules. DFT calculatioins resulted in similar findings.[25] The "minimal-pentagon-adjacency" rule for the stability of carbon fullerenes cannot be applied to phosphorous fullerene, since small phosphorus cages favor a pentagon fusion. In addition, the planar stabilized carbon rings of the carbon fullerenes may also be different in the phosphorus case, where pyramidal and puckered structures are more popular. Thus stable phosphorus fullerenes were ruled out.

However, Karttunen et al.[26] attempted to improve the stability for phosphorus fullerene by constructing three different types phosphorus fullerenes with tetrahedral, octahedral, and icosahedral structures (Fig. 3.5b–d). Starting from creating triangular faces cut from phosphorene, each polyhedral fullerene can be uniquely defined by a lattice index (h, k). The tetrahedral structures are made

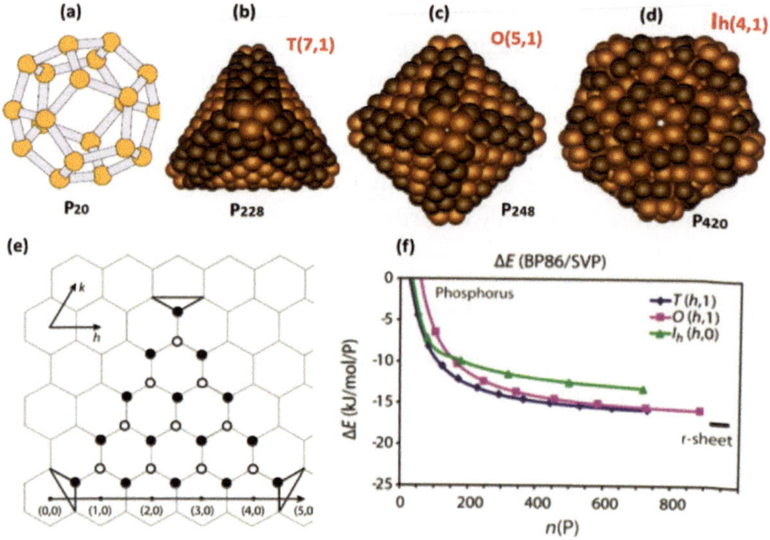

Figure 3.5 (a) Atomic structure of dodecahedral P_{20} fullerene-like cluster. (b–d) Representative polyhedral fullerenes of tetrahedral (b), octahedral (c), and icosahedral (d) family. (e) Lattice (h, k) index defines how the face of a polyhedron is cut from phosphorene. (f) Comparison of the total energies (relative to the tetrahedral P_4 cages) of the three families of phosphorus fullerenes: tetrahedral (blue diamonds), octahedral (violet squares), and icosahedral (green triangles). The value at *r* sheet denotes the relative energy of phosphorene monolayer. Reprinted with permission from Karttunen et al.[27] Copyright (2011) Springer.

of 4 triangular faces, and 4 vertices are formed from joining each of the 3 faces. The octahedral structures consist of 8 triangular faces, with 4 faces joining together at every vertex. The icosahedral fullerenes are composed of 20 triangular faces with 5 faces joining together at every vertex. These fullerenes consist of triangular faces with puckered hexagons and vertices with phosphorus triangles, squares, and hexagons. There are great flexibilities in these structures due to the puckered structure and fullerenes with larger sizes become increasingly stable, approaching phosphorene. Among the above three families of fullerene nanostructures, the tetrahedral structures are the most thermodynamically favorable, followed by octahedral andicosahedral structures (Fig. 3.5f). It is expected that by minimizing the strain arising, the vertices and edges adjoining the triangular faces could lead to potential phosphorous fullerenes. Via DFT and ab initio MP2 calculations, the structural stability of these structures, with respect to dissociation into P4 molecules, is found to increase as a function of their sizes.[26] These phosphorus fullerenes are found to be semiconductors with the HOMO–LUMO gaps between 1.6 and 2.9 eV for icosahedral structures,[27] comparable to the HOMO–LUMO gap (~1.7eV) in carbon fullerenes C_{60}.[28]

3.2.3 Phosphorus Nanotubes

Besides the phosphorus fullerene, phosphorus nanotubes with similar cylindrical structureas carbon nanotubes are proposed. Such 1D phosphorus structures can be of great interests and may be potentially synthesized as sucess was demonstrated in recent synthesis of phosphorus nanorods with diameters around 100 nm and lengths of several micrometers.[30] A phosphorene nanotube (PNT) can be intuitively regarded as single- or multi-wall, formed by rolling a phosphorene layer along specific or chiral angles. Therefore, a great variation of the types of PNT lies at the different rolling angle and diameter. Similar to the carbon nanotubes, a PNT can be uniquely defined form two indices (n_1, n_2) which correspond to the rollup vector $\mathbf{R} = n_1\mathbf{a}_1 + n_2\mathbf{a}_2$ in a 2D phosphorene layer (Fig. 3.6a).[29] Therefore, armchair (AC) and zigzag (ZZ) PNTs corresponds to $(0, n)$ and $(n, 0)$ types (Fig. 3.6b,c).

To examine the stability of the PNT, strain energies of the AC and ZZ PNTs with different diameters, defined as the energy difference between the PNT and the phosphorene layer, were calculated.[31] It was found that the ZZ PNTs have much larger strain energies than AC PNTs for the same diameter. Thus, smaller energy costs are needed for the formation of AC PNTs. Indeed, ZZ PNTs are unstable even at room temperature, while AC PNTs show a good structural integrity even at 600 K.[31] Simulations show a larger strain energy of PNTs than carbon nanotubes for the same diameter, which can be attributed to the promoted repulsion of the lone-pair electrons at the phosphorus atoms owing to the curvature effect.[29]

To reduce the strain energy, PNTs with faceted wall using a variety of stable planar phases of phosphorus allotropes were proposed.[32] Structures of AC and ZZ PNTs constructed from α-P,

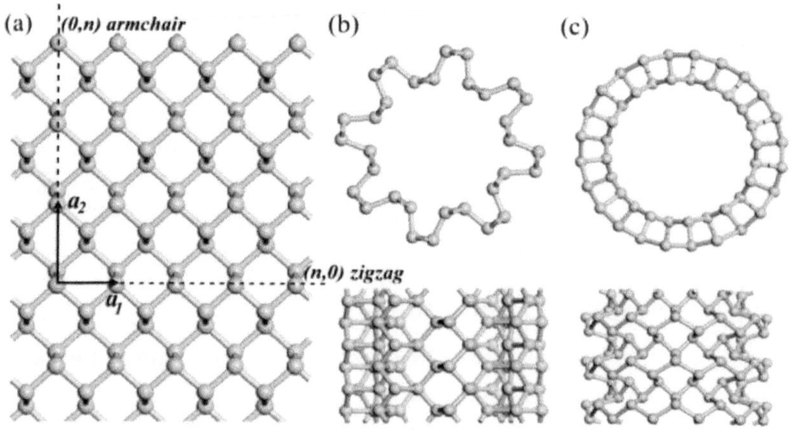

Figure 3.6 (a) Construction and classification of single-wall phosphorene nanotubes. Representative structures of (b) arm-chair PNT and (c) zigzag PNT. Reprinted with permission from Guo et al.[31] Copyright (2014) American Chemical Society.

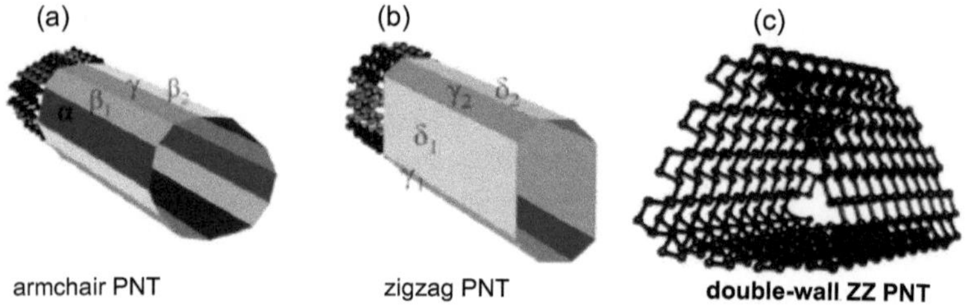

Figure 3.7 (a) Armchair and zigzag (b) faceted PNT built from different phosphorus allotropes distinguished by color and shading. (c) Double-wall zigzag PNT. Reprinted with permission from Guan.[32] Copyright (2014) American Physical Society.

β-P, γ-P, and δ-P allotropes of phosphorus are plotted in Fig. 3.7. As the lattice deformation at the faceted wall is minimized, the faceted PNTs are much more stable than the above-mentioned PNTs built from phosphorene. The faceted PNTs can be even stable up to 1000 K.

Concerning the electronic properties, both AC and ZZ PNTs are direct bandgap, which is strongly contrasted to carbon nanotubes where the nature of the bandgap strongly depends on the chirality of the tube. There is a direct relationship of the bandgap with the diameter of the PNTs due to the quantum confinement effect. For instance, the bandgap of AC (ZZ) PNTs increases from 0.28 to 0.61 eV (from 0.1 to 0.3 eV) for tubes with different diameters.[31] All these values are smaller compared to bandgaps of nanotubes of transition metal dichalcogenides (1.2–2.0 eV)[33] and boron nitride (~5 eV).[34]

The electronic structures of the PNTs can be tuned by applying external strain and electric fields.[31] For instance, a decrease (increase) in the bandgap of PNTs can occur when compressive (tensile) strain is applied along the tube direction. There is a direct-to-indirect bandgap transition for AC PNTs with tensile strain. Such consistent behavior in bandgap independent of the tube chirality is strongly different from carbon nanotubes which can increase or decrease with applied uniaxial strain depending on nanotube diameter and chirality. In addition, an order improvement of the holes mobility is found in AC PNTs by applying compressive strain.[35] The electronic structure of PNTs can also be modulated by applying electric field. The bandgap decreases with an increase in the magnitude of the electric field and there is a semiconductor–metal transition at a critical value of 0.5 eV/Å for AC PNTs (0, 8) and 0.1 eV/Å for ZZ PNTs (8, 0).[31]

3.3 Phononic Properties of Phosphorene

3.3.1 Single-Layer Phosphorene

Phosphorene has an orthorhombic layered structure coupled with van der Waals (vdW) interactions. Different from graphene, the sharing of neighboring puckered hexagons creates ridge structures. This feature induces new behaviors in phonons with respect to the interlayer coupling and strain modulation. For the monolayer phosphorene, each unit cell contains four atoms, and therefore there are 12 phonon branches in the phonon dispersion (Fig. 3.8): the in-plane transverse acoustic (TA) mode, the longitudinal acoustic (LA) mode and the out-of-plane acoustic (ZA) mode, and nine optical phonon branches. The irreducible representations of optical phonon modes at the zone center are six Raman-active modes (A_g^1, A_g^2, B_{1g},

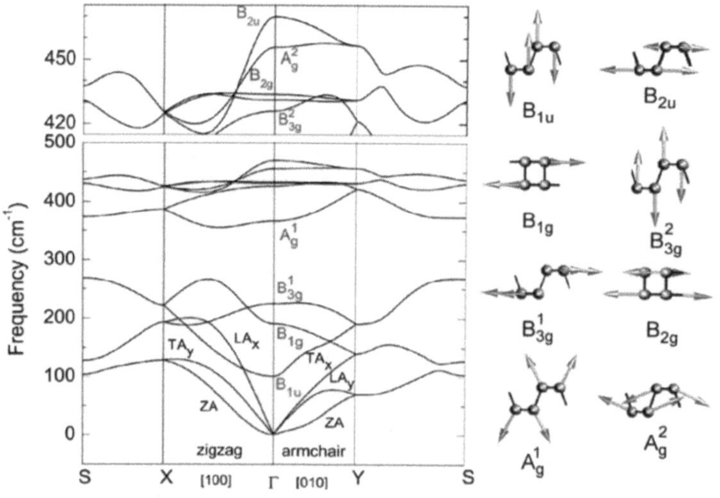

Figure 3.8 Phonon dispersion and IR and Raman-active modes of phosphorene. A suffix to the LA and TA acoustic branches denotes the polarization direction of the modes. Reprinted with permission from Cai et al.[3] Copyright (2015) John Wiley and Sons.

B_{2g}), the two IR-active modes, and one silent mode (A_u). In Table 3.2, the vibrational frequencies, and the degree of shift under strain of these optical modes at Γ are listed based on LDA and vdW included optB88 calculations.

In addition, the Grüneisen parameter, defined as $\gamma_{\mathbf{q}s} = -\frac{a}{2\omega_s(\mathbf{q})}\frac{d\omega_s(\mathbf{q})}{da}$, is also given for these modes. Compared with the experiment, the LDA calculations[3] predict very accurate frequencies for the Raman-active A_g^1 (366.1 cm^{-1}), B_{2g} (433.7 cm^{-1}), and A_g^2 (455.6 cm^{-1}) modes, in good agreement with the experimental values of 360.6, 438.9, and 469.0 cm^{-1}.[36,37] On the basis of the phonon dispersion, one can see a phosphorene as a 2D puckered honeycomb structure consisting of weakly coupled quasi-one-dimensional (1D) zigzag chains. For instance, the phonon curves for the LA, TA, and high-frequency optical modes are generally flatter and have a smaller group velocity in the armchair direction than along the zigzag direction. These less dispersive bands suggest that phosphorene consists of a weak coupling of atoms along the armchair direction. In contrast, the electronic carriers show a

Table 3.2 Frequency (ω), in cm^{-1}, Grüneisen parameter (γ), and the slope of strain–frequency relationship, $\Xi_\lambda = \Delta\omega/\varepsilon_\lambda$, in cm^{-1}, with $\lambda = x, y, z$, representing the direction of strains in monolayer phosphorene. Reprinted with permission from Cai et al.[3] Copyright (2015) John Wiley and Sons

	B_{1u}	B_{1g}	B^1_{3g}	A^1_g	B^2_{3g}	A_u	B_{2g}	A^2_g	B_{2u}
ω(LDA)	100.7	190.6	224.5	366.1	425.9	431.1	433.7	455.6	469.7
ω(optB88)	132.7	187.3	217.6	336.7	413.8	399.5	406.7	436.3	450.5
ω(Ref. 13)				360.6			438.9	469.0	
γ(LDA)	−8.63	−0.01	0.06	0.58	−0.54	1.07	1.01	0.32	0.92
Ξ_x	−566	61.7	−201	113	15.4	−641	−869	−535	−664
Ξ_y	899	−38.7	105	−234	151	58.7	169	205	70.9
Ξ_{xy}	1843	5.7	−27.6	−426	457	−927	−877	−296	−867

sharply opposite behavior. The effective mass along the armchair direction is much smaller than in the zigzag direction, resulting from the more dispersive electronic bands in the armchair direction.[38] Therefore, in phosphorene, there is a stronger intrachain interaction along the zigzag direction for phonons but a stronger interchain interaction for electrons/holes in the armchair direction. This dramatically opposite behavior of electrons and phonons is highly desired in thermoelectrics devices.[39]

To show the effect of anisotripic structure on the phononic properties, the anisotropy of the acoustic phonon and its strain modulation are discussed. Considering the dominant role of the in-plane thermal conduction and non-zero slope at the zone center, only the in-plane acoustic phonon modes are discussed. The 2D surfaces of these acoustic modes within the whole 2D Brillouin zone are plotted in Fig. 3.9a. Eigenvector analysis shows that for most of the phonons in the momentum space, a mixing character of transverse and longitudinal vibrations exists. Thus, quasi-LA (QLA) and quasi-TA (QTA) instead of LA and TA modes are used to respresent the different polarization of phonon modes. Pure LA and TA modes are only found in the zigzag and armchair directions. As shown in the phonon dispersion curves in Fig. 3.8, the transverse TA_x branch has a higher energy than the longitudinal LA_y mode, which is contrary to the common notion that a transverse sound wave has a smaller group velocity. The contour plot in Fig. 3.9a reveals a strong anisotropy of the phonon dispersion with strong orientational dependence.

The angular depedent group velocity (V) for the long-wavelength QTA and QLA modes is shown in Fig. 3.9b. The V of phonons propagating along x direction with polarization of y is represented by V_{xy}. There is a strong anisotropy of the V for acoustic waves. For instance, the V_{xx} for the QLA branch is 8762 m/s, which is twice the V_{yy} value of 4211 m/s. The stress-free boundary condition requires $V_{xy} = V_{yx}$, around 4700 m/s. There is generally a higher sound velocity along the Γ–X (zigzag) direction than along the Γ–Y (armchair) direction, in good agreement with the more dispersive acoustic curves shown in Fig. 3.9b.

The angular profile of the V evolves greatly with the type and degree of strain applied in phosphorene (Fig. 3.9b). For the QLA

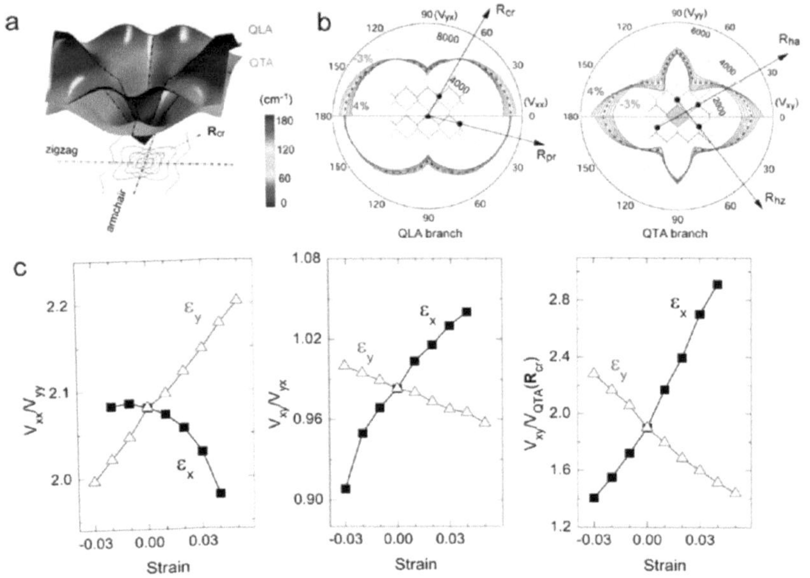

Figure 3.9 Angular dependence of the acoustic phonons in phosphorene. (a) 2D phonon dispersion surfaces for the in-plane acoustic branches. (b) Orientation dependence of the group velocity V (m/s) at the zone center a function of ε_x (upper part) and ε_y (lower part) strains from −3% to 4% for the QLA (left) and QTA branches (right). (c) Anisotropy of group velocity as a function of strain. Reprinted with permission from Cai et al.[3] Copyright (2015) John Wiley and Sons.

mode, a greate change of V occurs for the pure longitudinal LA_x (LA_y) modes, with V_{xx} and V_{yy} increasing with compressive uniaxial strain along ridge direction (ε_x) and along accordion direction (ε_y), respectively. In the case of the QTA mode, a large change of the V occurs in the zigzag direction for both uniaxial strains. An anomaly is found for V_{xy} showing an unexpected increase with ε_x. Applying strain can modify the phononic anisotropy in phosphorene (Fig. 3.9c). For instance, for the LA mode, the tensile ε_x reduces the V_{xx}/V_{yy} ratio, while the tensile ε_y can further enhance it. In contrast, an opposite behavior is found for the V_{xy}/V_{yx} ratio for applying the ε_x and ε_y. Interestingly, there are several special directions, shown by the arrows in Fig. 3.9b, where V is almost unchanged with strain, implying significantly smaller phonon deformation potential and significantly less anharmonicity.

3.3.2 Multilayer Phosphorene

Similar to the MoS_2, some of the Raman-active phonons show strong thickness dependence. The A_g^2 mode stiffens when the number of layers decreases: it shifts to the 469.5 cm^{-1} for a few layer phosphorene from 467.9 cm^{-1} of bulk black phosphorus. In contrast, A_g^1 (363.3 cm^{-1}) and B_{2g} (440.4 cm^{-1}) are less dependent on the number of layers. In addition, resulting from the double resonance scattering, A_g^2 and A_g^1 shift toward each other with increasing thickness, which can be used as the fingerprint for differentiating few- and single-layer phosphorene.

For layered materials like graphene and MoS_2, there exist intriguing rigid interlayer shearing and breathing modes.[40,41] These modes generally originate from the interaction between the vdW's coupled layers. Measuring their frequency can help quantify the number of layers. In addition, these weakly coupled phonons generally show small frequencies, which can be excited at very low temperatures. They normally show strong phonon anharmonicy and can significantly scatter the transporting phonons and electrons, thus greatly affecting the thermal and electronic conductivity. Similarly, for phosphorene, there also exist rigid-layered phonons. Due to the anisotropic structure and orthorhombic lattice, different from the hexagonal graphene and MoS_2, there are new features in these interlayer phonons induced by the lifting of degeneracy.[3]

Multilayer phosphorene structures can be represented by NL, where N is the number of the phosphorene sheets. The vibrational patterns and the frequencies of the rigid layer-shearing and breathing modes for the 1L–4L phosphorene examined by the LDA method are shown in Fig. 3.10. For an NL phosphorene, there are $N - 1$ layer-breathing modes, and $2^*(N - 1)$ in-plane shearing modes. While these shearing modes are doubly degenerated in hexagonal 2D materials, their degeneracy is lifted in low-symmetry phosphorene. As shown in Fig. 3.10, the vibrating displacement patterns and respective frequencies of equivalent modes are different for shearing along zigzag and armchair directions. The

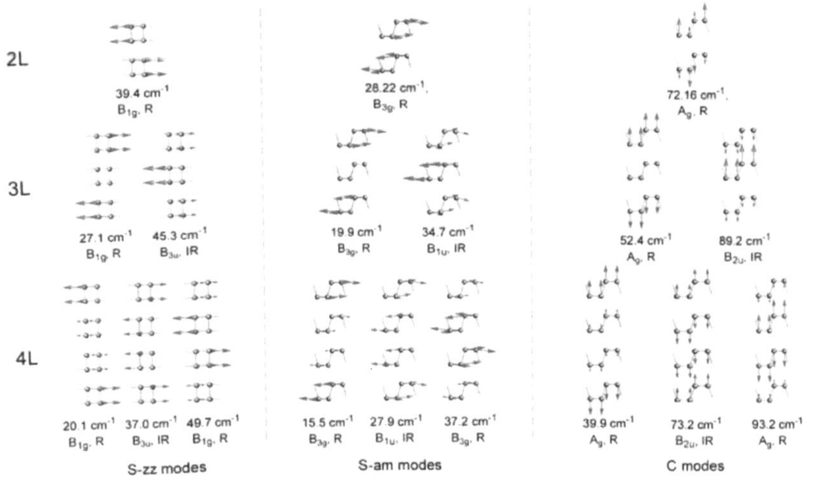

Figure 3.10 Interlayer shear and compressive modes for few-layer phosphorene. Reprinted with permission from Cai et al.[3] Copyright (2015) John Wiley and Sons.

hardest mode shows a relative shift of the innermost layers with out-of-phase vibration, while the outer layers are fixed. All of these ridid-layer modes can be Raman- or IR-active, suggesting that the number of phosphorene layers can be determined by measuring the frequencies of these modes.

A chain model can be used to quantitatively describe the relationship between the number of layers and frequencies. In this model, each phosphorene sheet is treated as a single pseudoatom with three freedom. Different from graphene and MoS_2, which are isotropic, phosphorene has a polarized-dependent character, and a different chain model is needed for three vibrating directions. The hardest shearing mode and softest compressing mode for NL phosphorene can be given by

$$\omega_{\lambda z}(N) = \frac{1}{\sqrt{2}\pi c}\sqrt{\frac{\alpha_{\lambda z}}{\mu}}\sqrt{1 \pm \cos\left(\frac{\pi}{N}\right)} \quad (3.1)$$

where the plus (minus) sign corresponds to the shear (compression) mode, $\mu = 1.46 \times 10^{-6}$ kg/m^2 is the mass density for the 1L phosphorene, c is speed of light in cm/s, λ is the polarization of the modes with $x(y)$ for shear mode along zigzag (armchair) direction and z for a compression mode, and α is the effective strength of the interlayer coupling for fitting.

The fitted value of α_{xz}, α_{yz}, and α_{zz} are 3.7×10^{19}, 2.1×10^{19}, and 1.4×10^{20} N/m^3, respectively. The effective shear strength α shows a strong angular anisotropy, about 1.8 times larger along the zigzag direction than that along the armchair direction. Compared with graphene, the shearing strengths α_{xz} and α_{yz} of phosphorene are 2.9 and 1.6 times larger, implying that few-layer phosphorene tends to be harder to shear in plane than graphene. The shear strengh of 2.7×10^{19} N/m^3,[41] of MoS_2 is found to be between the values along zigzag and armchair directions of phosphorene. However, the compressive strength of MoS_2 (4.2×10^{20} N/m^3,[41] is three times larger than phosphorene.

Based on the above value of interlayer coupling, the bulk modulus can be estimated as $C_{\lambda z} = \alpha_{\lambda z}t$, where $t = 0.51$ nm is the equilibrium distance between neighboring layers in bulk black phosphorus. The obtained value of C_{xz}, C_{yz}, and C_{zz} are 18.7, 10.6, and 70.8 GPa, respectively, consistent with experimental values of the shear moduli C_{44} (21.3 GPa), C_{55} (11.5 GPa), and stretching modulus of C_{33} (70.0 GPa) through ultrasonic[42,43] and neutron-scattering[44] measurements.

Some interesting information reflected in the anisotropic interlayer coupling is useful for phononic scattering and thermal conduction in phosphorene. It is now known that the low-frequency rigid-layer modes are highly anisotropic with a strong umklapp scattering and electron–phonon interaction.[41,40] Their low-frequency character suggests that they are fully activated at room temperature, thereby effectively affecting the electron and thermal transport. With a puckered honeycomb structure, the phononic properties of phosphorene are dramatically different from those of graphene. Under weak interalyer coupling approximation, the scattering rate (τ) for phonon modes propagating along direction can be derived as $\tau_{\lambda}^{-1} \propto D_{\lambda}(\omega)K_{\lambda}^{2}/\omega^{2}$, where $D_{\lambda}(\omega)$ is the density of states, and $K_{\lambda} = \alpha_{\lambda z}^{*}S$ is the effective interatomic force constant, S being the area of the in-plane unit cell of monolayer phosphorene. As the interlayer coupling is strongly anisotropic in the basal plane and much larger than that of graphene, it is expected that phosphorene

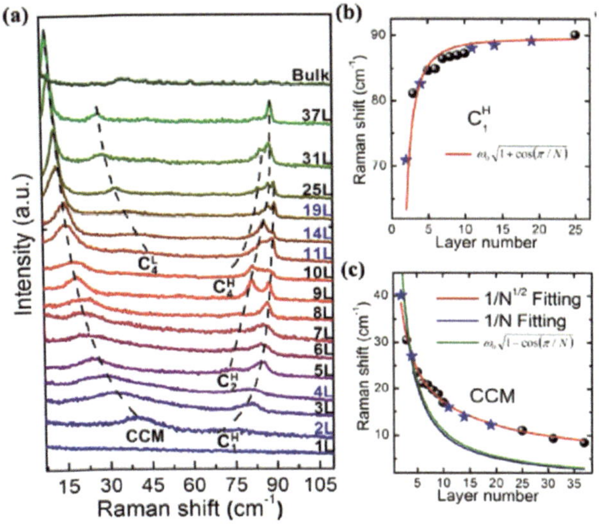

Figure 3.11 Low-frequency rigid-layer mode measured from few-layer to bulk black phosphorus samples. CCM and C_n^H (C_n^L) represent the collective compression mode, the nth-order compression modes from the higher (lower) branch with low (ultralow) frequencies, respectively. (b) The frequency of C_1^H mode extracted from the Raman spectra in (a) as a function of layer number. The red solid curve corresponds to the linear-chain fitting. (c) The frequency of CCM mode as a function of layer number. The green, red, and blue solid curves correspond to the fittings by the standard linear-chain model, $N^{-1/2}$ and N^{-1}. Reprinted with permission from Dong et al.[45] Copyright (2016) American Physical Society.

has a stronger interlayer interaction and a larger thermal leakage along the normal-surface direction, requiring a different strategy for thermal management compared with graphene.

Several experimental attempts have been made to measure these low-frequency rigid layer modes.[45,46] Raman measurements[45] revealed a strong interlayer coupling in black phosphorus through recording two distinct branches of collective compression mode (CCM) in few-layer samples: the low-frequency (LF) CCM around 100 cm^{-1} and the ultralow-frequency (ULF) CCM around 10 cm^{-1} (Fig. 3.11). Both of the two types of CCM show a clear layer-number frequency shift, consistent with the above-mentioned theoretical results.[3] The ULF and LF branches shift in opposite trends as the number of layers increases—the ULF mode moving to lower frequencies and the ULF mode shifting to higher frequencies. In addition, the width of these modes becomes more broadened as the layer number decreases, indicating the occurrence of a strong phononic anharmonicity in the thin samples.

Compared to bulk Raman-active phonon modes like the ($\sim$470 cm^{-1}) A_g^2 mode with a maximum frequency shift of $\sim$3 cm^{-1} between monolayer and bulk black phosphorus, the maximum frequency shifts for the ULF and LF modes are up to 32 and 19 cm^{-1}, respectively. Notably, the frequency shift of bulk A_g^2 mode disappears for thickness exceeding five layers, while the frequency shifts of LF and ULF still occur clearly for samples with more than ten layers. Similar findings were also reported by Ling et al.[46] This suggests that by measuring the ULF and LF Raman-active modes, it is possible to determine the thickness of phosphorene over a wide range up to tens of atomic layers, much higher than the previously used high-frequency Raman modes.

Quantitative analysis of thickness-dependent LF and ULF modes reveals that they follow a qualitatively different scaling behavior with layer numbers. For LF modes, it follows the standard linear-chain model with $\omega = \omega_0(1 + \cos(\pi/N))^{-1/2}$, where ω is the frequency. ω_0 is a fitted parameter that can be expressed as $\omega_0 = (1/(2^{-1/2})\pi c)(\alpha/\mu)^{-1/2}$, c is the speed of light, α is the interlayer force constant, and μ is the mass per unit cell area. However, for ULF modes, they do not follow the linear-chain model with $\omega = \omega_0(1 - \cos(\pi/N))^{-1/2}$. Instead, they scale as $N^{-1/2}$. Such unusual behavior can be attributed to the strong interlayer coupling to make all layers couple together, leading to an in-phase compression motion of all the phosphorene layers relative to the substrate.

3.3.3 Raman Measurement of Phonons

3.3.3.1 Principle of Raman scattering in phosphorene

Raman spectrostropy is a very powerful and nondestructive method to characterize the properties of layered materials such as the number of layers, the doping level, the type of defects, and

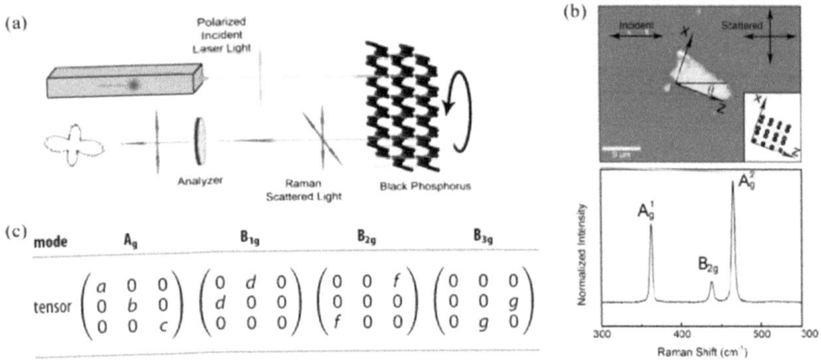

Figure 3.12 (a) Experimental setup of the Raman spectroscopy for the measurement of phosphorene. (b) The optical microscope image (upper panel) of the measured flake and crystallographic axes of phosphorene (inset in upper panel). The directions of the incident and scattered lights form an angle θ with the z axis (zigzag direction). Measured Raman spectrum for the phosphorene showing the A_g^1, A_g^2, and B_{2g} modes. (c) Raman tensors of the Raman-active modes in phosphorene. Reprinted with permission from Ribeiro et al.[47] Copyright (2015) American Chemical Society.

inherent strains within the samples. Figure 3.12a shows the measurement of the Raman spectrum in a backscattering geometry in phosphorene.[47,48] Following the definition of the crystallographic coordination with the y axis of the normal direction of the sheet, the x axis along the armchair direction, and the z axis along the zigzag direction, the incident light is polarized along the horizontal direction. The Cartesian coordinates coincide with the crystallographic axis. The analyzer is located before the entrance of the spectrometer for analyzing the scattered light with parallel and perpendicular polarizations relative to the incident light polarization. This setting allows the measurement of the orientational behavior of the Raman response by rotating the sample within the xz plane.

According to group theory, the representation of the Raman-active phonons modes are $2A_g+B_{1g}+B_{2g}+2B_{3g}$. The symmetry adapted Raman tensors for these phonon modes are shown in Fig. 3.12c. As only the A_g and B_{2g} modes contain the xz non-null components, the experimental setting shown in Fig. 3.12b, with the laser beam propagating along the y direction, only measures the A_g and B_{2g} modes. The typical Raman spectrum of bulk black phosphorus is shown in Fig. 3.12b. The measured values of the A_g^1, A_g^2 and B_{2g} modes are 360, 436, and 463 cm^{-1}, respectively.

Figure 3.13 shows the angular dependence of the Raman intensities of each mode in both parallel and cross-polarization configurations with obtained by 633 nm laser excitations. To understand the angular dependent behavior, it is useful to derive the Raman cross section S_k as:

$$S_k \propto |\mathbf{e}_i * R_k * \mathbf{e}_s| \quad (3.2)$$

where R_k is the Raman tensor (see Fig. 3.12c), $\mathbf{e}_i = (\sin\theta, 0, \cos\theta)$. $\mathbf{e}_s = (\sin\theta, 0, \cos\theta)$ and $(\cos\theta, 0, -\sin\theta)$ for the parallel and cross-polarization configurations, respectively. Based on the Raman tensor shown in Fig. 3.12c, the intensities for the A_g and B_{2g} modes in parallel and cross-polarization configurations are given by

$$S_{A_g}^{||} = (a\sin^2\theta + c\cos^2\theta)^2 \quad (3.3)$$

$$S_{A_g}^{\perp} = [(a-c)\cos\theta\sin\theta]^2 \quad (3.4)$$

$$S_{B_{2g}}^{||} = (2f\cos\theta\sin\theta)^2 \quad (3.5)$$

$$S_{B_{2g}}^{\perp} = [f\cos(2\theta)]^2 \quad (3.6)$$

The above expressions allow the fitting of the measured intensities which are shown in Fig. 3.13a, where the experimental and fitted data are represented by dotted and solid lines, respectively. It is noted that while the angular dependent behavior is well fitted for B_{2g} mode in both polarization configurations, there are large deviations between the fitted models and the experimental data for the two A_g modes. Especially for the A_g^2 mode in the parallel polarization configuration, there exhibits a secondary maximum at the 90° and 270°, which cannot be reproduced in the fitted curves.

The reason for the poor agreement is due to the assumption of a real component in the Raman tensor for obtaining the fitting curves in Fig. 3.13a. By assuming a complex value of the a, c, and f components in the Raman tensor for the A_g and B_{2g} modes in Fig. 3.12c with $a = |a|e^{i\phi_a}$, $b = |b|\, e^{i\phi_b}$, $c = |c|e^{i\phi_c}$, new expressions

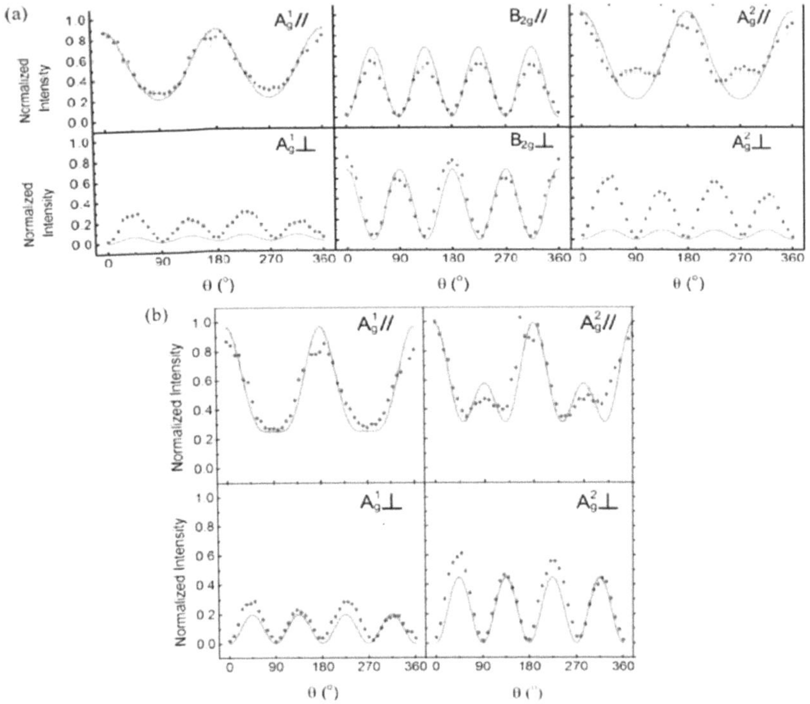

Figure 3.13 (a) Raman intensities of phonon modes measured in parallel and cross configurations using the 633 nm laser line. The dots and solid curves correspond to the experimental and fitted data by assuming real components of the Raman tensors of the modes, respectively. (b) Improved agreement with the measured value and fitted data by using Eqs. 3.7 and 3.8 by considering complex component of the Raman tensors of the A_g^1 and A_g^2 modes. Reprinted with permission from Ribeiro et al.[47] Copyright (2015) American Chemical Society.

of the intensities of each mode are given by

$$S_{A_g}^{\parallel} = (|a|\sin^2\theta + |c|\cos\phi_{ca}\cos^2\theta)^2 + |c|^2\sin^2\phi_{ca}\cos^4\theta \quad (3.7)$$

$$S_{A_g}^{\perp} = \left[(|a| - |c|\cos\phi_{ca})^2 + |c|^2\sin^2\phi_{ca}\right]\sin^2\theta\cos^2\theta^2 \quad (3.8)$$

$$S_{B_{2g}}^{\parallel} = (2|f|\cos\theta\sin\theta)^2 \quad (3.9)$$

$$S_{B_{2g}}^{\perp} = [|f|\cos(2\theta)]^2 \quad (3.10)$$

where ϕ_{ca} is the phase difference $\phi_c - \phi_a$. Figure 3.13b shows the excellent fitting by using the new formula which contains the complex value of the Ramen tensor. The necessity of using complex Raman tensor elements indicates that angular behavior of the polarized Raman spectra of phosphorene is very unusual. This abnormal behavior can be rooted in the light–phosphorene interaction, which is highly related to electron radiation and electron–phonon coupling.

3.3.3.2 Temperature dependence of phonons in phosphorene

For potential applications in nanoelectronic and optoelectronic devices, understanding the effect of temperature on the structural properties is critical for device fabrications. Actually, the phenomena related to electron–phonon coupling and atomic vibrations become more important with increasing temperature. In addition, by applying the external electric field in transistors, the device becomes self-heated, which can degrade the performance. Therefore, understanding the evolution of phononic properties is fundamental for examining structural stability and performance. Temperature-dependent Raman measurement allows uncovering the fine structure, thermostatic and thermokinetic properties such as thermal expansion and thermal conductivity by analyzing the evolution of the shapes, intensity, and half width at half maxima (fwhm) of Raman-active modes with temperature.

The evolution of the Raman spectrum with temperature for few-layer black phosphorus (Fig. 3.14a) was studied by D. J. Late.[49] In the experiment, it was found that the A_g^1, A_g^2, and B_{2g} modes soften as the temperature increases from 77 to 673 K. The changes in the peak position of these modes as a function of temperature are shown in Fig. 3.14b, where a linear relationship of the temperature frequency is observed. Within the temperature range from 77 to 673 K, the drop in frequency is 4.39, 8.14, and 8.63 cm^{-1} for A_g^1, B_{2g}, and A_g^2, respectively. By fitting with the Gruneisen model for each mode,

$$\omega(T) = \omega o + \chi T \tag{3.11}$$

where ω_o is the frequency at zero Kelvin and χ is the first-order temperature coefficient. The fitted temperature coefficients for the A_g^1, B_{2g}, and A_g^2 modes are -0.01, -0.013, and -0.014 cm^{-1}K^{-1}, respectively.

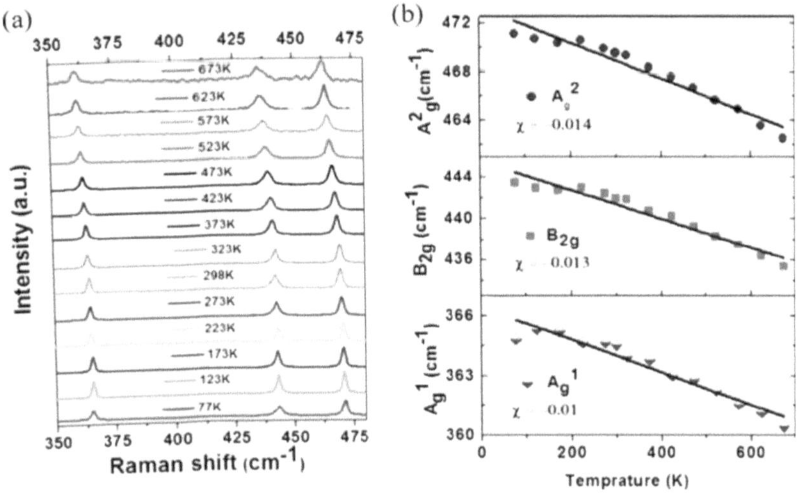

Figure 3.14 (a) Temperature dependence of the Raman spectra of the few-layer black phosphorus nanosheet. (b) The evolution of the Raman-active modes with temperature for the few-layer black phosphorus. Reprinted with permission from Late.[49] Copyright (2015) American Chemical Society.

The temperature-induced mode softening can be resulted from a double resonance process which also dominates the phononic behaviors in other 2D materials. Intrinsically, the mechanism behind the Raman measurement involves a four-phonon coupling. There is significant increase in fwhm for all the modes at elevated temperature due to phononic anharmonicity, which accounts for the thermal expansion and shifts of the frequencies. The observed softening of the A_g^1, B_{2g}, and A_g^2 modes could be due to a phonon decaying the lower-energy phonons and electron–phonon coupling-induced electron–hole pair. Overall, the changes in the mode positions in phosphorene result from the mode anharmonicity and the negative thermal expansion in the sample.

3.3.3.3 Strain effect on the vibrational frequencies

One of the advantages of 2D materials is their highly flexible structures, which can normally deform easily in bending or stretching configurations relatively. Strain engineering is an important technique to effectively tune the electronic, vibrational, optical, and chemical activities of 2D materials. The ability to quantify the strain subjected by the 2D materials is critical for strain engineering. Phosphorene is an ideal material for realizing strain engineering due to its good ability to sustain high strain up to 30%. Its unique puckered structure shows a strong anisotropic behavior under different deforming configurations.

The angular dependence of the strain effect on the lattice vibrational frequencies in few-layer phosphorene is investigated by in situ trained Raman spectroscopy.[50] It is found that the out-of-plane A_g^1 changes greatly when an uniaxial strain is applied along the near-armchair direction, while the in-plane B_{2g}, and A_g^2 modes are more sensitive to the near-zigzag direction strain. The near-armchair tensile strains from 0 to 2.2% can soften the A_g^1 mode with -1.3 cm^{-1}/%, while the other two in-plane modes are almost unchanged. For the near-zigzag tensile strains, the B_{2g} and A_g^2 modes show a linear dependence with -2.3 and -1.1 cm^{-1}/% red shift, respectively, while the A_g^1 mode is almost insensitive. These findings allow the possibility of intentional tuning of the in-plane and out-of-plane phonon modes by controlling the uniaxial strain.

3.3.3.4 Phonons at edges

Characterization of the edge structures of 2D materials is critically important for device fabrication and applications. This is due to the plenty of new phenomena induced by the break of the periodicity and the presence of dangling bonds and edge reconstructions. Compared with other 2D materials like graphene and MoS_2, the edge structures of phosphorene are still poorly understood. One effective way to uncover the atomic arrangements at edges is directly measuring the vibrational spectrum.

Polarized Raman measurement was performed on a 300 nm thick rectangular flake of black phosphorus with clear armchair (x direction) and zigzag (z direction) edge terminations (Fig. 3.15a,b).[51] New phonon modes related to the edge modes were observed which are forbidden in the bulk. Figure 3.15c shows the Raman spectra at the center and edge positions with different polarization configurations such as XX and XZ settings, where the

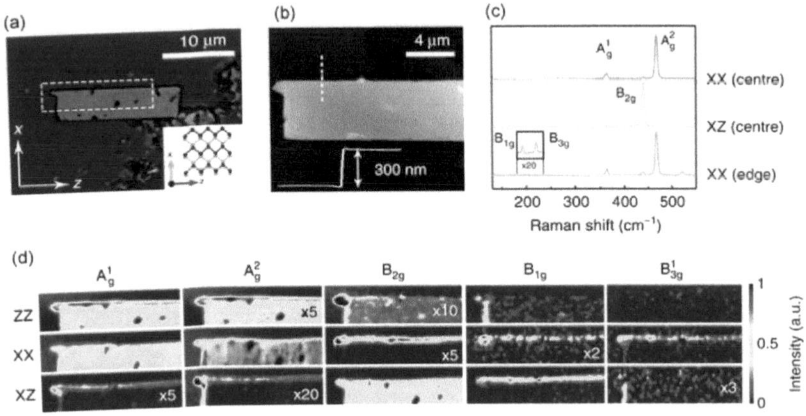

Figure 3.15 (a) Optical microscope image of the exfoliated phosphorene flake with zigzag and armchair edges. (b) Atomic force microscopy characterization of the flake. (c) The polarized Raman spectra at the center of the flake with XX (top) and XZ (middle) scattering configuration and the XX configuration at an armchair edge (bottom). (d) Hyperspectral Raman intensity images for all the observed modes with different scattering configurations. Reprinted with permission from Ribeiro et al.[51] Copyright (2016) Nature Publishing Group.

first and second index represents the polarization of the incident and scattered lights, respectively. Owing to the mode symmetry described before, the A_g^1, A_g^2, and B_{2g} modes appear at 360, 470, and 440 cm^{-1}, respectively in the XX and XZ spectrum measured in the center part of the sample flake. Interestingly, by measuring the armchair edge of the black phosphorus flake, new peaks assigned to B_{1g} and B_{3g}^1 modes were found at 190 and 230 cm^{-1}, respectively.

The hyperspectral images with spatially resolved intensity of all the modes are shown in Fig. 3.15d. The presence of the some high-intensity localized dots could originate from the defective or oxidized regions in the sample. Several new features are identified. For instance, for the bulk black phosphorus, A_g^1 and A_g^2 can only be observed in XX and ZZ polarized configurations. Surprisingly, A_g^1 and A_g^2 modes appear in the XZ crossed-polarization configuration at the armchair and zigzag edges. Another anomalous behavior was found for the B_{2g} mode, which can be observed at the zigzag edge for the XX configuration and at the armchair edge for the ZZ configuration.

In addition, there exist B_{1g} and B_{3g}^1 modes at the edges for all the scattering configurations. Normally, it is not expected to observe the B_{1g} and B_{3g}^1 modes in the bulk case for polarizations within the *xz* plane since they only contain *xy* and *yz* non-zero components of the Raman tensors, respectively. However, B_{1g} mode is clearly detected at the zigzag edge and B_{3g}^1 mode is observed in the armchair edge. It should be noted that the intensities associated with the edge phonons become weaker for thinner flakes.

3.4 Mechanical Deformation of Phosphorene

Mechanical properties of phosphorene are critical to its potential applications in flexible electronics. Owing to its intriguing puckered structure, phosphorene shows large structural flexibility. It can sustain tensile strain up to ~40% along the armchair direction. All its mechanical properties show a strong angular dependence. Interestingly, the unique negative Poisson's ratio is predicted in the normal direction for uniaxial strains applied in the direction parallel to the pucker. Young's modulus gives information about the resistance of a material toward uniaxial tensions, and Poisson's ratio describes the stability against shear strain. In the following section, both quantities are discussed.

3.4.1 Elastic Modulus and Fracture Strength

For a 2D system, the in-plane strain–stress relationship with orthogonal symmetry under plane stress conditions is expressed as

$$\begin{bmatrix} \sigma_x \\ \sigma_y \\ \tau_{xy} \end{bmatrix} = \frac{1}{1-\upsilon_{xy}\upsilon_{yx}} \begin{bmatrix} E_x & \upsilon_{yx}E_x & 0 \\ \upsilon_{xy}E_y & E_y & 0 \\ 0 & 0 & G_{xy}(1-\upsilon_{xy}\upsilon_{yx}) \end{bmatrix} \begin{bmatrix} \varepsilon_x \\ \varepsilon_y \\ \gamma_{xy} \end{bmatrix}$$
$$= \begin{bmatrix} C_{11} & C_{12} & 0 \\ C_{21} & C_{22} & 0 \\ 0 & 0 & C_{66} \end{bmatrix} \begin{bmatrix} \varepsilon_x \\ \varepsilon_y \\ \gamma_{xy} \end{bmatrix} \quad (3.12)$$

where $\upsilon_{ij} = -\frac{d\varepsilon_j}{d\varepsilon_i}$ is the Poisson ratio for a uniaxial strain applied along the *i* direction and a response strain in the *j* direction. σ_x,

σ_x, and τ_{xy} are the components of the stress tensor. γ_{xy} is the shear strain on the xy plane. E_i is Young's modulus along the ith axis, G_{xy} is the shear modulus on the xy plane, C_{ij} is the in-plane elastic stiffness constant, which is defined as $C_{ij} = \frac{1}{A_0}\frac{\partial E_s^2}{\partial \varepsilon_i \partial \varepsilon_j}$, where A_0 is the equilibrium area of different phosphorene phases, and E_s is the total energy of the system, defined as $E_s = a_1\varepsilon_x^2 + a_2\varepsilon_y^2 + a_3\varepsilon_x\varepsilon_y + a_4\gamma_{xy}^2$.

The mechanical properties of phosphorene are anisotropic due to an orthorhombic symmetry as examined by first-principles calculations.[52–54] Young's modulus is predicted to about two times larger in the zigzag direction (92.3–106.4 GPa) than the armchair direction (22.15–23.0 GPa). This behavior is rooted in the puckered structure, which is highly stretchable in the armchair direction but more rigid in the zigzag direction. The shear modulus is found to be around 22.3 GPa. The in-plane Poisson's ratio along armchair and zigzag directions is 0.175 and 0.703, respectively. Comparing with other 2D materials, the largest effective Young's modulus without the thickness of phosphorene (56.3 N m^{-1}) is only half that of monolayer MoS_2 (127.0 N m^{-1}),[55] and much smaller than that of 2D hexagonal boron nitride (292.1N m^{-1})[56] and graphene (335.0 N m^{-1}).[57] The much softer structure of phosphorene is due to its weaker P–P bond and puckered structure. Mechanical properties of the phosphorene are found to be tunable by chemical functionalization. For instance, via sodiation with decorating sodium atoms above the sheets, Young's modulus reduces owing to the softening of the P–P bonds.[58] The exact angular (θ) dependences of the Possoin's ratio and Young's modulus are plotted in Fig. 3.16.[54]

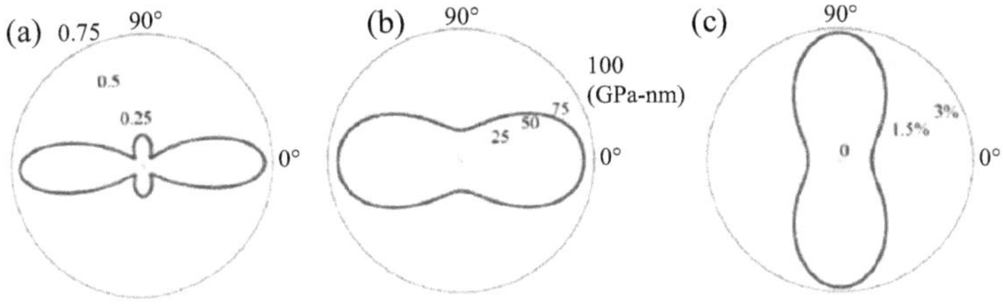

Figure 3.16 Orientation-dependent (a) Poisson's ratio, (b) Young's modulus of phosphorene, and (c) orientation-dependent strain in phosphorene corresponding to the constant stress. Reprinted with permission from Wang et al.[54] Copyright (2015) Royal Society of Chemistry.

Table 3.3 Elastic constants (C_{ij}, in GPa) and bulk modulus (B in GPa) of black phosphorus

	C_{11}	C_{22}	C_{33}	C_{44}	C_{55}	C_{66}	C_{12}	C_{13}	C_{23}	B
PBE-Grimme[59]	52.3	191.9	73.0	25.5	8.8	63.6	40.8	0.6	8.3	30.7
optB88[3]			70.8	18.7	10.6					
Exp[42–44,60]	55.1	178.6	70.0, 53.6	21.3	5.5, 11.5					32.5

For 3D bulk black phosphorus, the elastic constant and bulk modulus are compiled in Table 3.3. Calculation[59] shows that all the components of elastic constants satisfying Born's mechanical stability criteria and thus bulk black phosphorus is in a mechanical stable state. It can be seen that the C_{22} elastic constant is much larger than the C_{11} and C_{33} parameters, implying that bulk black phosphorus is stiffer against strain along the zigzag axis than along the normal and armchair directions. Based on fitting the frequency of the multilayer phosphorene, the obtained values of the shear moduli C_{44} and C_{55}, and stretching modulus of C_{33} are 18.7, 10.6, and 70.8 GPa, respectively, consistent with experimental values of 21.3, 11.5, and 70.0 GPa obtained through ultrasonic[42,43] and neutron scattering[44] measurements.

Based on the Voigt–Reuss–Hill approximations, the bulk and shear moduli of bulk black phosphorus can be derived. In the Hill approximation, the bulk moduli (B_{H}), the shear moduli (G_{H}), Young's modulus (E_{H}), and Poisson's ratio are 38.5, 29.4, and 70.3 GPa, and 0.30, respectively. Based on the bulk and shear moduli, Pugh's ratio, reflecting the brittle-ductile nature of a material, can be obtained as $B_{\mathrm{H}}/G_{\mathrm{H}}$. A ratio below (above) 1.75 tends to be brittle (ductile). For black phosphorus, the value of 1.31 shows a brittle behavior. In addition, the Poisson's ratio of a bulk material can indicate the volume change in a material subjected to a uniaxial strain. The Poisson's ratio smaller than 0.5 suggests a large volume change. The ratio of 0.30 for bulk black phosphorus indicates that a great change of volume occurs for even small mechanical deformations.

Concerning the tensile strength, the largest tensile strain that phosphorene can retain also shows strong anisotropy in different directions. Simulations found that phosphorene can be stretched around 0.48 strain along the armchair direction and only 0.11 strain along the zigzag direction.[61] For other 2D materials, MoS_2, graphene, and BN can sustain about 0.23, 0.25, and 0.29 strains. The large ductility of phosphorene along the armchair direction can be explained by bond rotation for unfolding the puckers instead of direct P–P bond extension. By first-principles calculations, Wei et al. predicted that the largest stress that phosphorene can sustain is up to 18 GPa (8 GPa) along the armchair (zigzag) direction, much smaller than the intrinsic strength of h-BN (102 GPa) and graphene (130 GPa).[52]

Mechanical properties of phosphorene have been measured by using reflection difference microscopy (RDM) and atomic force microscopy (AFM).[62] To identify the lattice orientation, RDM was used, which has the advantage of more rapid and direct visualization of the crystal origentations in a nondestructive and in situ way compared to the widely adopted Raman measurement. In the experiment, suspended phosphorene strips were used. The detailed procedures for their preparation are illustrated in Fig. 3.17a. The length (width) of the BP strips was from 0.80 to 1.92 μm (from 0.48 to 1.47μm). First, a phosphorene flake was transferred onto a Si/SiO_2 substrate, followed by spincoating a layer of PMMA, which was patterned by EBL into an L shape with two principal lattice axes aligned in the two arms respectively. Then the L-shaped pattern was transferred to the phosphorene flake and etched by Ar plasma and by removal of the PMMA. Finally, metal electrodes were deposited at the ends of each strip, releasing a sacrificial layer in BOE solution.

The AFM bending method was then used to measure the Young's modulus, as shown in Fig. 3.17d–f. A schematic of the test of bending is shown in Fig. 3.17d. During the experiment, the AFM cantilever tip was initially set at the middle of the suspended phosphorene strip. To bend the strip, the tip moved downward at a speed of 0.5 μm/s, followed by an upward motion to release the strip. The induced vertical deformation δ_{BP} can be calculated by $\delta_{BP} = \Delta_Z - D_{cant}$, where Δ_Z is the displacement given by the Z scanner, and D_{cant} is the deflection of the cantilever given by the AFM photodiode system. Therefore, the loading force (F) applied to the strip is expected to be $F = k_{cant} D_{cant}$, where k_{cant} is the effective spring constant of the

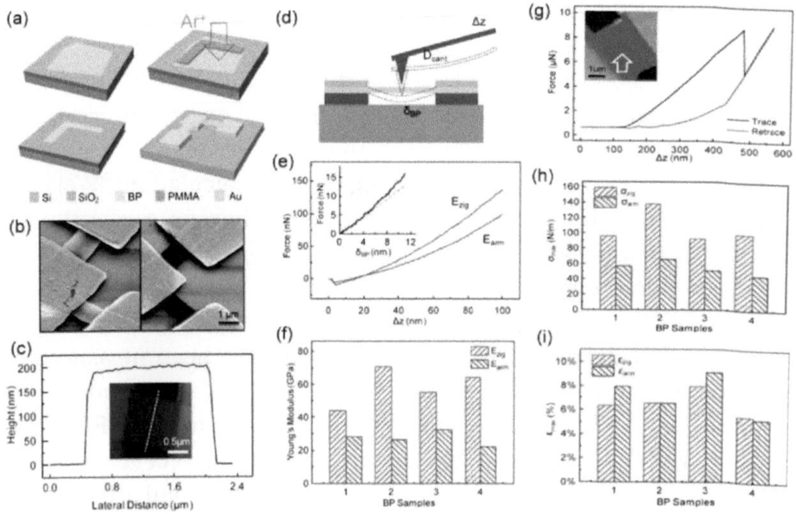

Figure 3.17 (a) Fabrication of suspended phosphorene strips. (b) SEM image of suspended BP strips (left) and collapsed strips damaged due to a high surface tension of water (right). (c) AFM image of a suspended BP strip and scanned topology profile. (d) Schematic of the AFM bending measurement. (e) Load force vs. displacement curves of the AFM tip. The inset shows the relationship between the deformation of the phosphorene strip and the loading force. (f) Young's modulus of suspended phosphorene strips. (g) Loading force–displacement curves of the AFM tip. Breaking stress (h) and strain (i) of the suspended strips. Reprinted with permission from Tao et al.[62] Copyright (2015) American Chemical Society.

AFM cantilever. The F–Δ_Z curves of the AFM tip measured on the armchair and zigzag strips are plotted in Fig. 3.17e. An almost linear curve was found below 5 nm, and the presence of nonlinear behavior at larger displacement was recorded. It should be noted that there always exists pre-tension in the 2D thin films which can affect the spring constant. Here the pre-tension in phosphorene flakes should be small due to the following double reasons. First, the trimming of the phosphorene flakes by Ar plasma greatly releases the tension within the interior region of the material. Second, the suspended structures were fabricated by release of the sacrificial layer instead of the approach of transferring to a pre-etched trench. The measured Young's moduli shown in Fig. 3.17f were found to be 27.2 $\pm$ 4.1

and 58.6 ± 11.7 GPa in armchair and zigzag directions, respectively. The ratio of around 2.21 between the zigzag and armchair directions is consistent with the theoretical results.

The breaking stress was measured by steadily increasing the forces until the phosphorene strips were broken (Fig. 3.17g–i). The loading $F-\Delta_Z$ curves are shown in Fig. 3.17g, where a sudden drop reflects the breaking of the strip. A pinhole was formed at the center of the broken strip, and it is shown in the inset. The breaking stress was found to be 54.8 N/m and 106.2 N/m along the armchair and zigzag directions, respectively (Fig. 3.17h), indicating a highly anisotropic behavior of the strength due to the anisotropic Young's modulus. Similarly, the breaking strain (Fig. 3.17i) was obtained based on the breaking stress and Young's modulus. Surprisingly, the average strain was found to be nearly isotropic with 7.2% and 6.5% in the armchair and zigzag directions, respectively, largely due to the cancellation of the anisotropic elastic modulus and stress.

3.4.2 Negative Poisson's Ratio

One of the most interesting mechanical properties of phosphorene is its negative Poisson's ratio. Defined as the negative ratio of transverse strain to an axial strain, it is a common notion that most materials show a positive Poisson's ratio, with structure becoming compressed in the normal direction of the axial direction of a uniaxial strain. So far, some cases reporting a negative Poisson's ratio have been related to bulk-scale and engineered structures.

For phosphorene, a positive Poisson's ratio along the zigzag direction ($\nu = 0.4$) and the normal surface direction ($\nu = 0.046$) is predicted when a strain is applied in the armchair direction.[63] There is a strong nonlinear dependence of ν on the uniaxial strain along the armchair direction. For uniaxial strain along the zigzag direction, the ν along the armchair direction is between 0.7 and 0.93,[61] around 2 times larger than that along the zigzag direction. Interestingly, the ν along the normal surface direction becomes negative ($\nu = -0.027$) when a uniaxial strain is applied along the zigzag direction. The underlying reason for this abnormal behavior is related to its unique structure with two coupling orthogonal hinges. The correlated deformations of the two hinges with one

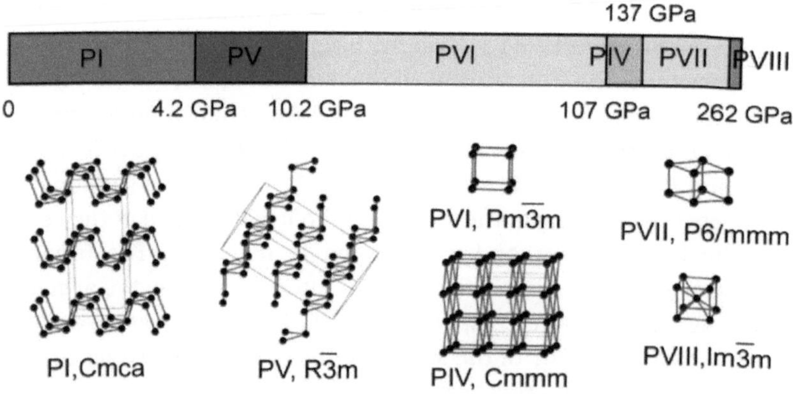

Figure 3.18 Sequences of phase transition starting from black phosphorus (PI) phase at ambient condition upon continually increasing the pressure.

hinge opening while the other hinge simultaneously closes lead to a negative Poisson's ratio.

3.4.3 Structural Evolution of Phosphorene under High Pressure

Black phosphorus (also named as the A17 phase and PI structure) is the ground state of phosphorus at ambient conditions. This orthorhombic phase transforms sequentially to other phases, continually increasing the pressure (see Fig. 3.18). The exact transformation sequence initiating from black phosphorus is PI (Cmca) $\rightarrow$ PV(R$\bar{3}$m) $\rightarrow$ PVI (Pm$\bar{3}$m) $\rightarrow$ PIV(Cmmm(00γ)) $\rightarrow$ PVII(P6/mmm) $\rightarrow$ PVIII(Im$\bar{3}$m).

At around 4.2 GPa, black phosphorus transforms to the PV phase (also named as A7 phase), which is a rhombohedral structure and R$\bar{3}$m space group. The PV phase is a semimetallic phase which also has a layered structure, and the phosphorus atoms are also threefold coordinated. However, the six-membered ring layers consist of pairs of chair-type rings which are connected in a trans-decalin way. During the transition, it was found that the electric conductivity of the material is very much dependent on the pressure, and two anomalies have been recorded at 1.7 and 4.2 GPa.[64] Magnetoresistance studies also found there is an electronic topological transition with the formation of a 3D Dirac semimetal where band crossover occurs near the Z point in the Brillouin zone with linear dispersion.[65]

At around 10.2 GPa, the PV phase starts to transform to the PVI structure (sc), which is a cubic structure with space group of $Pm\bar{3}m$, a subspace group of the $R\bar{3}m$. The PVI phase remains stable until 107 GPa, where it begins to transform to the PIV phase (ic), which is a structure incommensurate with the orthorhombic structure of Cmmm (00γ) as revealed from X-ray diffraction and first-principles calculations.[67] At 137 GPa, the PIV phase transforms to the PVII structure with the simple hexagonal phase (sh, P6/mmm). The transformations are first-order and both structures are metallic. At 262 GPa, the PVII phase transforms to the body-centered cubic phase PVIII (bcc, Im3m).

The above high-pressured phases of phosphorus are significantly different from other VA group elements like As, Sb, and Bi. In case of phosphorus, the phases tend to have simple packed structures such as the cubic PVI and simple hexagonal PVII. However, its other counterparts adopt more complex host–guest structures. Such difference can be attributed to the s–d band mixing induced at high pressures.[67] Displacive atomic mechanisms are proposed for the low-pressure A7 structure to the highest-pressure bcc structure, where two different structural paths are proposed for explaining the host–guest structures found in As, Sb, and Bi. The mechanism for yielding the incommensurate P IV structure was disclosed recently in P.[68]

3.5 Intrinsic Defects in Phosphorene

The presence of the intrinsic defects in nanomaterials is unavoidable and greatly affects their electronic, mechanical, thermal, and chemical properties. The energetics and dynamical behavior of these defects govern various microscopic and mesoscopic processes under heating or electronic irradiation such as hole evolution, mechanical deformation, and chemical stability, especially in atomically thin-layered materials. For phosphorene, it is well known that structural transformation occurs rapidly under ambient conditions.

While surface protection by oxides or other 2D layers improves the environment-related stability of phosphorene, intrinsic stability associated with the highly populated and dynamic vacancies becomes more important for applications at elevated temperatures. The knowledge of defect formation, healing, and migration of phosphorene is helpful for understanding the thermally activated process and protecting the structural integrity.

3.5.1 Review of Carbon Vacancies in Graphene

Owing to the same structural topology with hexagonal honeycomb structure between graphene and phosphorene, it is useful to compare their intrinsic structural defects, especially the vacancies which are the most popular defects in 2D materials. For the monovacancy (MV) in graphene, the loss of the carbon atom creates an intermediate structure with D_{3h} symmetry and three dangling carbon atoms. After a John–Teller reconstruction, the above structure evolves into the ground structure with pentagon–nonagon (59) rings and the site symmetry breaking down to C_{2v} (Fig. 3.19). Pairing of two of the three dangling atom creates a new bond and a 59 structure. DFT calculations predict that the formation energy (E_f) of this MV structure is in the range of 7.6–7.9 eV and the bond length of the new bond is between 1.8 and 2.0 Å,[69] which is consistent with the TEM measured value of 1.9 Å.[70] Apparently, there are three possible equivalent states of the 59 vacancy, and DFT calculations predict that the barrier for the transition between these equivalent states is only around 0.13 eV and the superposition of three equivalent states is observed in TEM observations.[71] The

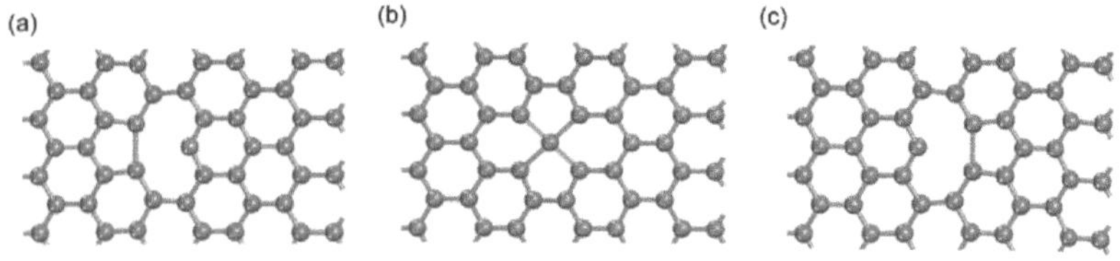

Figure 3.19 Migration of a MV in graphene. (a) Initial ground state of a 59 MV. (b) Transition state. (c) Final ground state of a 59 MV.

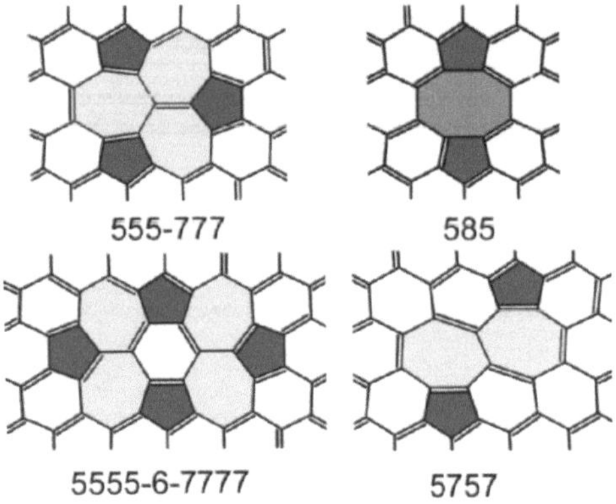

Figure 3.20 Atomic structures of the four different types of DV in graphene.

barrier of migration of a vacancy in graphene via the path shown in Fig. 3.19 is calculated to be between 1.0 and 1.4 eV.[72]

Under thermal activation and ion irradiation, coalescence of two MV in graphene into one divacancy (DV) occurs frequently. There are four different types of DV (Fig. 3.20) with a fully compensated structure satisfying a coordination number of three: the well-studied pentagon–octagon–pentagon (585), triple pentagon–triple heptagon (555-777), and quadruple pentagon–hexagon–quadruple heptagon (5555-6-7777), and pentagon–heptagon–pentagon–heptagon (5757) types. These DV structures were observed under electron irradiation in HRTEM.[74] DFT calculations[72] show that the energy cost of the formation of the 555-777 DV is the lowest, followed by the 5555-6-7777, 585, and 5757 DVs. The order of E_f follows 555-777 (8.09 eV) < 5555-6-7777 (8.69 eV) < 585 (8.99 eV) < 5757 (12.4 eV).

These DVs can be transformed into each other. The DFT simulations reveal the barriers and transition states for transformation between the 555-777 and 5555-6-7777, 585 and 555-777, and 585-5757 states, which are shown in Fig. 3.21. The very high barriers reaching above 5 eV suggest that three states of 585, 555-777 and 5555-6-7777 DV are stable at room temperature. Transformation between the 555-777, 585, 5555-6-7777 with relatively lower formation energy than the 5757 DV was recorded under electron irradiation in HRTEM.[75–77]

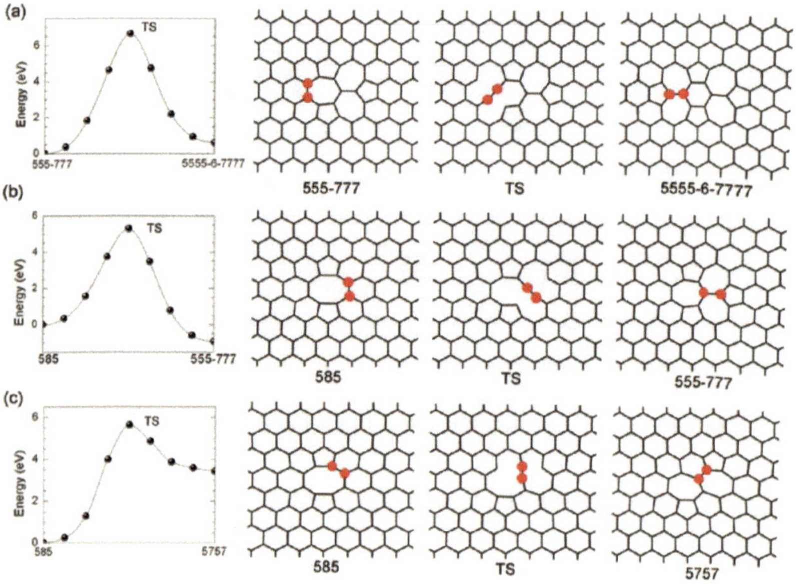

Figure 3.21 Transformations between various DVs in graphene. The activation energy and transition state (TS) for atomic process from 555-777 to 5555-6-7777 (a), from 585 to 555-777 (b) and from 585 to 5757 (c). All the energies of the profiles are relative to the initial structures. The transitions are realized through a SW transformation of the red dimmer. Reprinted with permission from Cai et al.[72] Copyright (2016) American Chemical Society.

Migration of these DVs is found to also involve a sequence of the above Stone–Wales (SW)-type transformations. Contrary to MV migration, the diffusion barrier for the migration of DV is considerably large. DFT calculations and analysis of HRTEM images show that the migration of the 585 DV takes place through two intermediate 5757 DV states. According to Cai et al., the barrier for transformation from the 585 to 5757 DV is 5.62 eV. For the migration of the 555-777 DV, a transformation to the 585 state is needed with a barrier of 6.20 eV.[78] Thus, thermally activated migration of DV at room temperature is impossible. In experiment, electron-irradiation-triggered motion of DV is found to occur via transitions between the 585, 555-777, and 5555-6-7777 states.[76] The HRTEM experiment also showed that the migration of the 585 DV under the

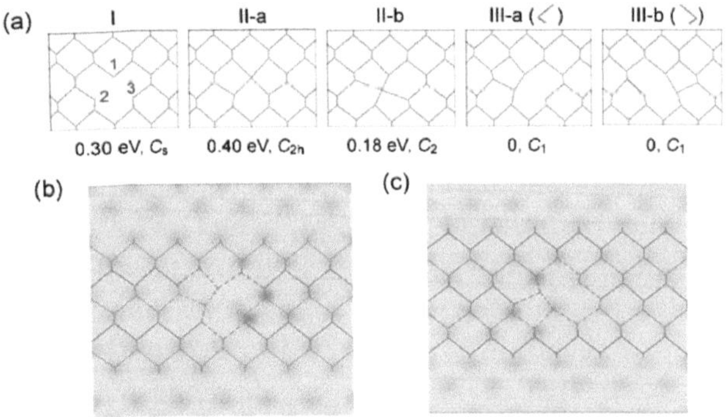

Figure 3.22 MV in phosphorene. (a) Various states of an MV: type I, II-a, and II-b structures are metastable structure, and type III-a and b are two equivalent structures and the ground state of a MV, similar to the 59 structure of the MV in graphene. (b) and (c) correspond to simulated STM images at +1 V, with the MV appearing at the top ridge close to and the bottom ridge far from the STM tip, respectively. Reprinted with permission from Cai et al.[72] Copyright (2016) American Chemical Society.

action of electron irradiation involves the intermediate metastable dislocation dipole 5757 state.[77]

3.5.2 Itinerant Vacancies in Phosphorene

Similar to graphene, an MV in phosphorene is created by the loss of a phosphorus atom and three dangling atoms (atoms 1, 2, and 3 in Fig. 3.22a). Through detailed first-principles calculations, Cai et al.[72] found that without any reconstruction, this structure denoted as type I is metastable. Driven by the Jahn–Teller distortion, different possible reconstructed structures can be formed by pairing any two of these dangling atoms: type II-a and II-b by adjusting atoms 1 and 2, respectively, and the ground state of the type III MV with two equivalent structures denoted as III-a and III-b. Similar to the 59 MV in graphene case, the type III MV in phosphorene also adopts a 59 structure but in a highly puckered manner. The puckered 59 structure resembles ticks such as "<" and ">" embedding the host lattice. In Fig. 3.22b,c, the STM image of MV is plotted. The two different images are shown to reflect the absence of in-plane mirror symmetry in phosphorene, where the MV can locate at either the top or the bottom ridge.

The DFT calculation shows that the E_f of type III MV is 1.65 eV and the type I, II-a, and II-b structures are 0.30, 0.40 and 0.18 eV higher in energy.[72] Compared with graphene, the E_f of phosphorene is much smaller than that of MV in graphene (7.57 eV). Similarly, the formation energy of both V_N (from 4.90 to 7.60 eV) and V_B (from 7.50 to 10.20 eV) in h-BN is greatly higher than that in phosphorene. Only the E_f of V_S in MoS_2 (from 1.22 to 2.25 eV) is less than or comparable with that in phosphorene, depending on the chemical potential of sulfur. The small energy cost of creating a vacancy in phosphorene can be attributed to the much softer P–P bonds than the C–C and B–N bonds in graphene and BN. An exceedingly larger density of MV in phosphorene will be expected than graphene and BN. Also, the small E_f implies that introduction of an MV in phosphorene can be realized even with solely moderate thermal stimuli which is different from graphene, where carbon vacancy is required to be induced by impinging energetic electrons or ions.

For the kinetics of a MV in phosphorene, owing to its anisotropic structure, there exist two different diffusing directions with hopping along zigzag and armchair directions, as shown in Fig. 3.23a. Via nudged-elastic band (NEB) calculation, the diffusion barrier (E_b) for a MV to an adjacent lattice position along the zigzag direction is found to be 0.30 eV, where several intermediates states (S1–S6) exist, as shown in Fig. 3.23c. For armchair-direction motion (Fig. 3.23d), the barrier is about 0.40 eV. It is interesting that the motion of a MV along zigzag direction involves continual rotation of 59 rings together with lattice shifts. Similar motion of a MV in graphene was earlier captured by TEM, and the atomic scale characteristic of this process was not understood.[79] This mode of motion was overlooked due to the somehow random diffusion of a carbon vacancy and its migration was generally believed to be through the path along Fig. 3.23. Due to the presence of structural anisotropy, it will be easier and highly interesting to experimentally observe the above rotation-plus-shift mode of diffusion of a MV in phosphorene.

Figure 3.23b compares the energetics and kinetics of the MV of other 2D materials. For phosphorene, E_b is the smallest with the

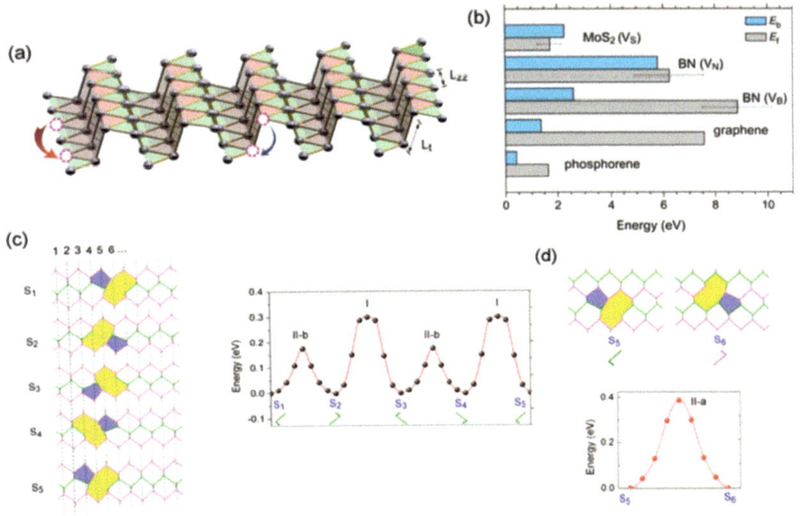

Figure 3.23 (a) Atomic structures showing a MV (dashed circle) diffusing along the armchair (blue arrow) and zigzag (red arrow) directions in phosphorene. (b) Comparison of the E_f of a MV and its diffusion barrier for various 2D materials. (c) Intermediate replicas and energetic profile for the migration of a MV along the zigzag direction by NEB calculation. (d) Intermediate replicas and energetic profile for the migration of a MV along the armchair direction. Reprinted with permission from Cai et al.[72] Copyright (2016) American Chemical Society.

following sequence: phosphorene (0.40 eV) < graphene (1.39 eV) < MoS_2 (V_S, 2.27 eV) < BN (V_B, 2.60 eV) < BN (V_N, 5.80 eV).[72] The hopping rate (v) of a vacancy can be calculated by the Arrhenius formula: $v = v_s \exp(-E_b/kT)$, where k is the Boltzman constant, v_s is a prefactor normally around 10^{13} Hz, and T is the temperature. Based on the computed E_b, the v of a vacancy in phosphorene is predicted to reach about 2.5×10^9 per second, which is around 16 orders larger than that of graphene at room temperature. The nearly itinerant hopping behavior of the MV implies a high mobility. This suggests a difficulty in observing them. As each MV hops around every 400 ps at room temperature, direct observation of them is impossible, because electron microscopy requires imaging timescales of the order of second. Cai et. al. suggest that it is needed to reduce the temperature to liquid nitrogen (around

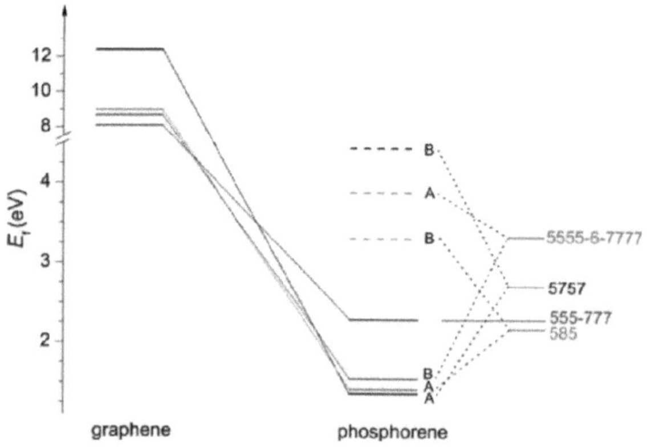

Figure 3.24 Comparison of the formation energy of DV in phosphorene. Note an opposite hierarchy of stability of DV between graphene and phosphorene. Reprinted with permission from Cai et al.[72] Copyright (2016) American Chemical Society.

70 K) to suppress the hopping even occurring around every 2 seconds. Moreover, the motion of a MV is highly anisotropic and the diffusion rate in the zigzag direction is three orders higher than that along the armchair direction. This anisotropy may account for the experimentally observed "eye"-shaped cracks during the thermal decomposition of phosphorene under vacuum.[6]

In the case of a DV, as shown in Fig. 3.24, the 585, 5757, and 5555-6-7777 DVs each evolve into two topologically equivalent structures due to the structural anisotropy.[72] Following to the nomenclature of Cai et al., a DV defect is denoted as "A" type if its characteristic geometric profile (such as pentagon–nonhexagon group) lying along the diagonal direction, and its sister conformer along the orthogonal (armchair and zigzag) direction is termed "B" type. Such lifting of structural degeneracy is not found in the 555-777 DV due to its pseudo-C_3 symmetry.

According to the previous discussion of the formation energy DVs in graphene, the order of the E_f of DV follows 555-777 (8.09 eV) < 5555-6-7777 (8.69 eV) < 585 (8.99 eV) < 5757 (12.4 eV). Surprisingly, such ordering is completely reversed in phosphorene with 555-777 (2.28 eV) > 5555-6-7777-B (1.54 eV) > 585-A (1.41 eV) > 5757-A (1.35 eV). The energy cost of forming a DV in phosphorene is greatly smaller than that of graphene. The 5757 structure is the lowest-energy configuration of DV, which is absolutely different from graphene, where the 5757 DV is the highest-energy configuration among the various DVs.

The path of transformation between the various DVs in phosphorene is shown in Fig. 3.25, where the continual transformations of 5555-6-7777, 555-777, 585, and 5757 DVs in graphene are also shown. In phosphorene, each equivalent pathway in graphene is split into two paths associated with the "A"- and "B"-type DVs. Compared to the large barriers in graphene, the barriers for transitions of DVs in phosphorene are much smaller, ranging from 0.44 to 1.83 eV (estimated rate from 2.0×10^9 to 6.8×10^{-3} per second at 600 K). This implies a high thermal activity of DV in phosphorene even at finite temperature, which is generally different from graphene, BN, and MoS_2 cases with an extremely low mobility of the DVs. The low barrier of the DV transformation in graphene is rooted in the small energy expense for bond rotation and torsion.

3.5.3 Linear Defects

Another important type of defect that commonly occurs in 2D materials is the linear defect, including the grain boundaries and edges, which can be produced during growth and fabrication of devices. Owing to the presence of a large amount of bond breaking and reconstruction, the linear defects normally induce a significant amount of new localized states. For instance, grain-boundary-induced defective states in MoS_2 are produced by the formation of new homoelemental Mo–Mo and S–S bonds that are absent in perfect lattice. Similar deep localized states are accompanied with homoelemental B–B or N–N bonds in a hexagonal BN sheet. These deep-lying states can significantly deteriorate the electronic performance by acting as trapping centers for charge carriers and electron–hole recombination centers.

Similarly, the presence of edges can exert more significant effects on the electronic properties due to the breakup of the lattice periodicity and the presence of dangling bonds. One well-known example is related to the edge states in graphene, forming the graphene nanoribbons. The edge states in graphene significantly

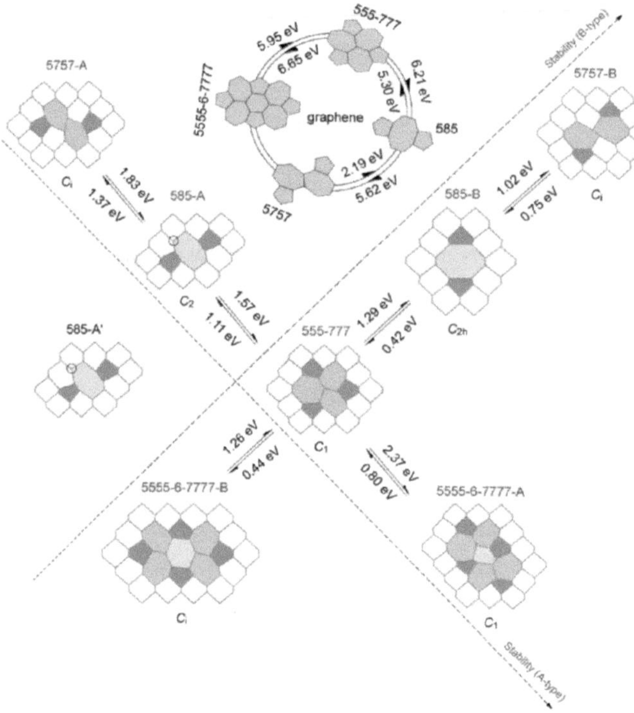

Figure 3.25 Transformation between different DV structures in graphene and phosphorene. The numbers beside the arrows are the diffusion barriers. Reprinted with permission from Cai et al.[72] Copyright (2016) American Chemical Society.

scatter the transport of carriers and degrade their mobility. In MoS_2, the presence of edges is accompanied with highly localized states which are very sensitive to the edge reconstructions.

In principle, formation of similar homoelemental bonds associated with these linear defects and accordingly its accompanying deep defective states can be avoided for elemental 2D materials like graphene and phosphorene. Calculations show that the grain boundaries are electronically benign and inactive due to the homoelemental bonding. Figure 3.26a shows the core structure of the dislocation with a (0, 1) Burgers vector in phosphorene which contains a buckled pentagon and heptagon pair. The basic units of dislocation can also be found in other honeycomb

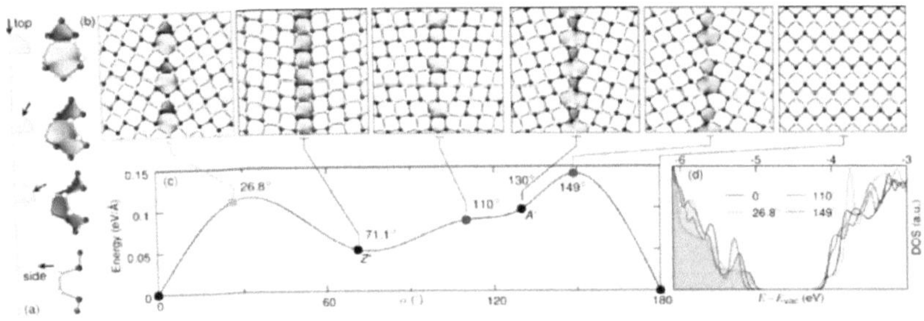

Figure 3.26 (a) Structures of grain boundaries in phosphorene at various misorientation angles. (b) Energy of grain boundary as a function of the misorientation angle. Reprinted with permission from Liu et al.[80] Copyright (2014) American Chemical Society.

structures such as graphene, BN, and MoS_2. The most intriguing characteristic of the dislocation of phosphorene is related to its same atomic coordination number and same chemical bonding of each phosphorus atoms as that in the perfect lattice. Such structural properties are not present in the dislocation cores or in the grain boundaries of MoS_2 and BN due to the presence of heteroelemental bonds at the defects.

Figure 3.26b shows the grain boundaries constructed from linear dislocation arrays with different misorientation angles (α). The formation energy of grain boundaries with the lowest energy is presented in Fig. 3.26c. Although most of the grain boundaries can be made of buckled pentagon–heptagon pairs, the 130° grain boundary consists of quadrilateral–octagon pair, which was previously observed in BN and MoS_2. It is found that the formation energies of the grain boundaries in phosphorene are smaller than those of graphene, MoS_2, and BN, indicating a higher chemical stability of these linear defects in phosphorene than its 2D counterparts. Figure 3.26d shows the electronic density of states for several grain boundaries embedding in a 2D phosphorene sheet. It is found that all these structures do not show deep gap states, which are sharply different from MoS_2, which always show deep-lying localized states in a bandgap.

In the next part, the characteristics of edges in phosphorene are introduced. Resembling the situation of a surface cutting from a bulk material, different edges can be constructed by cutting a 2D material. Normally, the electronic structures at edges can be sharply different from those of interior bulk. One well-known example is the spin-dependent gapless chiral edge state in graphene edges. For phosphorene, similar phenomena are also predicted and the presence of edges triggers great effect on the electronic properties of phosphorene. For instance, pristine armchair phosphorene nanoribbons are shown to be semiconducting, while zigzag ribbons are metallic. Both types of edges become semiconducting upon edge hydrogenation and the bandgap is inversely proportional to the ribbon width. Phosphorene nanoribbons also show intriguing properties under external field. A giant Stark effect is found for phosphorene nanoribbons, allowing applications in field-effect transistors working with a low bias. In addition, a high Seebeck coefficient is predicted which greatly favors thermoelectric apppli-cations.

Phosphorene has an anisotropic honeycomb structure with the absence of C_3 symmetry. Therefore, contrary to graphene, which has a perfect honeycomb lattice, phosphorene has additional edges in its structure besides the well-known armchair and zigzag directions. As shown in Fig. 3.27a, cutting along the phosphorene sheet along the armchair (zigzag) directions forms an angle other than 0 or 90 with the puckered ridges, leading to another set of armchair and zigzag directions, named as skewed-zigzag (sZZ) and skewed-armchair(sAC) nanoribbons.

Owing to the presence of dangling bonds and different electronic environments for the edge atoms, the lattice and atomic coordination shown in the bulk interior cannot be maintained in the edges. Normally, a pristine edge undergoes significant bond reorganization and atomic reconstruction for a better saturation of undercoordinated atoms. For instance, in graphene, the armchair edge is stablized by forming triple bonds at the edge. In contrast, the reconstruction of the zigzag edge involves a newborn pentagon and a heptagon with the replacement of two hexagons. This leads to the well-known haeckelite zz57 edge, which greatly reduces the edge energy by around 2.1 eV/nm. Similarly, for transition metal dichalcogenides, there is a (2×1) reconstruction for metal-terminated edges with an outward motion of second-row halogen atoms. This leads to a drop in edge energy of $\sim$0.4 eV/nm for MoS_2,

~1.0 eV/nm for $MoSe_2$, ~1.2 eV/nm for WS_2, and ~1.4 eV nm^{-1} for WSe_2.

Various possibilities of edge reconstruction of phosphorene nanoribbons have been discussed by using first-principles calculations.[81] Atomic nanoribbons are created to examine the energetics and stabilities of possible edges. A form of reconstructed zigzag edge forming a phosphorus nanotube attached at the edge front through the self-rolling of edge atoms is identified. This nanotube-terminated edge is the most stable edge among all the edge types explored. Its edge energy is around 1.5 eV/nm, much lower than that (2.4 eV/nm) of the pristine unreconstructed zigzag edge, indicating that this type of edge is highly favored thermostatically and possible in real phosphorene samples. The underlying reason for this stabilization is that all the edge atoms have a coordination number of 3, which is similar to the bulk atoms. This allows a full saturation of all the atoms without any dangling bonds. Notably, similar tube-decorated edge structures were proposed in graphene nanoribbon–nanotube junctions. However, the barrier for forming such nanotube-terminated junctions (see below) is much lower than that of graphene systems due to much flexible phosphorus–phosphorus bonds than the rigid carbon–carbon bonds. In principle, the diameter of the edge-connected nanotube could be any value with a proper rolling of the edge atoms. However, in phosphorene, the energy cost for the bending and rolling process for a large-diameter tube could outweigh that of energy saving during the saturation of dangling atoms. Therefore, the edge tube with a larger diameter is less unlikely.

Concerning the kinetics of edge reconstruction to form nanotube-terminated edges, Gao et al. performed detailed nudged elastic band calculations of the barrier for the rolling process.[81] It was found that this transformation from pristine zigzag edge to nanotube-terminated edge initiates with the rolling of a pair of two neighboring edge atoms (Fig. 3.27b), followed by another pair of atoms in a similar pair. This stepwise atomical scale process continues until the complete formation of an edge tube. The barrier for this rolling process is only around 0.47 eV, which can easily occur under finite temperature.

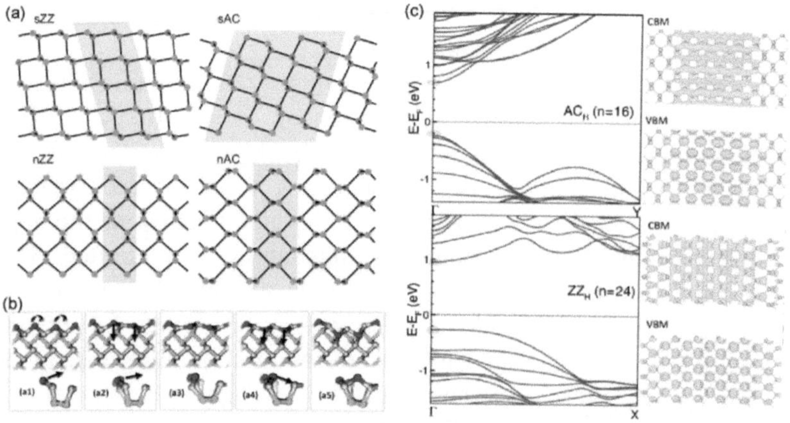

Figure 3.27 Edges in phosphorene. (a) Skewed-zigzag (sZZ), skewed-armchair (sAC), normal-zigzag (nZZ), and normal-armchair (nAC) nanoribbons in phosphorene. Reprinted with permission from Grujić et al.[82] Copyright (2016) American Physical Society. (b) Nanotube-terminated zigzag edges of phosphorene formed by self-rolling reconstruction. Reprinted with permission from Gao et al.[81] Copyright (2016) Royal Society of Chemistry. (c) Band structure and partial charges of the valence band maximum (VBM) and the conduction band minimum (CBM) of AC and ZZ nanoribbons of phosphorene. Reprinted with permission from Ramasubramaniam and Muniz.[83] Copyright (2014) American Physical Society.

Figure 3.27c shows the electronic band structures of hydrogen-terminated armchair (AC_{H}) and zigzag (ZZ_{H}) nanoribbons. All the hydrogenated phosphorene nanoribbons show a direct bandgap which is larger than their pristine counterparts. The gaps decrease with increasing ribbon width due to a quantum confinement effect. Interestingly, in phosphorene, most of the edge states are distributed within the phosphorene sheet instead of being localized at the edges.

3.5.4 Effect of Highly Mobile Vacancies on Properties and Applications

The ultrahigh mobility of the vacancies in phosphorene may allow the resistive switching where thermally and/or electrically activated motion of defects is critical. Conventional memresistive phenomena were realized based on a metal–oxide–metal architecture.[84] More recently, resistors were demonstrated based on anion-deficient 2D layered transition metal dichalcogenides like MoS_2[85,86] and transition metal oxides.[87] The advantage of using 2D sheets includes the tunability of the memristor's properties, the set voltage by electrostatic gating, and the scaled-down device dimensions. Similar to the 3D oxide case, the switching mechanism in an anion-based memristor is rooted in the motion of an anion. As shown in Fig. 3.28, three-terminal memristor devices are fabricated based on single-layer MoS_2 obtained by chemical vapor deposition on Si substrates oxidized with 300 nm thick SiO_2. There are two top gold electrodes defining the memristor channel with a Si back-gate electrode being used to control the set voltage. As the MoS_2 monolayers are polycrystalline, there are plenty of grain boundaries (GBs) with different orientations. It is found that the direction and the alignment of these grain boundaries relative to the Au electrodes play a dominating role in realizing the resistive switching. Positively charged oxygen vacancies accumulating at the GBs and drifting toward (away) the interface create conducting (insulating) channels that control the electronic barrier, resulting in the different resistance state. Sulphur vacancy migration between the GB region and the region with low sulphur vacancy concentration is directly responsible for the switching behaviors. Therefore, realization of the resistive switching requires an introduction of properly aligned conducting filaments or grain boundaries connecting the electrodes, which complicates the design due to these intrinsically randomly formed channels. It is expected that the ultramobile and highly anisotropic propagating behaviors of vacancies in phosphorene may initiate new design of resistive switching devices.

The energetics and kinetics of vacancies greatly affect the mechanical properties of low-dimensional materials. For instance, in graphene and carbon nanotubes (CNTs), the adjustment of the pentagon–heptagon (57) dislocation dipole under strain and at high temperature leads to the superplasic behavior in sp^2 carbon materials associated with the bond flips and mass loss.[88] Similar mechanisms should also exist in phosphorene. The small cost of energy for losing an atom and the extremely low barrier for the

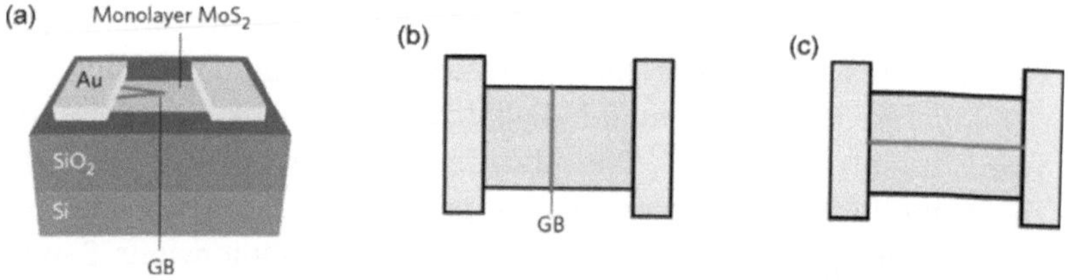

Figure 3.28 Schematics of memristor based on 2D MoS_2. (a) An interstecting-GB memristor where a GB only contacts with one of the two top electrodes. (b) A bisecting-GB memristor with a GB aligned perpendicular to the channel and isolated from the electrodes. (c) A bridging-GB memristor with a GB being parallel to the channel and connecting the two electrodes. Reprinted with permission from Yuan and Lou.[86] Copyright (2015) Nature Publishing Group.

atomic diffusion and bond flip in phosphorene allow the climb and glide of the 57 dislocation core. This suggests a similar superplasic deformation in phosphorene with even a much lower temperature than in CNTs.

The coalescence of a large number of vacancies is well known to lead to the formation of amorphous or glass structures.[89,90] Understanding the crystal-glass transformation is a long interest in the solid state field, where two competing models, random network theory and the crystallite theory, were proposed.[91] Since it is impossible to image a conventional bulk glass, graphene was believed to be an ideal material system to uncover this mystery as the rings can be imaged directly.[91] Unfortunately, as the melting temperature of graphene is amazingly high up to 12,000 K,[92] the carbon glass developed from graphene is normally obtained via continual high-energy electronic irradiation to induce the vacancies and trigger their migration and coalescence owing to their exceedingly high formation energy and migration barrier. Therefore, the displayed disorder and its evolution are not an intrinsic behavior of the material, but rather a tailored structure resulting from interactions with external particles. Disorder in graphene created in this approach occurs suddenly and shows a mixing character of the above two mechanisms highly depending on the power, the spinging time, and the size of the electron beam.[89] In contrast, with the small

E_f and high mobility of vacancies even at RT, phosphorene can lead to a new 2D glass produced through traditional methods via solely thermal heating and quenching without the need of irradiation. Indeed, a recent experiment observed the onset of amorphization of phosphorene by heating up to around 700 K in vacuum.[6] Plenty of new phenomena in phosphorene may be observed through in situ detection of the structural evolution at elevated temperatures, which is believed to enrich the understanding of the glass behavior in 2D.

The highly mobile vacancy in phosphorene implies attention needed to be paid for any attempts of vapor growth of phosphorene. A proper control of temperature and pressure should be required to enable the addition of atoms at the growing frontier while suppressing the disorder due to the fast vacancy combination. In addition, it is well known that phosphorene can be easily affected by the environment, showing a structural degradation in air, which is commonly ascribed to the adsorption of oxygen and water molecules. According to the itinerant vacancy uncovered by recent work,[72] the structural instability is more like to be an intrinsic origin. A promoted chemical activity associated with the defect centers may favor gas adsorption. It is likely that the structural degradation of phosphorene initiates at the edges, where the vacancies tend to agglomerate. Identifying the methods to protect the edges or the vacancies should effectively increase the stability of phosphorene.

3.6 Conclusions

Phosphorene shows a strong anisotropy in structure, phonons, and kinetics of defects. These highly anisotropic and strain-sensitive features of phononic properties suggest that phosphorene is a promising material for fabricating novel phononics devices, flexible nanoelectronics, and sensors. Its enhanced ability to manipulate the motion of encoding particles under various external stimuli leads to great progress in the design and fabrication of novel nanoelectronics devices nowadays. For instance, reversible cycling of insertion and transport of nucleus in host materials is at the heart of the fabrication of high-performance batteries. The remarkable motion of electrons/holes triggers great interest in graphene. Normally, transporting particles show itinerant or localized characters according to their moving capability under various external excitations. One exception is the atomic vacancies in bulk solids, which tend to be heavy and absent of itinerant behavior within an intact lattice through the application of solely limited thermal excitation. Mobile vacancies in bulk semiconductors, responsible for resistive switching, predominantly occur in filament or grain boundaries because of the synergistic effect of Joule heating and a high electric field. Owing to the reduced neighbors and the structural flexibility to afford the accompanying strain, one-dimensional (1D) and two-dimensional (2D) materials such as carbon nanotubes, graphene, MoS_2, and BN are believed to be more feasible for the production and control of vacancies. However, for both the native and the physically induced vacancies in these low-dimensional materials, thermally activated massive motion of vacancies around room temperature (300 $\sim$ 600 K) is still impossible and has to resort to continual electronic irradiation.

To this end, the itinerant features of vacancies in phosphroene, even at room temperature, resembling the ultrahigh mobile electrons/holes of its renowned predecessor, graphene, enable phosphorene to be a bright new opportunity to broaden the knowledge of the evolution of vacancies and their effects in the 2D limit. Deep understanding of the thermal-structural-mechanical relationships may inspire its potential applications in phase-change devices, and control of the anisotropic motion of these exceedingly mobile vacancies in phosphorene under external stimuli is highly promising for resistive switching devices. The salient features of vacancies in phosphorene enrich the understanding of the thermally activated behaviors of vacancies in low-dimension systems which are previously mainly acquired based on carbon materials and may eventually lead to a new paradigm for defect engineering of low-dimensional materials.

References

1. Corbridge DEC, (1978). *Phosphorus: An Outline of Its Chemistry Biochemistry and Technology.* Elsevier, Amsterdam, p. 1.
2. Corbridge DEC, (1974). *The Structural Chemistry of Phosphorus.* Elsevier, Amsterdam, p. 13.
3. Cai Y, Ke Q, Zhang G, Feng YP, Shenoy VB, Zhang Y-W, (2015). Giant phononic anisotropy and unusual anharmonicity of phosphorene: interlayer coupling and strain engineering. *Adv. Funct. Mater.*, **25**:2230–2236.
4. Brown A, Rundqvist S, (1965). Refinement of the crystal structure of black phosphorus. *Acta Crystallogr.*, **19**:684.
5. Zhang CD, Lian JC, Yi W, Jiang YH, Liu LW, Hu H, Xiao WD, Du SX, Sun LL, Gao HJ, (2009). Surface structures of black phosphorus investigated with scanning tunneling microscopy. *J. Phys. Chem. C*, **113**:18823–18826.
6. Liu X, Wood JD, Chen K-S, Cho E, Hersam MC, (2015). In situ thermal decomposition of exfoliated two-dimensional black phosphorus. *J. Phys. Chem. Lett.*, **6**: 773–778.
7. Douthwaite RE, Green MLH, Heyes SJ, Rosseinsky MJ, Turner JF, (1994). Synthesis and characterisation of the inclusion complex {(P4)2C60}. *J. Chem. Soc., Chem. Commun.*, 1367.
8. Ruck M, Hoppe D, Wahl B, Simon P, Wang Y, Seifert G, (2005). Fibrous red phosphorus. *Angew. Chem. Int. Ed.*, **44**: 7616.
9. Thurn H, Krebs H, (1969). Über struktur und eigenschaften der halbmetalle. XXII. Die kristallstruktur des hittorfschen phosphors. *Acta Crystallogr. B*, **25**:125–135.
10. Boecker S, Haeser M, Anorg Z, (1995). *Allg. Chem.*, **621**:358–386.
11. Pfitzner A, Braeu MF, Zweck J, Brunklaus G, Eckert H, (2004). Phosphorus nanorods - two allotropic modifications of a long-known element. *Angew. Chem. Int. Ed.*, **43**:4228.
12. Kim Y, Park Y, Choi A, Choi N-S, Kim J, Lee J, Ryu JH, Oh SM, Lee KT, (2013). An amorphous red phosphorus/carbon composite as a promising anode material for sodium ion batteries. *Adv. Mater.*, **25**:3045–3049.
13. Hittorf W, (1865). *Ann. Phys. Chem.*, **126**:193.
14. Martin TP, (1986). *Z. Phys. D*, **3**:211.

15. Bréchignac C, Cahuzac Ph, Carlier F, de Frutos M, Leygnier J, Roux JPh, (1995). Coulombic fission and evaporation of antimony cluster ions. *J. Chem. Phys.*, **102**:763.
16. Häser M, Schneider U, Ahlrichs R, (1992). Clusters of phosphorus: a theoretical investigation. *J. Am. Chem. Soc.*, **114**:9551–9559.
17. Jones RO, Hohl D, (1990). Structure of phosphorus clusters using simulated annealing-P_2 to P_8. *J. Chem. Phys.*, **92**:6710.
18. Mu L, Yang S, Bao X, Yin H, Kong X, (2015). Medium-sized phosphorus cluster cations P^+_{2m+1} ($6 \leq m \leq 32$) studied by collision-induced dissociation mass spectrometry. *J. Mass Spectrom.*, **50**:1352–1357.
19. Nava P, Ahlrichs R, (2008). Theoretical investigation of clusters of phosphorus and arsenic: fascination and temptation of high symmetries. *Chem. Eur. J.*, **14**:4039–4045.
20. Häser M, (1994). Structural rules of phosphorus. *J. Am. Chem. Soc.*, **116**:69256926.
21. Chen MD, Li JT, Huang RB, Zheng LS, Au CT, (1999). Structure prediction of large cationic phosphorus clusters. *Chem. Phys. Lett.*, **305**:439.
22. Jones RO, Ganteför G, Hunsicker S, Pieperhoff P, (1995). Structure and spectroscopy of phosphorus cluster anions: theory (simulated annealing) and experiment (photoelectron detachment). *J. Chem. Phys.*, **103**:9549.
23. Bulgakov AV, Bobrenok OF, Kosyakov VI, (2000). Laser ablation synthesis of phosphorus clusters. *Chem. Phys. Lett.*, **320**:19–25.
24. Seifert G, Heine T, Fowler PW, (2001). Inorganic nanotubes and fullerenes: structure and properties of hypothetical phosphorus fullerenes. *Eur. Phys. J. D*, **16**:341–343.
25. Han JG, Morales JA, (2004). A theoretical investigation on fullerene-like phosphorus clusters. *Chem. Phys. Lett.*, **396**:27–33.
26. Karttunen AJ, Linnolahti M., Pakkanen TA, (2008). Structural principles of polyhedral allotropes of phosphorus. *ChemPhysChem*, **9**:2550–2558.
27. Karttunen AJ, Linnolahti M., Pakkanen TA, (2011). Structural and electronic trends among group 15 polyhedral fullerenes. *Theor. Chem. Acc.*, **129**:413–422.
28. Yang Y, Arias F, Echegoyen L, Felipe Chibante LP, Flanagan S, Robertson A, Wilson LJ, (1995). Reversible fullerene electrochemistry: correlation with the HOMO-LUMO energy difference for C60, C70, C76, C78, and C84. *J. Am. Chem. Soc.*, **117**:7801–7804.

29. Seifert G, Hernández E, (2000).Theoretical prediction of phosphorus nanotubes. *Chem. Phys. Lett.*, **318**:355–360.

30. Winchester RAL, Whitby M, Shaffer MSP, (2009). Synthesis of pure phosphorus nanostructures. *Angew. Chem. Int. Ed.*, **48**:3616–3621.

31. Guo H, Lu N, Dai J, Wu X, Zeng XC, (2014). Phosphorene nanoribbons, phosphorus nanotubes, and van der Waals multilayers. *J. Phys. Chem. C*, **118**:14051–14059.

32. Guan J, (2014). High stability of faceted nanotubes and fullerenes of multiphase layered phosphorus: a computational study. *Phys. Rev. Lett.*, **113**:226801.

33. Ghorbani-Asl M, Zibouche N, Wahiduzzaman M, Oliveira AF, Kuc A, Heine T, (2013). Electromechanics in MoS_2 and WS_2: nanotubes vs. monolayers. *Sci. Rep.*, **3**:2961.

34. Golberg D, Bando Y, Huang Y, Terao T, Mitome M, Tang C, Zhi C, (2010). Boron nitride nanotubes and nanosheets. *ACS Nano*, **4**:2979–2993.

35. YuS, Zhu H, Eshun K, Arab A, Badwan A, Li Q, (2015). A computational study of the electronic properties of one-dimensional armchair phosphorene nanotubes. *J. Appl. Phys.*, **118**:164306.

36. Liu H, Neal AT, Zhu Z, Luo Z, Xu X, Tománek D, Ye PD, (2014). Phosphorene: an unexplored 2D semiconductor with a high hole mobility. *ACS Nano*, **8**:4033–4041.

37. Lu W, Nan H, Hong J, Chen Y, Zhu C, Liang Z, Ma X, Ni Z, Jin C, Zhang Z, (2014). Plasma-assisted fabrication of monolayer phosphorene and its Raman characterization. *Nano Res.*, **7**:853–859.

38. Qiao J, Kong X, Hu Z-X, Yang F, Ji W, (2014). High-mobility transport anisotropy and linear dichroism in few-layer black phosphorus. *Nat. Commun.*, **5**:4475.

39. Ong Z-Y, Cai Y, Zhang G, Zhang Y-W, (2014). Strong thermal transport anisotropy and strain modulation in single-layer phosphorene. *J. Phys. Chem. C*, **118**:25272–25277.

40. Tan PH, Han WP, Zhao WJ, Wu ZH, Chang K, Wang H, Wang YF, Bonini N, Marzari N, Pugno N, Savini G, Lombardo A, Ferrari AC, (2012). The shear mode of multilayer graphene. *Nat. Mater.*, **11**:294–300.

41. Boukhicha M, Calandra M, Measson M-A, Lancry O, Shukla A, (2013). Anharmonic phonons in few-layer MoS_2: Raman spectroscopy of ultralow energy compression and shear modes. *Phys. Rev. B*, **87**:195316.

42. Yoshizawa M, Shirotani I, Fujimura T, (1986). Thermal and elastic properties of black phosphorus. *J. Phys. Soc. Jpn.*, **55**:1196–1202.

43. Kôzuki Y, Hanayama Y, Kimura M, Nishitake T, Endo S, (1991). Measurement of ultrasound velocity in the single crystal of black phosphorus up to 3.3 GPa gas pressure. *J. Phys. Soc. Jpn.*, **60**:1612–1618.
44. Yoshihara A, Fujimura T, Oka Y, Fujisaki H, Shirotani I, (1987). Brillouin scattering studies on low frequency surface dynamics in black phosphorus. *J. Phys. Soc. Jpn.*, **56**:1223–1232.
45. Dong S, Zhang A, Liu K, Ji J, Ye YG, Luo XG, Chen XH, Ma X, Jie Y, Chen C, Wang X, Zhang Q, (2016). Ultralow-frequency collective compression mode and strong interlayer coupling in multilayer black phosphorus. *Phys. Rev. Lett.*, **116**:087401.
46. Ling X, Liang L, Huang S, Puretzky AA, Geohegan DB, Sumpter BG, Kong J, Meunier V, Dresselhaus MS, (2015). Low-frequency interlayer breathing modes in few-layer black phosphorus. *Nano Lett.*, **15**:4080–4088.
47. Ribeiro HB, Pimenta MA, de Matos CJS, Moreira RL, Rodin AS, Zapata JD, de Souza EAT, Castro Neto AH, (2015). Unusual angular dependence of the Raman response in black phosphorus. *ACS Nano*, **9**:4270–4276.
48. Kim J, Lee J-U, Lee J, Park HJ, Lee Z, Lee C, Cheong H, (2015). Anomalous polarization dependence of Raman scattering and crystallographic orientation of black phosphorus. *Nanoscale*, **7**:18708.
49. Late DJ, (2015). Temperature dependent phonon shifts in few-layer black phosphorus, *ACS Appl. Mater. Interfaces*, **7**:5857–5862.
50. Wang Y, Cong C, Fei R, Yang W, Chen Y, Cao B, Yang L, Yu T, (2015). Remarkable anisotropic phonon response in uniaxially strained few-layer black phosphorus. *Nano Res.*, **8**:3944–3953.
51. Ribeiro HB, Villegas CEP, Bahamon DA, Muraca D, Castro Neto AH, de Souza EAT, Rocha AR, Pimenta MA, de Matos CJS, (2016). Edge phonons in black phosphorus. *Nat. Commun.*, **7**:12191.
52. Wei Q, Peng X, (2014). Superior mechanical flexibility of phosphorene and few-layer black phosphorus. *Appl. Phys. Lett.*, **104**:251915.
53. Hu T, Dong J, (2015). Structural phase transitions of phosphorene induced by applied strains. *Phys. Rev. B*, **92**:064114.
54. Wang L, Kutana A, Zou X, Yakobson BI, (2015). Electro-mechanical anisotropy of phosphorene. *Nanoscale*, **7**:9746.
55. Cai Y, Zhang G, Zhang Y-W, (2014). Polarity-reversed robust carrier mobility in monolayer MoS_2 nanoribbons. *J. Am. Chem. Soc.*, **136**:6269–6275.
56. Song L, Ci L, Lu H, Sorokin PB, Jin C, Ni J, Kvashnin AG, Kvashnin DG, Lou J, Yakobson BI, Ajayan PM, (2010). Large scale growth and

characterization of atomic hexagonal boron nitride layers. *Nano Lett.*, **10**:3209–3215.

57. Lee C, Wei X, Li Q, Carpick R, Kysar JW, Hone J, (2009). Elastic and frictional properties of graphene. *Phys. Status Solidi B*, **246**:2562–2567.
58. Kulish VV, Malyi OI, Persson C, Wu P, (2015). Evaluation of phosphorene as anode material for Na-ion batteries from first principles. *Phys. Chem. Chem. Phys.*, **17**:13921–13928.
59. Appalakondaiah S, Vaitheeswaran G, Lebègue S, Christensen NE, and Svane A, (2012). Effect of van der Waals interactions on the structural and elastic properties of black phosphorus. *Phys. Rev. B*, **86**:035105.
60. Akai T, Endo S, Akahama Y, Koto K, Maruyama Y, (1989). The crystal structure and oriented transformation of black phosphorus under high pressure. *High Press. Res.*, **1**:115.
61. Jiang J-W, Park HS, (2014). Negative poisson's ratio in single-layer black phosphorus. *Nat. Commun.*, **5**:4727.
62. Tao J, Shen W, Wu S, Liu L, Feng Z, Wang C, Hu C, Yao P, Zhang H, Pang W, Duan X, Liu J, Zhou C, Zhang D, (2015). Mechanical and electrical anisotropy of few-layer black phosphorus. *ACS Nano*, **9**:11362–11370.
63. Peng X, Wei Q, Copple A, (2014). Strain-engineered direct-indirect band gap transition and its mechanism in two-dimensional phosphorene. *Phys. Rev. B*, **90**:085402.
64. Okajima M, Endo S, Akahama Y, Narita S-i, (1984). Electrical investigation of phase transition in black phosphorus under high pressure. *Jpn. J. Appl. Phys.*, **23**:15.
65. Xiang ZJ, Ye GJ, Shang C, Lei B, Wang NZ, Yang KS, Liu DY, Meng FB, Luo XG, Zou LJ, Sun Z, Zhang Y, Chen XH, (2015). Pressure-induced electronic transition in black phosphorus. *Phys. Rev. Lett.*, **115**:186403.
66. Marqués M, Ackland GJ, Lundegaard LF, Falconi S, Hejny C, McMahon MI, Contreras-García J, Hanand M, (2008). Origin of incommensurate modulations in the high-pressure phosphorus IV phase. *Phys. Rev. B*, **78**:054120.
67. Hässermann U, (2003). High-pressure structural trends of group 15 elements: simple packed structures versus complex host-guest arrangements. *Chem. Eur. J.*, **9**:1472.
68. Katzke H, Tolédano P, (2008). Displacive mechanisms and order-parameter symmetries for the A7-incommensurate-bcc sequences of high-pressure reconstructive phase transitions in Group Va elements. *Phys. Rev. B*, **77**:024109.

69. Skowron ST, Lebedeva IV, Popov AM, Bichoutskaia E, (2015). Energetics of atomic scale structure changes in graphene. *Chem. Soc. Rev.*, **44**:3143–3176.
70. Robertson AW, Montanari B, He K, Allen CS, Wu YA, Harrison NM, Kirkland AI, Warner JH, (2013). Structural reconstruction of the graphene monovacancy. *ACS Nano*, **7**:4495–4502.
71. El-Barbary AA, Telling RH, Ewels CP, Heggie MI, Briddon PR, (2003). Structure and energetics of the vacancy in graphite. *Phys. Rev. B*, **68**:144107.
72. CaiY, Ke Q, Zhang G, Yakobson BI, Zhang Y-W, (2016). Highly itinerant atomic vacancies in phosphorene. *J. Am. Chem. Soc.*, **138**:10199–10206.
73. Trevethan T, Latham CD, Heggie MI, Briddon PR, Rayson MJ, (2014). Vacancy diffusion and coalescence in graphene directed by defect strain fields. *Nanoscale*, **6**:2978–2986.
74. Kim Y, Ihm J, Yoon E, Lee G-D, (2011). Dynamics and stability of divacancy defects in graphene. *Phys. Rev. B*, **84**:075445.
75. Robertson AW, Allen CS, Wu YA, He K, Olivier J, Neethling J, Kirkland AI, Warner JH, (2012). Spatial control of defect creation in graphene at the nanoscale. *Nat. Commun.*, **3**:1144–1151.
76. Kotakoski J, Meyer JC, Kurasch S, Santos-Cottin D, Kaiser U, Krasheninnikov AV, (2011). Stone-Wales-type transformations in carbon nanostructures driven by electron irradiation. *Phys. Rev. B*, **83**:245420.
77. Kotakoski J, Krasheninnikov AV, Kaiser U, Meyer JC, (2011). From point defects in graphene to two-dimensional amorphous carbon. *Phys. Rev. Lett.*, **106**:105505.
78. Kim Y, Ihm J, Yoon E, Lee G-D, (2011). Dynamics and stability of divacancy defects in graphene. *Phys. Rev. B*, **84**:075445.
79. Robertson AW, Montanari B, He K, Allen CS, Wu YA, Harrison NM, Kirkland AI, Warner JH, (2013). Structural reconstruction of the graphene monovacancy. *ACS Nano*, **7**:4495–4502.
80. Liu Y, Xu F, Zhang Z, Penev ES, Yakobson BI, (2014). Two-dimensional mono-elemental semiconductor with electronically inactive defects: the case of phosphorus. *Nano Lett.*, **14**:6782–6786.
81. Gao J, Liu X, Zhang G, Zhang Y-W, (2016). Nanotube-terminated zigzag edges of phosphorene formed by self-rolling reconstruction. *Nanoscale*, **8**:17940.
82. Grujić MM, Ezawa M, Tadić MŽ, Peeters FM, (2016). Tunable skewed edges in puckered structures. *Phys. Rev. B*, **93**:245413.

83. Ramasubramaniam A, Muniz AR, (2014). Ab initio studies of thermodynamic and electronic properties of phosphorene nanoribbons. *Phys. Rev. B*, **90**:085424.

84. Yang JJ, Strukov DB, Stewart DR, (2013). Memristive devices for computing. *Nat. Nanotechnol.*, **8**:13–24.

85. Sangwan VK, Jariwala D, Kim IS, Chen K-S, Marks TJ, Lauhon LJ, Hersam MC, (2015). Gate-tunable memristive phenomena mediated by grain boundaries in single-layer MoS_2. *Nat. Nanotechnol.*, **10**:403–406.

86. Yuan J, Lou J, (2015). 2D materials: memristor goes two-dimensional. *Nat. Nanotechnol.*, **10**:389–390.

87. Bessonov AA, Kirikova MN, Petukhov DI, Allen M, Ryhänen T, Bailey MJA, (2015). Layered memristive and memcapacitive switches for printable electronics. *Nat. Mater.*, **14**:199–204.

88. Ding F, Jiao K, Wu M, Yakobson BI, (2007). Pseudoclimb and dislocation dynamics in superplastic nanotubes. *Phys. Rev. Lett.*, **98**:075503.

89. Eder FR, Kotakoski J, Kaiser U, Meyer JC, (2014). A journey from order to disorder-atom by atom transformation from graphene to a 2D carbon glass. *Sci. Rep.*, **4**:4060.

90. Pan C-T, Hinks JA, Ramasse QM, Greaves G, Bangert U, Donnelly SE, Haig SJ, (2014). In-situ observation and atomic resolution imaging of the ion irradiation induce damorphisation of graphene. *Sci. Rep.*, **4**:6334.

91. Wright AC, Thorpe MF, (2013). Eighty years of random networks. *Phys. Status Solidi B*, **250**:931–936.

92. Kumar A, Wilson M, Thorpe, MF, (2012). Amorphous graphene: a realization of Zachariasen's glass. *J. Phys. Condens. Matter*, **24**:485003.

Chapter 4

Thermal Properties of Phosphorene

Gang Zhang and Yong-Wei Zhang
Institute of High Performance Computing, 1 Fusionopolis Way, #16-16 Connexis North, Singapore 138632, Singapore
zhangg@ihpc.a-star.edu.sg, zhangyw@ihpc.a-star.edu.sg

In this chapter, we give a review on the recent advances in the study of thermal properties of monolayer and few-layer phosphorene. We first discuss some of the interesting features in the thermal conduction of monolayer and multilayer phosphorene and their underlying mechanisms. Next, we present recent understandings in developing phosphorene phononic crystal. Finally, we examine recent experimental works on the thermal conductivity of few-layer phosphorene.

Due to its unusual structural and electronic properties, phosphorene holds great promise for its applications in nanoelectronic and optoelectronic devices. However, effective thermal management of strong localized Joule heating (or hot spots) in those nanoscale devices remains challenging. Hence, use of highly thermally conducting materials in or around hot spots is of great important for dissipating heat and preventing device performance from degradation and breakdown. Recent studies have presented several interesting findings about the thermal properties of phosphorene. Thus, it would be of great value to critically review these findings and

Phosphorene: Physical Properties, Synthesis, and Fabrication
Edited by Yongqing Cai, Gang Zhang, and Yong-Wei Zhang

ISBN 978-981-4774-64-2 (Hardcover), 978-0-203-71061-6 (eBook)
www.jennystanford.com

evaluate the possibility of integrating this interesting material into nanoscale electronic devices. In this chapter, we would like to review the thermal conduction of monolayer and multilayer phosphorene, focusing on some of the most important characteristics, including anisotropic thermal conduction in monolayer phosphorene, thermal conduction in multilayer phosphorene, phosphorene phononic crystal, and recent experimental progresses.

4.1 Anisotropic Thermal Conduction of Monolayer Phosphorene

Using first-principles calculations and the nonequilibrium Green's function method, Ong et al. explored the ballistic thermal conductance of monolayer phosphorene.[1] Figure 4.1a shows the thermal conductance in the zigzag (G_{zigzag}) and armchair (G_{armchair}) direction at different values of biaxial tensile strain as a function of temperature. At room temperature, G_{zigzag} and G_{armchair} for phosphorene (without strain) are equal to 1.08 and 0.76 nW/K/nm^2, respectively. The value of G_{zigzag} is comparable to that of MoS_2 ($\sim$1.28 nW/K/nm^2).[2] Such a low thermal conductance may be due to the non-planar structure of single-layer phosphorene. Interestingly, in stark contrast to the isotropic thermal conductance observed in monolayer graphene and MoS_2, monolayer phosphorene exhibits a strong anisotropy in its thermal conductance. Figure 4.1b shows the thermal conductance ratio $G_{\mathrm{zigzag}}/G_{\mathrm{armchair}}$ as a function of temperature. It is seen that at room temperature, $G_{\mathrm{zigzag}}/G_{\mathrm{armchair}} \approx$ 1.4. The anisotropy ratio increases as the temperature decreases, attaining a peak value of 1.6 at around 28 K at zero strain.

The strong anisotropy is certainly one of the most unique characteristics in the thermal conduction of phosphorene. More interestingly, this anisotropy ratio can be modulated by strain. Figure 4.1a shows the decrease in both G_{zigzag} and G_{armchair} as a function of strain. Careful examination shows, however, the decrease in G_{armchair} is proportionally larger than that in G_{zigzag}. Figure 4.1b shows the variation of thermal conductance ratio $G_{\mathrm{zigzag}}/G_{\mathrm{armchair}}$ as a function of temperature. It is seen that at 300 K, the ratio

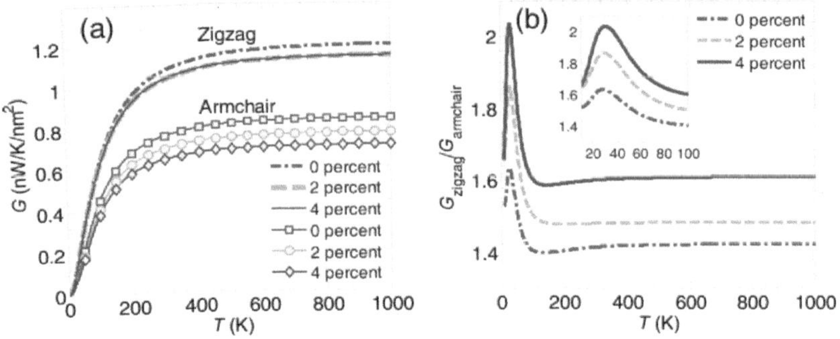

Figure 4.1 (a) Thermal conductance in the zigzag (G_{zigzag}) and armchair (G_{armchair}) direction as a function of temperature at different values of biaxial strain. (b) Plot of their ratio ($G_{\text{zigzag}}/G_{\text{armchair}}$) as a function of temperature. The inset shows the ratio at low temperature. Reprinted with permission from Ong et al.[1] Copyright (2014) by the American Chemical Society.

$G_{\text{zigzag}}/G_{\text{armchair}} \approx 1.4$ under zero strain. As the strain increases to 4%, the ratio rises to 1.6. Therefore, the conductance anisotropy can be enhanced by applying a biaxial strain. In particular, at the low-temperature range ($T < 100$ K), the anisotropy peaks at around 28 K, with a value of 1.6 under zero strain and 2.0 under 4% strain.

The anisotropy in thermal conduction has implications for heat dissipation in 2D crystal-based nanoscale devices, where Joule heating in the confined volume of the atomically thin crystal can create localized hot spots and the management of waste heat is expected to be a bottleneck to the large-scale functional circuit integration. In monolayer phosphorene, the carrier mobility is considerably higher along the armchair direction than that along the zigzag direction. Thus in a phosphorene field effect transistor, the source–drain electrical current should be aligned in the armchair direction, which is unfortunately also the direction with the inferior thermal conductance, around 20% of that of graphene. Thus, new design strategies for the thermal management of phosphorene-based electrical and optoelectronic devices are needed. The relatively higher thermal conductance along the zigzag direction suggests that heat dissipation should be channeled transversely rather than to metal source/drain electrodes. The cross-shape device structure

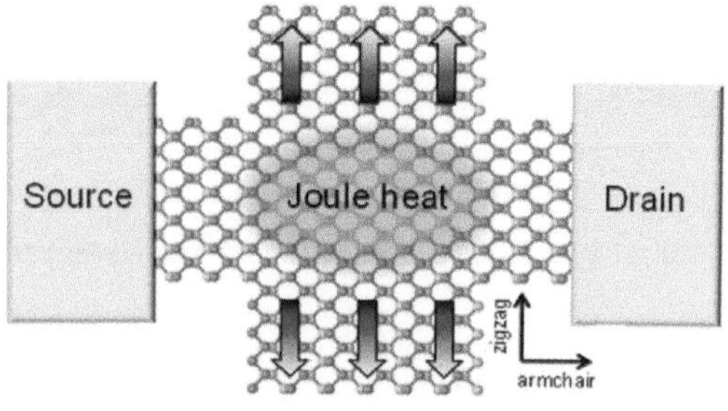

Figure 4.2 Schematic of the cross-shape device structure for dissipating heat away in the direction transverse to the electrical current. Reprinted with permission from Ong et al.[1] Copyright (2014) by the American Chemical Society.

as shown in Fig. 4.2 allows the electrical current to flow along armchair direction while enabling heat to be dissipated transversely, taking advantage of the superior thermal conductance in the zigzag direction.

Figure 4.3 shows the thermal conductance in the zigzag and armchair direction as a function of temperature for strain applied in the zigzag and armchair direction. G_{armchair} decreases with the

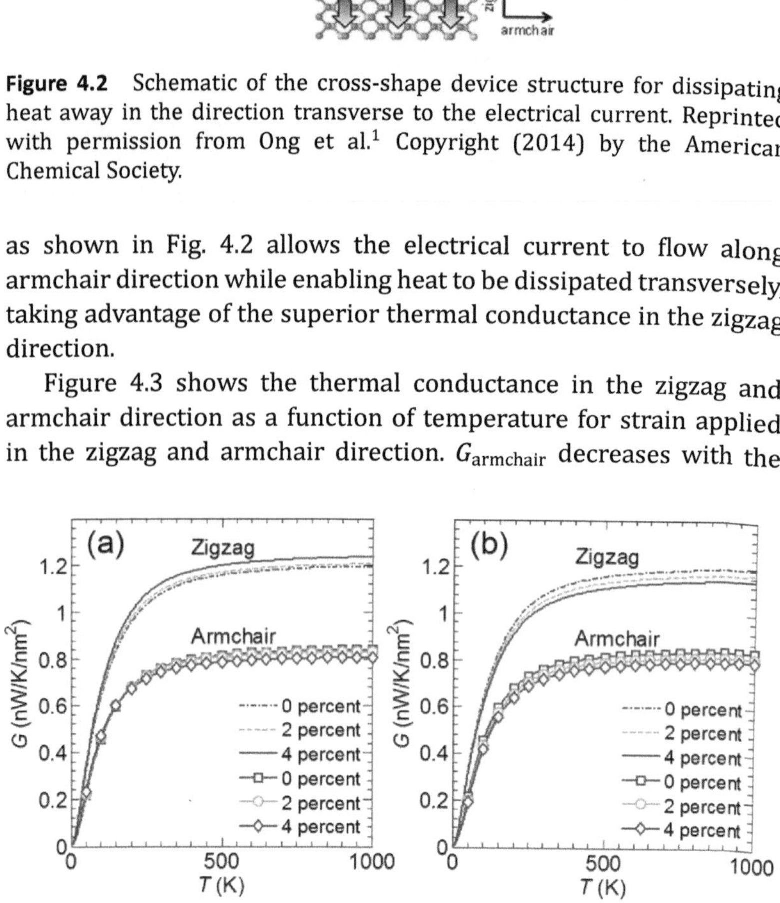

Figure 4.3 Thermal conductance in the zigzag (G_{zigzag}) and armchair (G_{armchair}) direction as a function of temperature at different values of uniaxial strain applied in the (a) zigzag and (b) armchair direction. Reprinted with permission from Ong et al.[1] Copyright (2014) by the American Chemical Society.

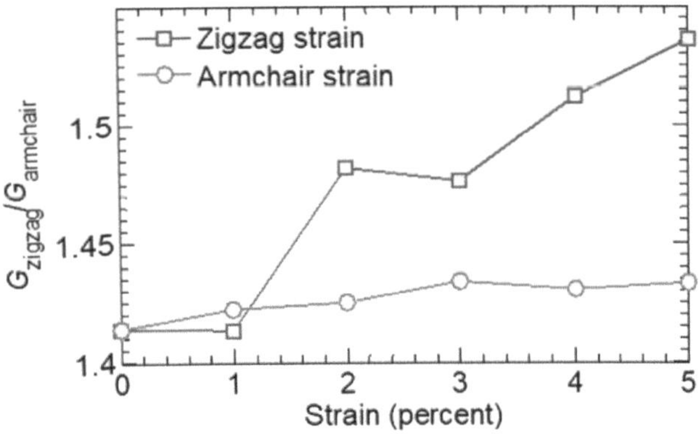

Figure 4.4 Thermal conductance anisotropy ($G_{\text{zigzag}}/G_{\text{armchair}}$) at different values of uniaxial strain applied in the zigzag and armchair direction at 300 K. Reprinted with permission from Ong et al.[1] Copyright (2014) by the American Chemical Society.

tensile strain in either the armchair or the zigzag direction. The decrease in G_{armchair} under tensile strain is similar to the trend observed in thermal transport simulations of bulk materials, which has been explained as originating from the lower phonon group velocities when strain is applied.

On the other hand, it is noted that G_{zigzag} exhibits an anomalous orientation-dependent change. When the strain is applied in the armchair direction, G_{zigzag} decreases like G_{armchair}. However, when the tensile strain is applied in the zigzag direction, G_{zigzag} actually increases by a small but perceptible amount. This anomaly in the change in G_{zigzag} and G_{armchair} implies that the anisotropy of thermal conductance can be modulated by changing the direction of the applied strain. Figure 4.4 shows the anisotropy ratio ($G_{\text{zigzag}}/G_{\text{armchair}}$) as a function of the applied strain in the zigzag and armchair direction. The anisotropy ratio is $\sim$1.4 for zero strain. But with the increase in the applied strain, the anisotropy ratio increases. This increase in the anisotropy ratio is substantially greater for strain in the zigzag direction than in the armchair direction.

The phonon transmittance along the zigzag direction ($\Xi_{\text{zigzag}}(\omega)$, Fig. 4.5a) and armchair direction ($\Xi_{\text{armchair}}(\omega)$, Fig. 4.5b) are explored to understand the nontrivial strain effects on thermal

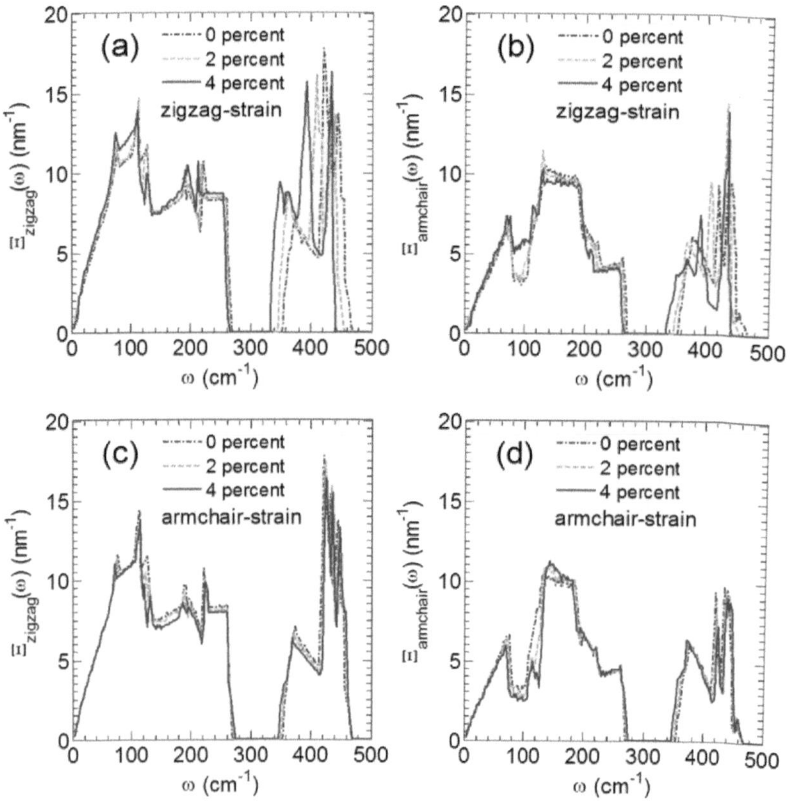

Figure 4.5 The zigzag- and armchair-oriented cross-sectional phonon transmission (Ξ_{zigzag} and $\Xi_{armchair}$) as a function of frequency at different values of uniaxial strain applied in the (a, b) zigzag and (c, d) armchair direction. Reprinted with permission from Ong et al.[1] Copyright (2014) by the American Chemical Society.

conductance. In the plot of the transmittances, the spectrum gives the frequency-dependent relative contribution to the thermal conductance. It is clear that the strain in the zigzag direction results in marked phonon softening. Another consequence of the strain is that the contribution from the low frequency modes ($\omega < 120$ cm^{-1}) Ξ_{zigzag} is enhanced. In contrast, when the strain is applied in the armchair direction, the frequency redshift in $\Xi_{zigzag}(\omega)$ (Fig. 4.5c) and $\Xi_{armchair}(\omega)$ (Fig. 4.5d) are much smaller. There is also a small

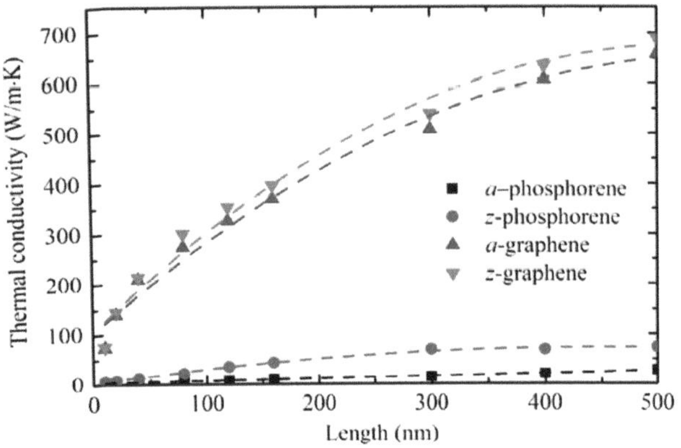

Figure 4.6 Thermal conductivities of phosphorene and graphene along the armchair or zigzag directions versus length. Reprinted with permission from Hong et al.[4] Copyright (2015) by the Royal Society of Chemistry.

but visible reduction in the phonon transmittances, which explains the decrease in G_{zigzag} and G_{armchair}.

The obvious direction-dependent thermal conduction is confirmed by other independent theoretical works,[3–10] including molecular dynamics simulation and first-principle-based Boltzmann transport calculations. Hong et al. studied the length dependence of thermal conductivities of the phosphorene sheet in the armchair and zigzag directions and compared them with those of graphene sheet.[4] Figure 4.6 shows that the thermal conductivities of phosphorene sheet are around one order of magnitude lower than those of graphene sheet. The results shown in Fig. 4.6 indicate that the anisotropic thermal transport in phosphorene sheet is much more significant than that in graphene sheet. This high anisotropy is partially attributed to the direction-dependent group velocities and anisotropic phonon dispersion in phosphorene. Undoubtedly, the distinct pucker structures in the armchair and zigzag directions also contribute to the strong anisotropic thermal conductivities in the phosphorene sheet.

Liu and Chang studied the heat transport in other crystal directions in monolayer phosphorene.[9] As shown in Fig. 4.7, the thermal conductivity of monolayer phosphorene shows a smooth, monotonic decreasing trend with increasing the crystal chirality.

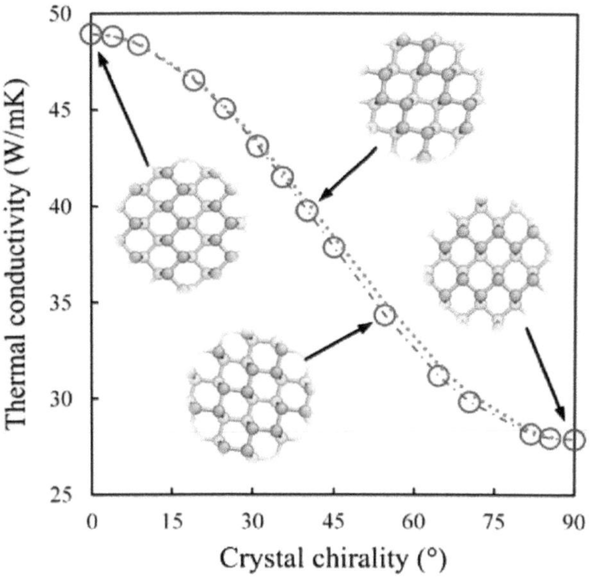

Figure 4.7 Orientation-dependent thermal conductivity of phosphorene sheets. The gray dot line was fitted by cosine function with a period of 2π. Reprinted with permission from Liu and Chang.[9] Copyright (2015) by the Royal Society of Chemistry.

Figure 4.8 shows the temperature-dependent thermal conductivity. It is interesting to see that at high temperatures, the thermal conductivities are exactly in T^{-1} dependence for phosphorene sheets while the temperature exceeds 200 K. In addition, the 90° chirality has the highest thermal conductivity at the very low temperatures, yet has the lowest value with increasing temperature because of its smaller phonon group velocities and larger Grüneisen parameters.

Furthermore, the thermoelectric properties of phosphorene also have been investigated.[11–15] Although a high thermoelectric figure of merit *ZT* was expected, unfortunately, a theoretical value of only $ZT = 0.14$ is found in p-type impurity-free phosphorene at 500 K.[14] By analyzing the various effects on *ZT*, Zhou et al. proposed the possible means to enhance the figure of merit *ZT*.[15] In weakly doped condition, the Seebeck coefficient of monolayer phosphorene

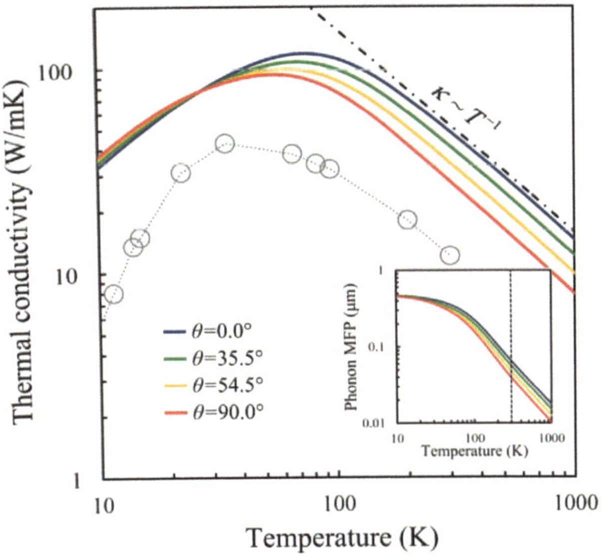

Figure 4.8 Temperature-dependent thermal conductivity of phosphorene sheets. The blue, green, yellow, and red lines represent the thermal conductivity along the 0.0°, 35.5°, 54.5°, and 90.0° chiralities, respectively. The inset shows the phonon MFP as a function of temperature. Reprinted with permission from Liu and Chang.[9] Copyright (2015) by the Royal Society of Chemistry.

is high. Thus the electrical/thermal conductance ratio is the major factor that affects its thermoelectric performance, due to the low electron conduction and high thermal conduction. In this regime, the phonon thermal conductance plays a major role in comparison with electronic thermal conductance; therefore, reduction of lattice thermal conductivity will significantly enhance *ZT*. Possible means include creation of superlattice to localize the long-wavelength phonons or introduction of mass disorders and/or edge roughness to suppress the phonon transmissions. On the other hand, in highly doped condition, the reduction of Seebeck coefficient and the rapid increase of electronic thermal conduction are the major factors that limit *ZT*. In this regime, controlling electronic structure is critical for boosting *ZT*. Possible methods include electromagnetic field modulation and/or strain engineering.

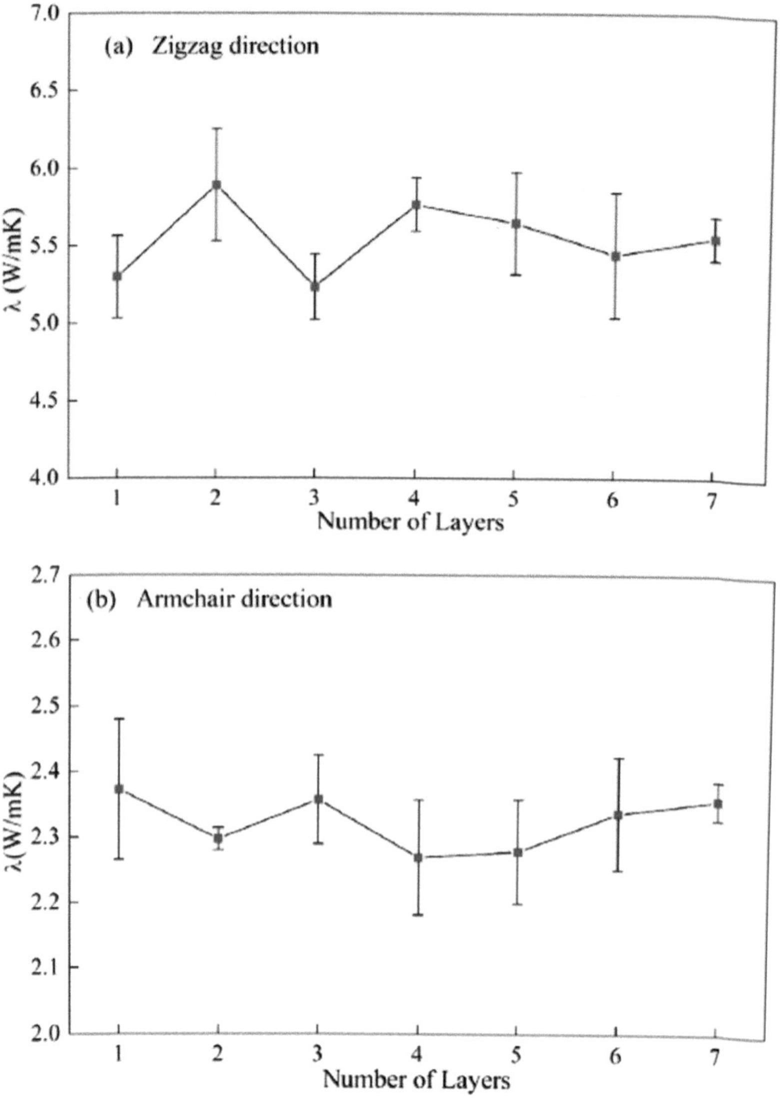

Figure 4.9 The in-plane thermal conductivity versus the number of layers in multilayer phosphorene along the zigzag (a) and armchair (b) directions. Reprinted with permission from Zhang et al.[5] Copyright (2016) by the Royal Society of Chemistry.

4.2 Thermal Conductivity of Multilayer Phosphorene

It is worth noting that in addition to monolayer phosphorene, multilayer phosphorene was also studied experimentally. Clearly, it is interesting to explore the thickness dependence of thermal conductivity. Zhang et al. studied the thermal conductivity of multilayer phosphorene by varying the number of layers from 2 to 7 using molecular dynamics simulations.[5] The multilayer phosphorene samples have in-plane dimensions of 20 $\times$ 20 nm^2 and the layers are stacked in the stable A–B stacking sequence. Figure 4.9 shows the in-plane thermal conductivities of multilayer phosphorene along both the zigzag and armchair directions with different numbers of layers. It is clear that the thermal conductivities along these two in-plane directions are insensitive to the number of layers.

This layer number-independent thermal conductivity is in sharp contrast to that of graphene. Unlike graphene with only one-atom thickness, the atoms in single-layer phosphorene are arranged in two sub-layers and form a puckered geometry. This puckered structure in single-layer phosphorene hinders the out-of-plane (flexural) phonon mode and thus diminishes the layer effect in multilayer phosphorene. This is the underlying physical origin for this layer-independent thermal conductivity of multilayer phosphorene. Figure 4.10 shows the VDOS for both single-layer phosphorene and single-layer graphene. Figure 4.10a demonstrates that the out-of-plane vibration mode (flexural mode) in single-layer graphene makes predominant contribution to the heat conduction in the low frequency range. Thus for multilayer graphene, the interlayer vdW interactions restrict the flexural mode and thus the thermal conductivity of multilayer graphene decreases with increasing the layer number. In contrast to single-layer graphene, as shown in Fig. 4.10b, the flexural mode in mono-layer phosphorene contributes only slightly to the low-frequency thermal conduction. Therefore, the increase in the layer thickness and the associated interlayer vdW interactions exert a negligible effect on the thermal conductivity, resulting in the observed layer-independent characteristic in multilayer phosphorene.

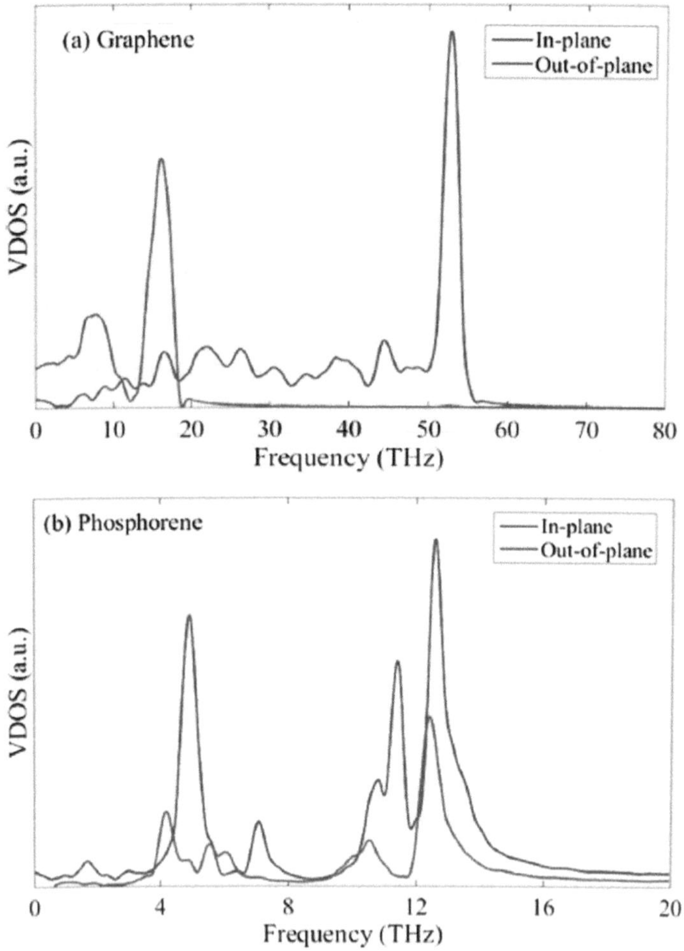

Figure 4.10 In-plane and out-of-plane VDOS of single-layer graphene (a) and single-layer phosphorene (b). Reprinted with permission from Zhang et al.[5] Copyright (2016) by the Royal Society of Chemistry.

4.3 Phosphorene Phononic Crystal

Phononic crystal structure has drawn intensive interest recently as such structure is able to modulate thermal conductivity but exert a weak influence on electrical conductivity. Therefore, if a

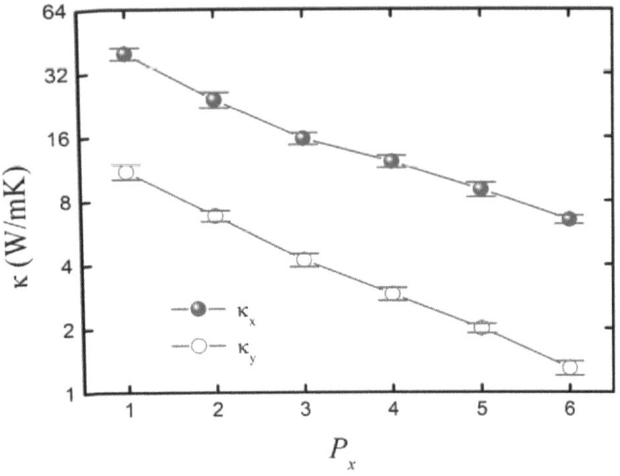

Figure 4.11 The room-temperature themal conductivity of phoshphoren phononic crystal versus the size of pores. Reprinted with permission from Xu and Zhang.[16] Copyright (2016) by the Institute of Physics.

similar trend holds for phosphorene, the phosphorene phononic crystal would have much higher ZT with respect to the pristine phosphorene monolayer in weakly doping regime.

Using molecular dynamics simulation, Xu and Zhang[16] studied the thermal properties of a phosphorene phononic crystal, which is a nanomesh structure with periodically arranged pores. In arranging a rectangular pore into each rectangular supercell of phosphorene, the dimension of each pore and each supercell is of $P_x \times P_y$ and $U_x \times U_y$ primitive unit cells, respectively. As shown in Fig. 4.11, there is a remarkable reduction in thermal conductivity in the phosphorene phononic crystal compared with that in pristine phosphorene, with κ_y being about 11.0 W/mK when $P_x = 1$, which is only about 33% of thermal conductivity along armchair direction in pristine phosphorene. When P_x increases to 6, the thermal conductivity further decreases to only 4%.

Two empirical models, that is, Eucken model and Maxwell–Garnett model, are commonly used to explore the thermal conductivity of porosity structures. In the Eucken model, $\kappa/\kappa_0 = (1 - \phi)/(1 + \frac{2}{3}\phi)$, and in the Maxwell–Garnett model, $\kappa/\kappa_0 = (1 - \phi)/(1 + \phi)$, where κ_0 is the thermal conductivity of the intact material and ϕ is the porosity (which is equal to $P_x P_y / U_x U_y$ in this work). When $P_x = 4$, ϕ is $P_x^2/100 = 0.16$, the Eucken model and the Maxwell–Garnett model predict $\frac{\kappa_x}{\kappa_{zig}}$ to be 76% and 72%, respectively. However, according to the molecular dynamics simulation, the calculated κ_x is only 8%. It is obvious that this remarkable reduction cannot be understood by the Eucken model and Maxwell–Garnett model.

In nanostructures, phonons usually experience intensive phonon-boundary scattering. In general, there are two types of phonon-boundary scattering processes: The particle-like incoherent scattering and the wave-like coherent scattering. The latter leads to Brillouin zone folding and new phonon dispersion due to wave interference. To realize the coherent scattering, the periodicity of the phononic crystal must be comparable or smaller than the phonon mean free path. The phonon mean free paths of phosphorene along the two directions are 141.4 nm (zigzag direction) and 43.4 nm (armchair direction),[3] which are larger than the periodic length considered here. Thus, the coherent picture is a reasonable approximation to describe the frequent phonon scattering at the high-density pore boundaries. In this condition, according to the flattened acoustic phonon branches and increased number of optical phonon branches, the zone folding greatly suppresses the phonon group velocities, and the suppression becomes even stronger with the increase of pore size. Moreover, since the zone folding in the phononic crystal shifts the phonon dispersion, the phonon lifetime is reduced in the phononic crystal as compared to that in the pristine phosphorene. Therefore, the remarkable reduction of thermal conductivity in phosphorene phononic crystal is the result of a combination of the suppressed phonon group velocities and reduced phonon lifetimes.

4.4 Experimental Results

The strong anisotropy in thermal conductivity of few-layer black phosphorous was confirmed by recent experimental observations.[17,18] In experiment,[17] the anisotropic in-plane thermal conductivity of suspended few-layer BP was measured using

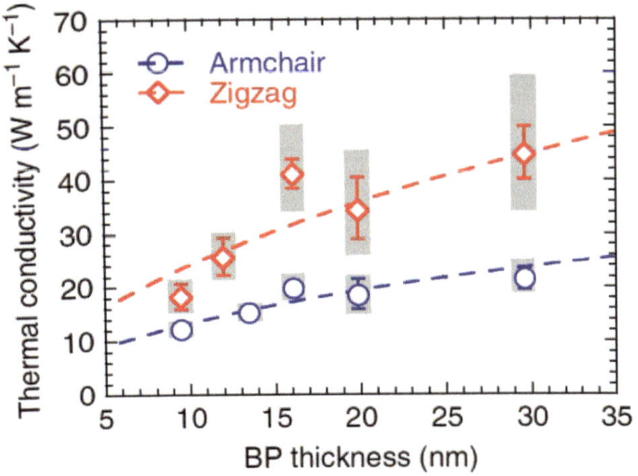

Figure 4.12 Experimentally measured armchair and zigzag in-plane thermal conductivities of multiple black phosphorous films. The grey error bars account for the uncertainty in the thermal conductivity of SiN substrate, whereas the blue/red error bars do not. Reprinted with permission from Luo et al.[17] Copyright (2015) by Nature Publication Group.

micro-Raman spectroscopy. Under laser illumination, the rise in local temperature leads to softened atomic bonds, resulting in a change in the optical phonon frequency and a red shift of the corresponding Raman peaks. Such laser-induced temperature-dependent Raman peak can serve as an optical thermometer, whereas the focused laser itself acts as a steady-state heat source. As shown in Fig. 4.12, the measured thermal conductivity ranges from 10 to 20 W/mK along the armchair direction, and about 20 to 40 W/mK along the zigzag direction. These measurements exhibit a strong thickness dependence and significant in-plane anisotropy, which are consistent with the aforementioned theoretical results.

4.5 Conclusions

In this chapter, we present an overview of the thermal properties of phosphorene. Our intent is to present the state-of-the-art view about this fascinating material with regard to its thermal properties, from both theoretical and experimental perspectives. It is hoped that the insights and understandings revealed and summarized here provide an important reference and guideline for the further study and applications of phosphorene in thermal management and heat-to-electricity conversion.

References

1. Ong Z-Y, Cai Y, Zhang G, Zhang Y-W, (2014). Strong thermal transport anisotropy and strain modulation in single-layer phosphorene. *J. Phys. Chem. C*, **118**:25272–25277.
2. Cai Y, Lan J, Zhang G, Zhang Y-W, (2014). Lattice vibrational modes and phonon thermal conductivity of monolayer MoS_2. *Phys. Rev. B*, **89**:035438.
3. Xu W, Zhu L, Cai Y, Zhang G, Li B, (2015). Direction dependent thermal conductivity of monolayer phosphorene: parameterization of Stillinger-Weber potential and molecular dynamics study. *J. Appl. Phys.*, **117**:214308.
4. Hong Y, Zhang JC, Huang X, Zeng XC, (2015). Thermal conductivity of a two-dimensional phosphorene sheet: a comparative study with graphene. *Nanoscale*, **7**:18716.
5. Zhang Y, Pei Q, Jiang J, Wei N, Zhang Y, (2016). Thermal conductivities of single-and multi-layer phosphorene: a molecular dynamics study. *Nanoscale*, **8**:483.
6. Zhu L, Zhang G, Li B, (2014). Coexistence of size-dependent and size-independent thermal conductivities in phosphorene. *Phys. Rev. B*, **90**:214302.
7. Jain A, McGaughey AJH, (2015). Strongly anisotropic in-plane thermal transport in single-layer black phosphorene. *Sci. Rep.*, **5**:8501.
8. Peng X, Chen K-Q, (2015). Comparison on thermal transport properties of graphene and phosphorene nanoribbons. *Sci. Rep.*, **5**:16215.
9. Liu T-H, Chang C-C, (2015). Anisotropic thermal transport in phosphorene: effects of crystal orientation. *Nanoscale*, **7**:10648.
10. Aierken Y, Cakir D, Sevik C, Peeters FM, (2015). Thermal properties of black and blue phosphorenes from a first-principles quasiharmonic approach. *Phys. Rev. B*, **92**:081408.
11. Qin G, Yan Q-B, Qin Z, Yue S-Y, Cui H-J, Zheng Q-R, Su G, (2014). Hinge-like structure induced unusual properties of black phosphorus and

new strategies to improve the thermoelectric performance. *Sci. Rep.*, **4**:6946.

12. Zhang J, Liu HJ, Cheng L, Wei J, Liang JH, Fan DD, Shi J, Tang XF, Zhang QJ, (2014). Phosphorene nanoribbon as a promising candidate for thermoelectric applications. *Sci. Rep.*, **4**:6452.
13. Fei R, Faghaninia A, Soklaski R, Yan J-A, Lo C, Yang L, (2014). Enhanced thermoelectric efficiency via orthogonal electrical and thermal conductances in phosphorene. *Nano Lett.*, **14**:6393.
14. Liao B, Zhou J, Qiu B, Dresselhaus MS, Chen G, (2015). Ab initio study of electron-phonon interaction in phosphorene. *Phys. Rev. B*, **91**:235419.
15. Zhou H, Cai Y, Zhang G, Zhang Y-W, (2016). Thermoelectric properties of phosphorene at the nanoscale. *J. Mater. Res.*, **31**:3179–3186.
16. Xu W, Zhang G, (2016). Remarkable reduction of thermal conductivity in phosphorene phononic crystal. *J. Phys. Condens. Matter*, **28**:175401.
17. Luo Z, Maassen J, Deng Y, et al. (2015). Anisotropic in-plane thermal conductivity observed in few-layer black phosphorus. *Nat. Commun.*, **6**:8572.
18. Lee S, Yang F, Suh J, et al. (2015). Anisotropic in-plane thermal conductivity of black phosphorus nanoribbons at temperatures higher than 100 K. *Nat. Commun.*, **6**:8573.

Chapter 5

Electronic and Optical Properties of Phosphorene

Yongqing Cai, Gang Zhang, and Yong-Wei Zhang

*Institute of High Performance Computing, The Agency for Science, Technology and Research (A*STAR), Singapore 138632, Singapore*

zhangg@ihpc.a-star.edu.sg, zhangyw@ihpc.a-star.edu.sg

Since the first isolation of single graphene layer from bulk graphite via Scotch tape, various other layered two-dimensional (2D) materials, including hexagonal boron nitride and transition metal dichalcogenides like MoS_2 and WS_2, were isolated through a similar method. Although the bulk form of these materials had been well studied for decades, the atomically thin 2D layers attracted new interest due to the new electronic, chemical, and optical characteristics compared with their bulk counterparts. The synthesis of phosphorene, a monolayer form of bulk black phosphorus, is another attempt in the exploration of post-graphene 2D materials.

Phosphorene is gaining great attention due to a finite bandgap and another elemental 2D material besides graphene. Field-effect transistors made from graphene show a high room-temperature mobility up to 1000 $cm^2V^{-1}s^{-1}$ and an ON/OFF ratio of 10^5. The bandgap of bulk black phosphorus is around 0.3 eV. The bandgap

Phosphorene: Physical Properties, Synthesis, and Fabrication
Edited by Yongqing Cai, Gang Zhang, and Yong-Wei Zhang

ISBN 978-981-4774-64-2 (Hardcover), 978-0-203-71061-6 (eBook)
www.jennystanford.com

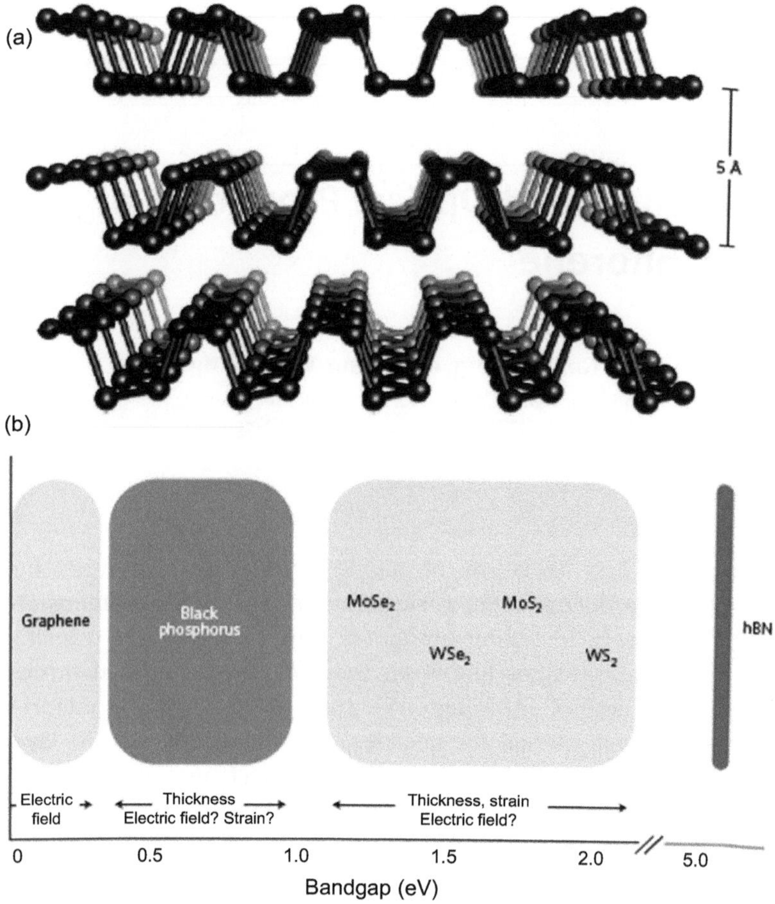

Figure 5.1 (a) Atomic structure of black phosphorus. (b) Comparison of the bandgap of black phosphorus with other common layered materials. The range of the bandgap for each type of material can be achieved through thickness control or strain engineering. Reprinted with permission from Churchill and Jarillo-Herrero.[1] Copyright (2014) Nature Publishing Group.

increases gradually to 1.5 eV as the thickness decreases to the monolayer. The large tunability of bandgap allows promising applications from infrared optoelectronics to high-mobility nanoelectronics. As shown in Fig. 5.1, the magnitude of the bandgap of phosphorene bridges those of the metallic graphene, semiconducting transition metal dichalcogenides, and insulator h-BN, thereby producing layered 2D materials covering a complete spectrum of bandgaps for applications in nanoelectronics. Importantly, in contrast to the semiconducting dichalcogenides with a direct bandgap in the monolayer sheet and an indirect bandgap in multilayer forms, phosphorene shows a robust direct bandgap for all thickness, which is highly desired for optoelectronic devices. Another strikingly different character associated with its anisotropic structure is that each individual layer is buckled. This intrinsic anisotropy leads to a strong anisotropy in physical properties like the electronic and phononic transports. In the following sections of this chapter, we briefly introduce the electronic and optical properties of phosphorene.

5.1 Electronic Structure of Phosphorene

Table 5.1 summarizes the fundamental properties of phosphorene in comparison with other 2D materials. It can be seen that phosphorene possesses dramatically different electronic and mechanical properties compared with other 2D materials. For example, it is well known that graphene has superior carrier mobility owing to an ideally zero mass of carriers and a Dirac cone band structure around the Fermi level. However, its zero bandgap leads to a very low ON/OFF ratio, which limits its application in transistors. Transition metal dichalcogenides like MoS_2 and WSe_2 exhibit a high ON/OFF ratio due to the presence of a finite bandgap. However, their low mobility hinders their widespread applications in nanoelectronics. Phosphorene, owing to its relatively high carrier mobility and satisfying bandgap, is a highly promising material for novel nanoelectronics and solar cell applications. Phosphorene has a relatively high carrier mobility ($\sim$000 $cm^2V^{-1}s^{-1}$ at 120 K) and outstanding ON/OFF ratio (10^3–10^5). Complex band structure analysis[2] shows that phosphorene has a smaller decay rate and a better transmission and electronic performance than those of the WTe_2 despite a similar bandgap value.

Another intriguing property is related to its asymmetric transport of electrons and phonons. For electronic transport, the favored

Table 5.1 Comparison of the fundamental properties of phosphorene to those of other common 2D materials*

	Graphene	Phosphorene	MoS_2	h-BN
Bandgap (eV)	0	0.3 – 2.0	1.2 – 1.8	5.9
Carrier mobility ($cm^2V^{-1}S^{-1}$)	$\sim 2 \times 10^5$	~1000	10 – 200	–
ON/OFF ratio	5.5 – 44	$10^3 - 10^5$	$10^6 - 10^8$	–
Thermal conductance ($W\,m^{-1}K^{-1}$)	5×10^3	10 – 36	34.5 – 52	250 – 360

*Data from Batmunkh et al.[3]

direction is the armchair direction, whereas for phonons the preferred transport direction is the zigzag direction, which is sharply different from those of electrons. This opposite transport, together with its low thermal conductivity and high electronic conductivity, allows promising thermoelectric applications.

5.1.1 Electronic Structures of Bulk Black Phosphorus and Few-Layer Phosphorene

The band structures of bulk black phosphorus and few-layer phosphorene were systematically studied by Cai et al.[4] The band structure of bulk black phosphorus calculated by hybrid functional (HSE06) is shown in Fig. 5.2, where a direct bandgap of 0.39 eV is obtained at the Z point, which is in good agreement with the experimental results.[5] In principle, a more accurate bandgap can be obtained by using *GW* approximation. A slightly smaller value of 0.3 eV is calculated with a non-self-consistent G_0W_0 scheme.[6] In addition, the adoption of the more reliable random phase approximation (RPA) within the G_0W_0 scheme leads to an even smaller value of 0.1 eV.[7]

The band structures for few-layer (1L to 5L) phosphorene calculated by HSE06 method is shown in Fig. 5.3a. It can be seen that all the phosphorene layers with different thicknesses have a direct bandgap at Γ point and share a similar topology. The bandgap decreases with increasing thickness owing to the quantum confinement effect. For instance, the bandgaps for 1L and

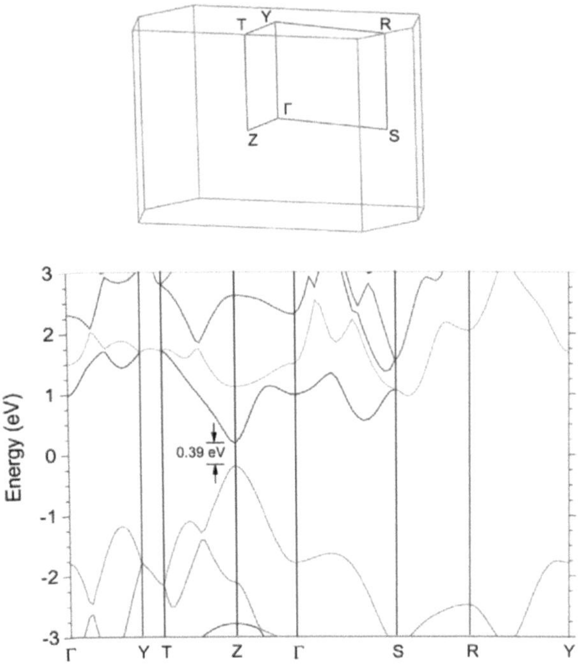

Figure 5.2 Brillouin zone of bulk black phosphorus (top panel) and the electronic band structure of bulk black phosphorus calculated by hybrid functional (HSE06) method (bottom panel). Reprinted with permission from Cai et al.[4] Copyright (2014) Nature Publishing Group.

2L phosphorene are 1.52 and 1.02 eV, respectively. This thickness-dependent gap is particularly useful for the construction of multi-junction solar cells, where a promoted efficiency can be obtained due to a full covering of the solar spectrum. This intriguing character was used in experiments, and it resulted in a fast response to the optical excitations from visible to 940 nm wavelength.[8] Most importantly, contrary to the 2D transition metal dichalcogenides materials, the 1L, few-layer, and bulk phosphorene has a direct bandgap, which is highly desirable for optoelectronic applications and efficient solar energy harvest. In addition, there are similar asymmetric band dispersions around the zone center along the Γ–X and Γ–Y directions for the few-layer phosphorene, implying a robust anisotropic behavior of the electronic structures.

On the basis of derived band structures, it is easy to analyze the effective mass by calculating $\hbar^2[\partial^2\varepsilon(k)/\partial k_\alpha \partial k_\beta]^{-1}$. Owing to the anisotropic structure and band dispersions, the effective mass for holes and electrons is also highly angularly dependent. The electron mass is smaller by one order along the armchair direction (Γ–Y) than along the zigzag direction (Γ–X) for all phosphorene layers with different thicknesses. For instance, the effective mass of electron is 0.14 m_e (1.23 m_e) along the armchair (zigzag) direction for monolayer phosphorene. The anisotropic behavior in effective mass is even larger for holes up to a two-order difference between the two directions. There is a large decrease in hole mass from 1L (13.09 m_e) to 2L (2.18 m_e) phosphorene along the zigzag direction, implying a large modification of band curvature due to interlayer coupling. It is expected that few-layer phosphorene should outperform monolayer phosphorene with a higher carrier mobility due to a higher carrier density, a smaller effective mass, and a weaker scattering because of the improved screening. In addition, such significant anisotropy in the effective mass of the conducting carriers in phosphorene is absent in other common 2D materials like MoS_2 and BN and may lead to novel applications.

This strongly orientationally different effective mass of electrons and holes can account for the large anisotropic conduction observed in experiment.[9] Angular dependence of the transport behavior was measured based on a peeled black phosphorus flake of around 10 nm, which was transferred to a 90 nm SiO_2-capped Si substrate. Metal electrodes were deposited above the flake with an angular increment of 45° (Fig. 5.3b,c). It was found that the drain current is angularly dependent, reflecting an anisotropic and orientationally dependent transport behavior in phosphorene. The maximum drain current roughly follows a sinusoidal trend with a minimum value of ~85 mA/mm at 45° and 225° and a maximum value of ~137 mA/mm at 135° and 315°. A similar oscillation trend was also observed in the transconductance, a quantity closely related to the carriers' mobility. The nearly 50% difference between the two orthogonal directions is not present in other conventional 2D materials. The dramatically different transport behavior can be rooted in uniquely puckered structure and the anisotropic electronic properties shown above.

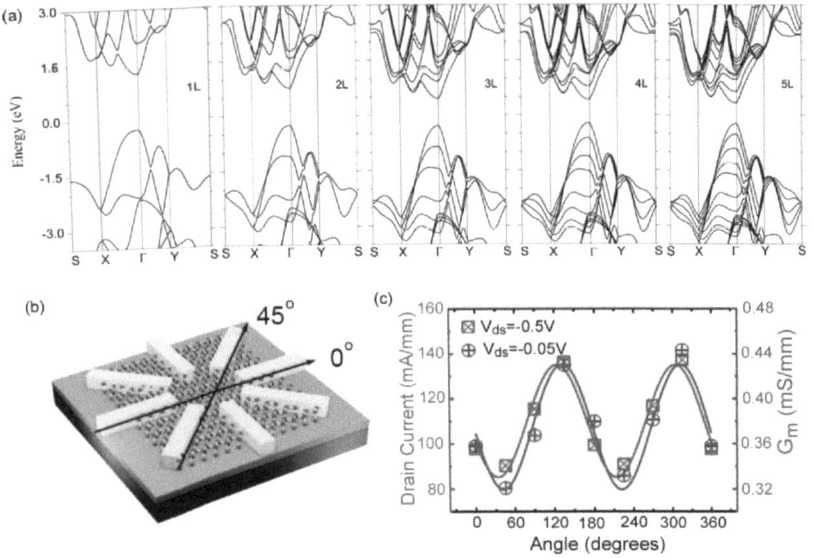

Figure 5.3 (a) Thickness-dependent band structures for few-layer phosphorene calculated by HSE06 method. Reprinted with permission from Cai et al.[4] Copyright (2014) Nature Publishing Group. (b) Experimental setup of electrodes for measuring the angle-dependent transport of phosphorene. (c) Angular-dependent drain current and the transconductance G_m of a phosphorene flake with a thickness of ~10 nm. Reprinted with permission from Liu et al.[9] Copyright (2014) American Chemical Society.

Figure 5.4a presents the thickness-dependent bandgap for multilayer phosphorene from 1L to 5L with different exchange-correlation functionals. It was found that the HSE06 method is more accurate than the PBE and optB88 methods for a better description of the screening effect and electron–electron exchange interaction. The bandgap decreases greatly from 1L to 3L predicted for all the different methods, followed by a slight drop with a further increase in the thickness. This quantum confinement effect can be quantitatively described by using a linear model[4] to describe the scaling behavior of the bandgap of few-layer phosphorene with a number of layers:

$$E_g = E_0 + C/n^{\beta} \tag{5.1}$$

where E_g and E_0 are the bandgap of few-layer and bulk phosphorene, respectively, and n represents the number of layers. By fitting

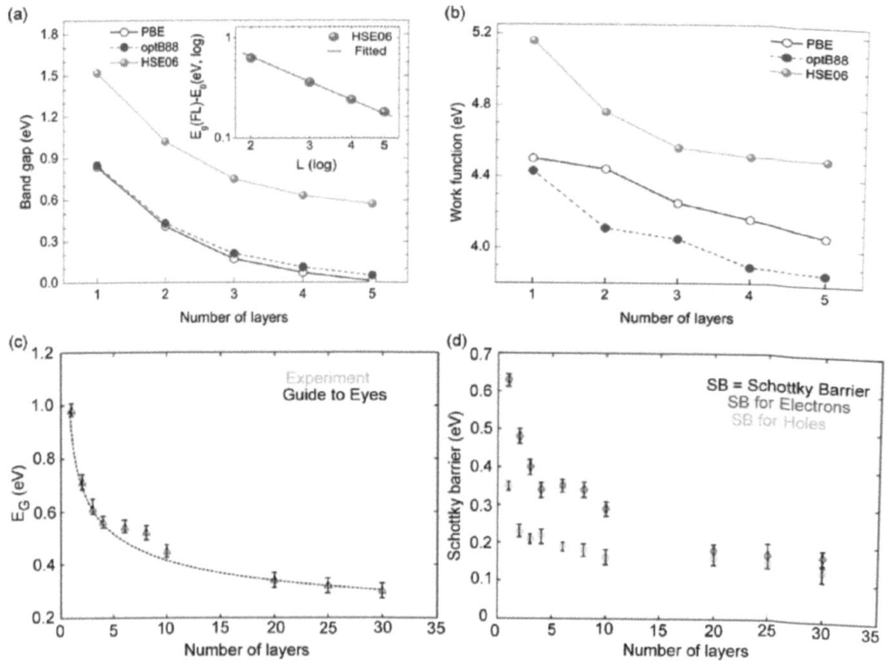

Figure 5.4 Thickness-dependent bandgap (a) and work function (b) of few-layer phosphorene. The inset of (a) shows the log–log scale fitting of the relationship of bandgap with the number of layers. Reprinted with permission from Cai et al.[4] Copyright (2014) Nature Publishing Group. Thickness-dependent transport gap (c) and Schottky barrier heights (d) for electrons and holes measured in experiment. Reprinted with permission from Das et al.[5] Copyright (2014) American Chemical Society.

the bandgap of phosphorene from 2L to 5L sheets, the values for β and C are found to be 1.4 and 1.62 eV, respectively. It should be noted that this model is only appropriate for few-layer phosphorene and does not apply to the monolayer case, as there is no interlayer coupling. The above empirical model can be used to estimate the bandgap of few-layer phosphorene, which is hard to be measured experimentally.

Similar thickness-dependent behavior is also found in the work function, which can be obtained theoretically by subtracting the Fermi energy from the vacuum potential (Fig. 5.4b). It shows that different methods of describing the exchange-correlation functional can lead to different values of work function. For instance, the value of work function predicted by the PBE, HSE06, and optB88 methods for 1L phosphorene is 4.50, 5.16, and 4.43 eV, respectively. All the methods predict a decrease in the work function by 0.43 eV (PBE), 0.66 eV (HSE06), and 0.57 eV (optB88) with the thickness increasing from 1L to 5L. Similar to the variation of the bandgap, there is a significant drop in the work function from 1L to 3L and only minor changes occur with a further increase in the thickness. The tunable work function with a number of layers allows reduction in the contact resistance and Schottky barrier through a proper control of the number of layers of phosphorene together with well-aligned contact metals.

The thickness-dependent electronic properties of phosphorene were demonstrated experimentally.[5] It was found that the transport gap of phosphorene (Fig. 5.4c) increases monotonically from 0.3 eV of bulk to 1.0 eV for a monolayer, leading to thickness-dependent ON/OFF ratios accordingly. Compared to the MoS_2 with bandgap changes from 1.8 eV in monolayer case to 1.1 eV of the bulk form, the nearly threefold change in the bandgap of phosphorene is unique, which implies an entirely different physical mechanism. In addition, due to the different work function and bandgap shown above, the Schottky barrier heights vary when the thickness is scaled, which leads to asymmetric electron and hole currents due to the differently aligned position of the Fermi level of the electrode with the phosphorene layers with different thickness. Figure 5.4d shows the extracted Schottky barrier heights for electron injection into the conduction band and hole injection into the valence band.

The band alignments between the multilayer phosphorene are shown in Fig. 5.5. The valence band maximum (VBM) and conduction band minimum (CBM) for 1L to 5L phosphorene are aligned and referenced to the vacuum level based on HSE06 calculation. Similar to the bandgap and work function shown before, both the VBM and CBM are thickness dependent. There is an upward (downward) shift of the VBM and CBM with a number of layers, with more thickness sensitivity for the VBM than for the CBM. For instance, the VBM for 1L, 3L, and 5L is located at −5.43, −4.85, and −4.70 eV, respectively, and the CBM for the 1L, 3L, and 5L is located at −3.91, −4.09, and −4.13 eV, respectively. This

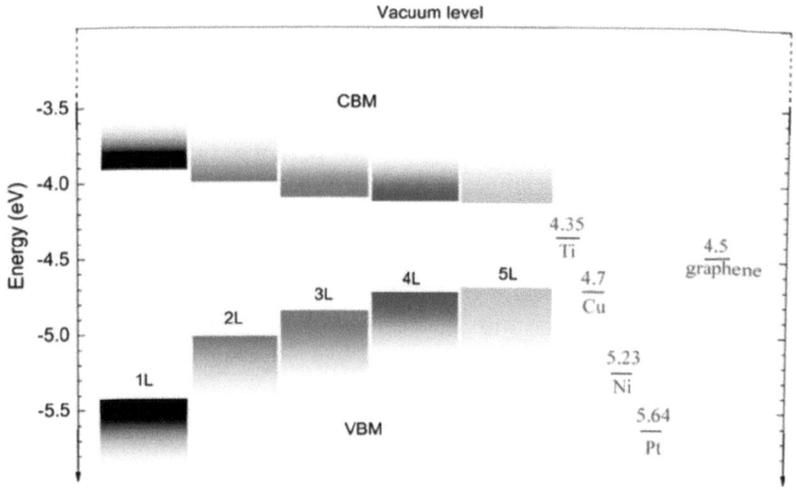

Figure 5.5 Band alignment of the conduction band minimum (CBM) and valence band maximum (VBM) of few-layer phosphorene with the number of layers (from 1L to 5L) based on HSE06 calculation. The VBM and CBM are referenced to the vacuum level. Reprinted with permission from Cai et al.[4] Copyright (2014) Nature Publishing Group.

strong layer number-dependent alignment of band edges allows a tunable Schottky barrier height for more efficient hole injection and transport across the interface. The highly tunable bandgap, work function, and positions of the CBM and VBM of phosphorene facilitate the match of the electronic states with other contacting materials.

5.1.2 Electron–Phonon Coupling

The interaction between the carriers and phonon governs the transporting behavior of electrons/holes and phonons such as the mobility, thermal conductivity, and superconductivity of a material. For phosphorene, electron–phonon coupling (EPC) directly accounts for phononic anharmonicity and superconductivity. Experimental and theoretical studies investigate the shift of phonons and the EPC in phosphorene by tracing the evolution of the Raman active phonons with doping.[10] The A_g^1 and A_g^2 modes show a softening and a broadening of linewidths with electron doping, while remaining unaffected with hole doping. An enhanced EPC is found for these two modes with electron doping, as reflected from the increased full width at half maximum (FWHM) of A_g^1 and A_g^2 modes with electron doping (Fig. 5.6a,b). However, for hole doping, the FWHM of the two modes are nearly unchanged. In contrast, the frequency and the linewidth of B_{2g} mode are insensitive to the doping. In principle, the strength (λ) of the EPC for a phonon mode ν at $\boldsymbol{q}$ with frequency $\omega_{\mathbf{q}\nu}$ is expressed as

$$\lambda_{\mathbf{q}\nu} = \frac{2}{N(\epsilon_F)\hbar\omega_{\mathbf{q}\nu}} \sum_{\mathbf{k}} \sum_{ij} \left| g_{\mathbf{k+q,k}}^{\mathbf{q}\nu,ij} \right|^2 \times \delta(\epsilon_{\mathbf{k+q,i}} - \epsilon_F)\delta(\epsilon_{\mathbf{k,j}} - \epsilon_F) \tag{5.2}$$

$$g_{\mathbf{k+q,k}}^{\mathbf{q}\nu,ij} = \left(\frac{\hbar}{2M\omega_{\mathbf{q}\nu}} \right)^{1/2} \langle \psi_{\mathbf{k+q,i}} \mid \Delta V_{\mathbf{q}\nu} \mid \psi_{\mathbf{k,j}} \rangle \tag{5.3}$$

where $N(\epsilon_F)$ is the density of states at Fermi energy (ϵ_F), g is the electron–phonon coupling matrix element, $\psi_{k,j}$ is electronic wavefunction of the jth band at $\boldsymbol{k}$ point, M is the effective mass associated with the phonon, $\Delta V_{\mathbf{q}\nu}$ is the change in the self-consistent potential induced with atomic displacements of phonon $\mathbf{q}\nu$. The calculated λ for the B_{2g}, A_g^1, and A_g^2 modes as a function of the doping concentration are shown in Fig. 5.6b. Consistent with the experimentally observed FWHM, the λ of A_g^1 and A_g^2 modes correlate linearly and the λ of the B_{2g} mode is nearly unchanged with electron concentration. The different behavior can be rooted in the symmetry of the phonon modes. Since the A_g modes reserve the symmetry of the lattice, the electron–phonon coupling matrix element g of the intraband transition ($i = j$) has a non-zero value. However, for the B_{2g} mode, the matrix component is zero due to a loss of the lattice symmetry. In addition, the asymmetry of the λ with hole and electron doping is related to the symmetry of the states at VBM (even under mirror operation) and CBM (odd under mirror operation) at Γ point.

Figures 5.6c–f present the scattering matrix elements for the coupling of the valence state (the hole of energy 50 meV) with the acoustic LA (Fig. 5.6c,e) and optical B_{3g}^1 (Fig. 5.6d,f) phonons as a function of the phonon wave vector $\boldsymbol{q}$ for monolayer phosphorene.[11] To explore the anisotropy of the EPC, the scattering matrix is examined with hole states along the armchair (Fig. 5.6c,d) and

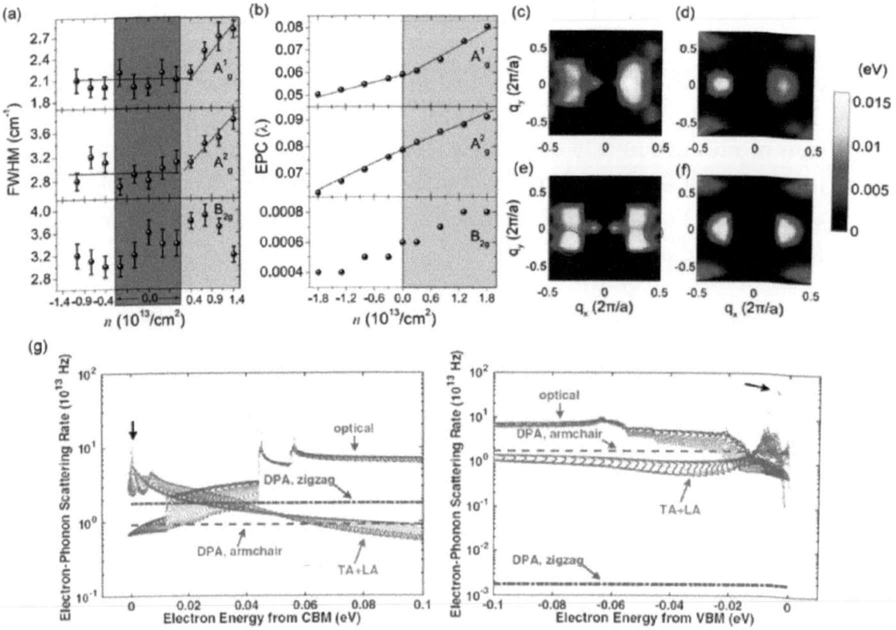

Figure 5.6 (a, b) Measured FWHM of Raman peaks and calculated EPC as a function of carrier density (n). Reprinted with permission from Chakraborty et al.[10] Copyright (2016) Institute of Physics. (c–f) The intraband hole–phonon scattering matrix elements for (c, e) LA and (d, f) B^1_{3g} mode phonons with momentum $\boldsymbol{q}$ in the whole 2D Brillouin zone. (c, d) The hole states transporting along the armchair direction. (e, f) The hole states transporting along the zigzag direction. Reprinted with permission from Ref. Jin et al.[11] Copyright (2016) American Institute of Physics. (g) Calculated electron–phonon scattering rates for electrons (left panel) and holes (right panel). Reprinted with permission from Ref. Liao et al.[12] Copyright (2015) American Physical Society.

zigzag (Fig. 5.6e,f) directions, suggesting that the strength of EPC depends on the direction of the momentum of the electronic state $\boldsymbol{k}$. The hole carrier transport along the zigzag direction should couple more strongly than that along the armchair direction. In addition, acoustic modes are more strongly scattered than optical phonons. By using the Fermi golden rule and the calculated scattering matrix for intraband transition, the scattering rate (τ) of electrons at band

i with momentum $\boldsymbol{k}$ is obtained as

$$\frac{1}{\tau_{\mathbf{k}}^{i}} = \frac{2\pi}{\hbar} \sum_{\mathbf{q},\nu,\pm} \left| g_{\mathbf{q},\mathbf{k}}^{\mathbf{i},\nu} \right|^{2} \left(N_{\mathbf{q},\nu} + \frac{1}{2} \pm \frac{1}{2} \right) \delta \left(E_{\mathbf{k}\mp\mathbf{q}}^{i} \pm \hbar\omega_{\nu,\mathbf{q}} - E_{\mathbf{k}}^{i} \right) \tag{5.4}$$

where $N_{\mathbf{q},\nu}$ is the occupation number of the phonon according to the Bose–Einstein distribution and $\pm$ represents the emission and absorption of a phonon. Based on the evaluated scattering rate, the intrinsic carrier transport properties, including the low-field mobility, can be calculated. It was found that the hole mobility along armchair direction (460 cm^2/Vs) is around 5 times that along the zigzag direction (90 cm^2/Vs) due to the directional dependence of the scattering rate.

A similar study on evaluating the transport and thermoelectric properties of phosphorene was carried out through the method of Wannier interpolation.[12] The energy-dependent scattering rate for electrons and holes is shown in Fig. 5.6g. Since the electrons with different $\boldsymbol{k}$ would experience different scattering although they have the same energy, the energy-dependent scattering rate is multivalued at each energy. Also, the deformation potential approximation was found to give a reasonable estimate of the electron–acoustic–phonon interactions. In particular, the scattering from optical phonons is not negligible and becomes significant for carriers slightly away from the band edges. Moreover, there exist peak structures (indicated by the arrows) in the plots of the scattering rate around the band edges. This can be due to the coupling of the states along armchair and zigzag directions via EPC. The high densities of states along the zigzag direction can serve as the final states for carriers transporting along the other direction to be scattered to.

5.1.3 Tight-Binding Model

Most of the theoretical investigation of phosphorene discussed above is mainly based on parameter-free first-principles calculations within the framework of DFT and its many-body extensions like the GW method. While these first-principles calculations can produce very accurate results of ground state electronic properties, their calculations can only be limited to small systems and can obscure the physics at their root. Another method which allows a large-scale simulation for edges, surfaces, and disorders is the tight-binding (TB). The low-energy electronic properties of graphene were extensively examined by a simple TB model involving only the hopping parameters. The TB model of phosphorene is more complex due to its relatively low symmetry. Another useful method for describing the low-energy electronic properties of phosphorene is the k.p method, which has been used to investigate the thickness-dependent optical properties based on the effective low-energy Hamiltonian[13] and underlying mechanism of the anisotropy of the valence band.[14] Based on the fitted effective mass of electrons and holes from DFT calculations, the k.p Hamiltonian is determined in reciprocal space to determine the conduction band (CB) and valence band (VB) edges. Therefore, this method is not designed for real-space problems.

Effective four-band TB models[7,15] for multilayer phosphorene have been established based on the GW Hamiltonian by projecting the reciprocal electronic states to the real-space Wannier function:

$$\mathrm{H} = \sum_{i} \varepsilon_{\mathrm{i}} n_{\mathrm{i}} + \sum_{i \neq j} t_{ij}^{||} c_{\mathrm{i}}^{\dagger} c_{j} + \sum_{i \neq j} t_{ij}^{\perp} c_{\mathrm{i}}^{\dagger} c_{j} \tag{5.5}$$

where the summation runs over the atomic sites i of single-layer phosphorene, ε_{i} is the energy of the electron at site i, $t_{ij}^{||}$ ($t_{ij}^{\perp}$) is the intralayer (interlayer) hopping parameter between the ith and jth sites, and $c_{\mathrm{i}}^{\dagger}$ (c_j) is the creation (annihilation) operator of electrons at site $i(j)$. The above Hamiltonian was obtained by the four maximally localized Wannier functions $|w_{\mathrm{i}}(\mathbf{r})\rangle$ of low-energy VB and CB states around 0.3 eV near the band edges. The $\varepsilon_{\mathrm{i}} t_{ij}^{||}$ and $t_{ij}^{\perp}$ was determined by the matrix elements of the original GW Hamiltonian transformed into the Wannier function basis $\langle w_{\mathrm{i}}(\mathbf{r}) \mid H \mid w_j(\mathbf{r}) \rangle$ $|w_{\mathrm{i}}(\mathbf{r})\rangle$. The hopping parameters were determined to atomic neighbors with a distance up to 4.23 Å. The obtained TB band structure is compared to the original GW bands in Fig. 5.7. It can be seen that the CB and VB bands match overall with the GW results within around 0.3 eV from band edges. It was found that the nearest-neighbor in-plane ($t_1^{||} = -1.22$ eV) and nearest-neighbor out-of-plane ($t_2^{||} = 3.67$ eV) play dominant roles in the TB matrix. The bandgap of monolayer phosphorene can be

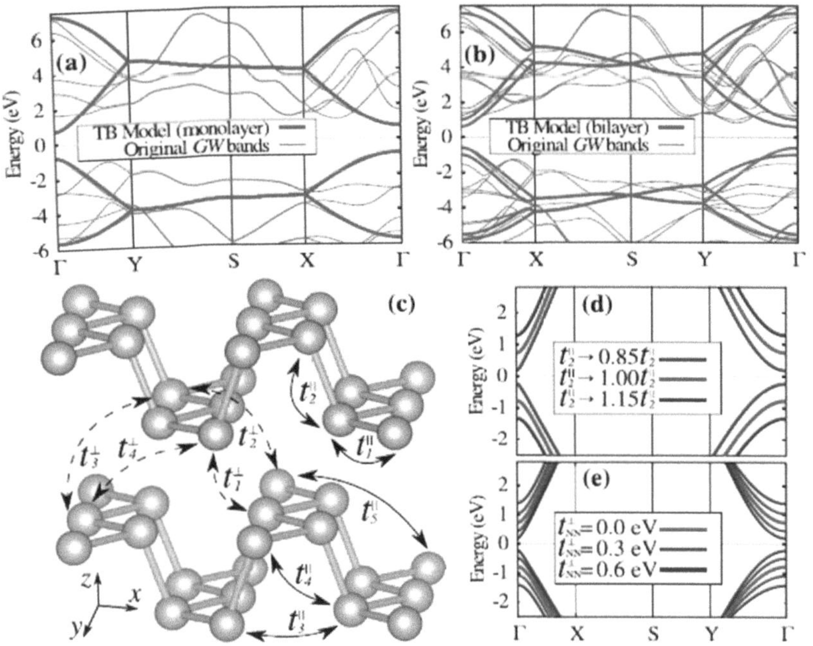

Figure 5.7 Comparison of band structures obtained by GW and TB for monolayer (a) and bilayer (b) phosphorene. (c) Hopping terms in the four-band TB model. The evolution of band structures on $t_2^{||}$ in the monolayer phosphorene (d) and nearest-neighbor interlayer $t_{NN}^{\perp}$ hopping parameters in the bilayer phosphorene. Reprinted with permission from Rudenko and Katsnelson.[7] Copyright (2014) American Physical Society.

expressed as $E_g \approx 2t_2^{||} - 4t_1^{||}$. Different from graphene where only the nearest-neighbor hopping parameter is enough to accurately describe the electronic properties, two additional parameters, $t_1^{||}$ and repulsive $t_2^{||}$, in phosphorene are required which are believed to have originated from the puckered honeycombs. By intentionally increasing (decreasing) the $t_1^{||}$ parameter, the CB and VB rigidly shift toward (apart from) each other (Fig. 5.7d), implying a significant role of the Coulomb repulsion between *p* electrons of atoms in the nearest-neighbored up and down ridges in phosphorene. The hopping parameters ($t_{\perp}$) describe the interlayer interactions. The repulsive nature of this term for atoms between the layers leads to a reduction in bandgap.

Starting from the electronic structures based on DFT, a tight-binding model was proposed using by using the Wannierization method to obtain an exact representation of the DFT Hamiltonian in a basis set of localized orbitals.[16] Different from the k.p method considering only the states close to the high symmetry points, this approach can fully capture the orbital weight of the states and the topology in the whole Brillouin zone. The tight-binding Hamiltonian for the bulk black phosphorus is used for ribbons structures by neglecting the edge reconstruction effect. The effect of the orbital magnetic field can be considered by using the Peierls substitution in the hopping matrix elements to calculate the Landau levels of phosphorene.

The above four-band TB model can only be used for low-energy electronic properties. A parametrization of the TB model for a full Hamiltonian of the black phosphorus was derived based on the quasiparticle dispersion relations of bulk black phosphorus.[17] The TB Hamiltonian in reciprocal space is expressed as follows:

$$H_{mn}(\mathbf{k}) = E_{\mathrm{m}}\delta_{mn} + \sum\nolimits_{\mathbf{R}\neq 0} e^{\mathrm{i}\mathbf{k}\cdot\mathbf{R}}\gamma_{mn}(\mathbf{R}) \quad (5.6)$$

where $\mathbf{k}$ is the wavevector of electrons, $\mathbf{R}$ the lattice vector, E_{m} the on-site energy, and γ_{mn} the hopping parameters, with m and n running over the different orbitals of all the atoms in the primitive cell. γ_{mn} are expressed with the Slater–Koster approximation. Considering the $3s$, $3p_x$, $3p_y$, and $3p_z$ orbitals in this four-atom primitive cell, the Hamiltonian is a 16×16 matrix with six fit parameters of E_{s}, E_{p}, $V_{\mathrm{ss}\sigma}$, $V_{\mathrm{sp}\sigma}$, $V_{\mathrm{pp}\sigma}$, and $V_{\mathrm{pp}\pi}$.

The above parameters are fit to angle-resolved photoemission spectroscopy (ARPES) data of pristine black phosphorus. The ARPES spectrum records the many-body effects, leading to a renormalized quasiparticle spectrum. To fully consider the description of the low-energy part of the conduction band, ARPES measurement was performed for the lithium-doped black phosphorus, where the Fermi level can be rigidly shifted within the conduction band. The parameterization of the conduction band allows the investigation of the evolution of the Fermi surface topology, which is critical for explaining the superconducting phenomenon observed in alkali metal–doped black phosphorus. In addition, the accurate description of the conduction band and valence band allows the

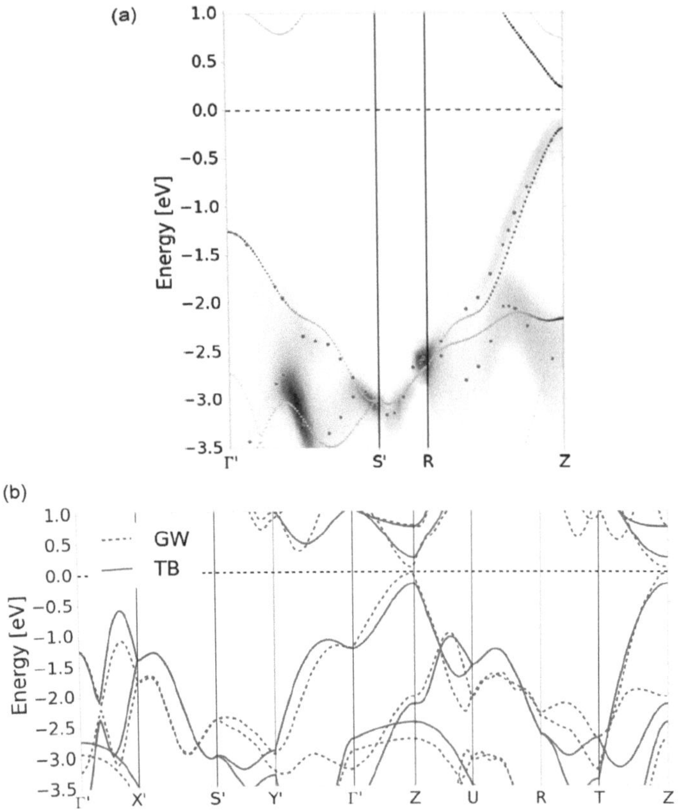

Figure 5.8 (a) Fitting of the TB with the electronic dispersions measured from ARPES measurements for bulk black phosphorus. (b) Fitting of the TB with the GW calculations for bulk black phosphorus. Reprinted with permission from Ehlen et al.[17] Copyright (2016) American Physical Society.

study of the thickness dependence of the quasiparticle bandgap of phosphorene.

The maxima of ARPES along the high-symmetry path of the Brillouin zone is fit with the above TB model by using a steepest-descent algorithm to minimize the residuals between the TB bands and measured curves. It was found that at least five intralayer and two interlayer matrix elements are needed to be taken into account for a good description of the experimental data (Fig. 5.8a). Based on the parameters obtained from bulk black phosphorene, the band dispersion can be obtained for few-layer phosphorene by using the zone-folding method. Starting from this parameterized TB model, the gating effect can be considered with a proper modification of the on-site energy to describe a different charge on each layer. The above TB model was also fit with the GW calculations. Figure 5.8b compares the band structures calculated with the TB and GW, respectively, which shows good agreement.

5.2 Strain Engineering on Electronic Properties

In contrast to the conventional 3D semiconductors which suffer from structural degradation on applying moderate strain, 2D black phosphorus is highly promising for strain engineering for modulating the electronic and optical properties due to its outstanding mechanical resilience. Black phosphorus has a very large compressibility in the armchair and normal-surface directions, which allows pronounced strain-induced evolution of electronic structure. While direct compression or extension of covalent P–P bonds requires a large energy cost, lattice distortion in black phosphorus can occur relatively easily by modifying the bond angles and the weak van der Waals gap between the layers. Therefore, in addition to the layer control approach related to quantum confinement effect, the strain engineering is a promising avenue for an effective modification of the electronic structure of black phosphorus.

Theoretical calculations showed that the phosphorene can withstand a tensile strain up to 30%.[18,19] The electronic properties are found to be sensitive to the direction and magnitude of the strain applied. For instance, phosphorene shows a direct-indirect-direct transition of the bandgap when uniaxial strains are applied along the zigzag direction. A direct-indirect bandgap transition occurs at −2% and 9% strains, whereas further increase in the strain at −10.2% and 11.3% leads to the indirect-direct gap transition again. Figure 5.9a,b shows the evolution of the band structures on applying uniaxial strains along the zigzag (ε_x) and armchair (ε_y) direction, respectively. At around 9% ε_x strain, the conduction band minimum (CBM) shifts from Γ to K1 (0, 0.3, 0) point due to the relative

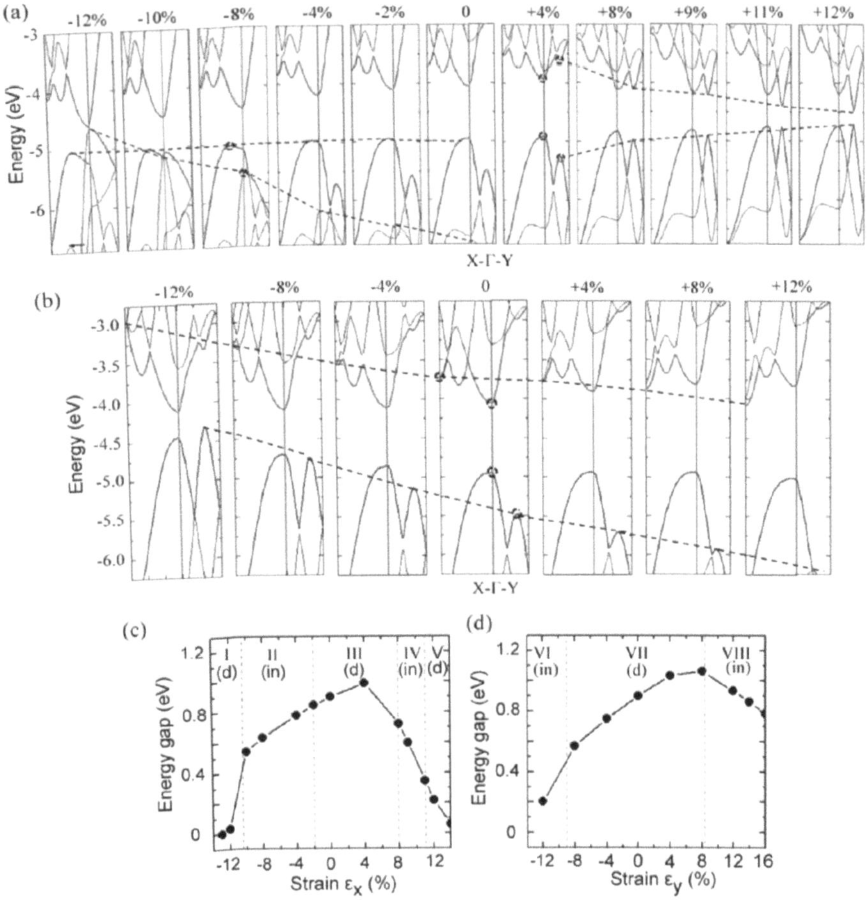

Figure 5.9 Variation of band structures of monolayer phosphorene with uniaxial strain along zigzag (a) and armchair (b) direction. (c, d) The variation of bandgap with strains along zigzag (c) and armchair (d) direction. The nature of bandgap ("d" for direct gap and "in" for indirect gap) is shown. Reprinted with permission from Peng et al.[18] Copyright (2014) American Physical Society.

shifting of the two lowest conduction bands, while the valence band maximum (VBM) still remains at the Γ, thus leading to an indirect bandgap. At around 12% ε_x strain, similar transition from indirect to direct gap occurs due to the shifting of the VBM from Γ to K1 (0, 0.3,

0). Concerning the compressive strain along the zigzag direction, the bandgap changes from direct to indirect at around −2% strain due to the VBM shifts from Γ to $K2$ (0.15, 0, 0), while the CBM remains at the zone center. At around −12% strain, recovery of the indirect to direct gap occurs to the Γ due to the exchange of the positions of the top two valence bands.

With respect to the effect of the strain along armchair direction (ε_y), a much larger compressive strain up to around −8% than the value of −2% compressive strain along the zigzag direction is required to trigger the direct-indirect gap transition. At around +8% strain, an indirect bandgap occurs due to the evolution of the conduction bands. Figure 5.9c shows the bandgap as functions of ε_x and ε_y strains, indicating the indirect/direct nature of the gap. For tensile ε_x strain, the bandgap increases linearly, reaching the maximum at the 4% strain and then drops rapidly upon further deformation, reducing to 0.22 eV at 12% strain. Driven by the downward shifting of the CBM, the drop in bandgap occurs for the compressive ε_x strain, reaching below to 0.04 eV at −10% strain and becoming metal upon further compressing.

Fei et al.[20] showed that the strain can effectively tune the anisotropic properties of the electronic transport by creating a different angular-dependent behavior compared with the strain-free sheet due to the exchanging of the first and second lowest-energy conduction bands. Under around 5% biaxial tensile strain, the effective mass of transporting electrons via the conduction bands shows a different angular preference: the angular profile of the electron effective mass with an approximate "8" shape rotates 90° with a small effective mass along the zigzag direction and a large mass along the armchair direction, sharply in contrast to the strain-free case. Interestingly, the electronic anisotropy seems to be less sensitive to the strain along the armchair direction.

For the compressive strain along the normal direction, theoretical calculation by Rodin et al. showed that the gap becomes narrow with even a semiconductor–semimetal–metal transition.[21] The theoretical value of the critical strain for triggering the metallicity in compressed black phosphorene is around 24 GPa based on the calculation of Hellman–Feynman forces. High-pressure experiment showed that a moderate hydrostatic pressure induces a

Lifshitz transition and narrows the bandgap of black phosphorus. The compressed black phosphorus evolves into a semimetal at around 1.2 GPa, reflecting a positive colossal magnetoresistance and a nonlinear field dependence of Hall resistivity.[22] There is a very complex Fermi surface topology of the semimetallic black phosphorene with the presence of a nontrivial Berry phase and Dirac-like dispersion, as revealed by the Shubnikov–de Haas oscillations. The coexistence of hole and electron pockets, low carrier densities, and linear dispersion in deformed bulk black phosphorus suggests a potential Dirac fermion system in this layered material. Figure 5.10a shows the temperature-dependent longitudinal resistivity at different pressures. The low-temperature resistive divergence gradually disappears with increasing pressure. The upturn trend completely vanishes at 1.25 GPa and a nonmetallic behavior was not found up to room temperature at 1.38 GPa.

The above observation of the topological Fermion surface in black phosphorus under pressure can be rooted in the gradual overlap of the atomic orbitals, especially the p_z orbitals of adjacent interlayer phosphorus atoms.[23] Figure 5.10b shows that the order of lattice compressibility follows: out-of-plane > armchair > zigzag directions. The pressure-dependent band structures for bulk black phosphorus calculated by the mBJ method are shown in Fig. 5.10c. The bandgap gradually decreases from 0.28 eV at 0 GPa to 0.17 eV at 0.5 GPa. The bandgap becomes 0 eV at 1.23 GPa, with the conduction and valence bands overlapping at the Z point. The predicted critical pressure of 1.23 GPa for semiconductor–semimetal transition is comparable to the experimental observation.[22] Further increasing the pressure induces anti-inverse and degenerate Dirac point at the Z point, suggesting that pressured black phosphorus is a 3D Dirac semimetal. This fourfold degenerated Dirac cone gradually evolves into four Dirac cones around the Z point with further increase in the pressure (Fig. 5.10d). Orbital analysis shows that the dominant weight of the Dirac cones is contributed by the $3p_z$ orbital, similar to the $2p_z$ orbital of graphene, which accounts for the six Dirac cones in graphene. It is likely that the overlapping of these $3p_z$ orbitals at high pressure leads to the linearization of the bands and the band crossover near the Z point in the Brillouin zone boundary.

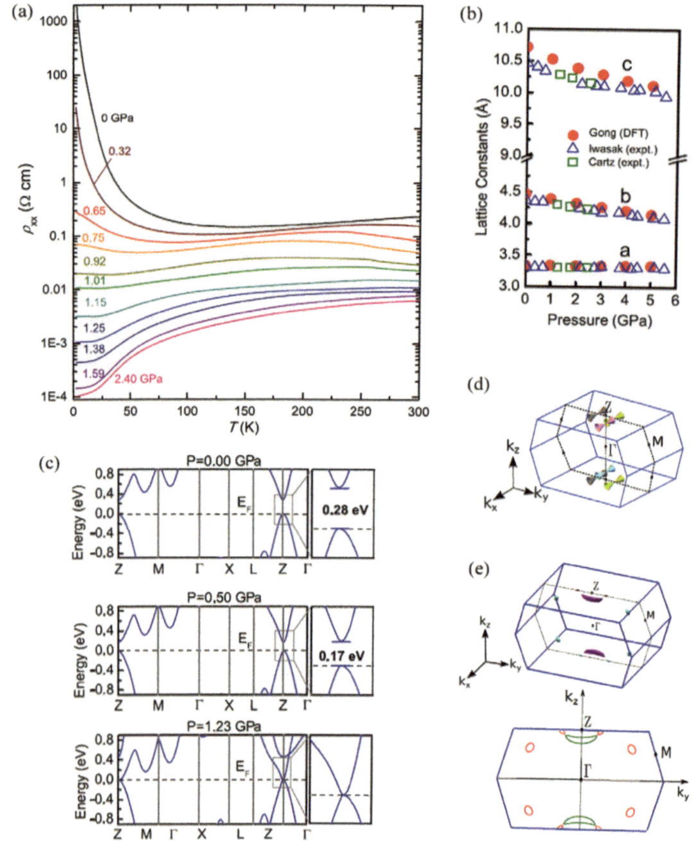

Figure 5.10 Pressure effect on bulk black phosphorus. (a) Experimental measured longitudinal resistivity (ρ_{xx}) of black phosphorus as a function of temperature (T) at different hydrostatic pressures (in GPa, indicated by the numbers beside the curves). Reprinted with permission from Xiang et al.[22] Copyright (2015) American Physical Society. (b) Predicted lattice constants of bulk black phosphorus as a function of pressures by first-principles calculations (red circles), compared with experimental data from Iwasaki et al.[24] and Cartz et al.[25,26] The a, b, and c lattices correspond to the zigzag, armchair, and out-of-plane directions, respectively. (c) Electronic band structure of black phosphorus under pressure from 0 to 1.23 GPa calculated by the mBJ method. (d) The first Brillouin zone of bulk black phosphorus under 2.0 GPa showing four Dirac cones at the zone boundary. (e) Fermi surface (top panel) of bulk black phosphorus under 2.0 GPa and the projection of the Fermi surface on the plane $(0, k_y, k_z)$ (bottom panel). The electron- and hole-related pockets are shown in red and black lines, respectively. Reprinted with permission from Gong et al.[23] Copyright (2016) American Physical Society.

The Fermi surface of black phosphorus under 2.0 GPa and its projection are shown in Fig. 5.10e. Three different kinds of Fermi pockets are found in the Fermi surface: two hole pockets in the Γ–Z line and two tiny electron pockets around the Dirac cones in the M–Z line. The left four electron-type pockets are found to be unrelated with the Dirac cones. It is clear that the topology of the Fermi surface of bulk black phosphorus undergoes dramatic changes and a *Lifshitz* transition under critical pressure of 1.23 GPa with simultaneously nearly compensated electron and hole pockets. The 3D Dirac semimetal thus obtained is highly promising for optoelectronic applications.

5.3 Defects, Doping, and Heterostructure

5.3.1 Defects

By using density functional theory with hybrid functional HSE06, the structural and electronic properties of intrinsic defects like vacancies (V_{P}) and self-interstitials (P_{i}) in monolayer and few-layer phosphorene have been investigated.[27] The calculations based on a 3 × 2 supercell show that a smaller energy cost of forming these defects is needed in the thinner layers. Both defects have a lower formation energy in the surface layer (denoted as X_{out}, where $X = V_{\mathrm{P}}$ or P_{i}) than the interior part (denoted as X_{in}) in the few-layer case. It was found that both vacancies and self-interstitials can serve as shallow acceptors, accounting for the p-type conductivity of phosphorene.

The formation energy and transition energy of charged defects as a function of the electronic chemical potential are plotted in Fig. 5.11a. The formation energy of a charged defect, compensated with a uniform background charge with opposite charge, was obtained via

$$\Delta E_D^f(\alpha, q) = E_{\mathrm{tot}}(\alpha, q) - E_{\mathrm{tot}}(\mathrm{host}, 0) - n_\alpha \mu_\alpha + q(\mu_{\mathrm{e}} + \epsilon_v) + E_{\mathrm{corr}}[q] \tag{5.7}$$

where $E_{\mathrm{tot}}(\alpha, q)$ and $E_{\mathrm{tot}}(host, 0)$ are the total energies of the α defective system in q charged state and corresponding perfect host system, respectively. n_α denotes the loss ($n_\alpha < 0$) or increase ($n_\alpha > 0$) of the number of atoms of species α of the perfect system

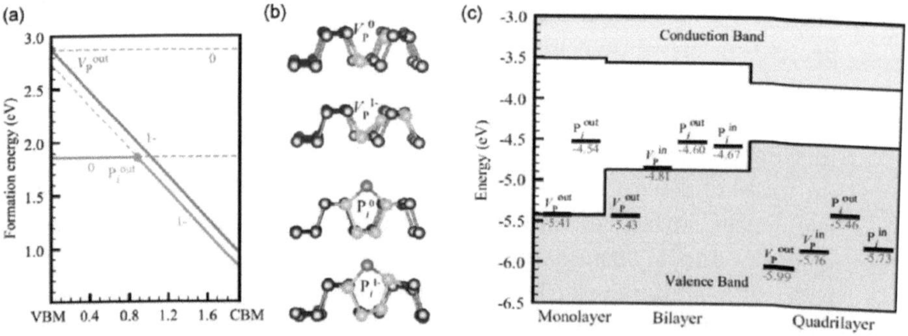

Figure 5.11 (a) Formation energy of charged V_P and P_i in monolayer phosphorene. (b) Geometry readjustment of charged defects relative to the neutral state. (c) Evolution of the transition levels with the thickness of phosphorene. Reprinted with permission from Wang et al.[27] Copyright (2015) American Physical Society.

to form the α defective system. μ_α and μ_e are the atomic and electronic chemical potential, respectively. Note that the μ_e is in reference to the top of the valence band (ϵ_v) and varies between 0 and the bandgap (E_g). The E_{corr} term corresponds to the correction of the electrostatic potential of the defective system with the perfect system and the finite-size effects for the charged systems. The transition energy level $\epsilon_\alpha(q/q')$ between two different charged states (q and q') for a defect can be evaluated by the following formula:

$$\epsilon_\alpha(q/q') = [\Delta E_D^f(\alpha, q) - \Delta E_D^f(\alpha, q')]/(q' - q) \tag{5.8}$$

Figure 5.11a shows that phosphorus vacancy is more stable in the -1 charged state, implying its acceptor role. The local atomic geometry of the defective center changes greatly with the charged states (Fig. 5.11b). P_i has around 1.0 eV smaller formation energy than V_P, suggesting that P_i should be more abundant in the real samples. Figure 5.11c shows the diagram of thickness-dependent transition levels of the native defects with reference to the vacuum level. It can be seen that the transition levels of V_P and P_i shift downwardly with increasing thickness, implying shallower acceptor levels in thicker layers.

5.3.2 Atomic Doping

Owing to the presence of lone-pair electrons on the phosphorus atoms, the surface of phosphorene tends to be very active, which can easily host foreign adatoms. The atomically thin nature of the material implies that its electronic, chemical, and structural properties are very sensitive to these surface dopants or impurities. Some typical anionic and cationic atomic dopants are introduced below.

Oxygen: Phosphorene exposed to air tends to be easily oxidized, which causes structural degradation and surface roughening even within several hours after exposure in air. Like graphene, there exist several different types of oxidized species in the phosphorene surface. First-principles calculations showed that there is an energy release of around 2 eV for each adsorption of oxygen atoms.[28] The lowest energy configuration of adsorbed oxygen atoms, with a binding energy of −2.08 eV, is the dangling oxygen which forms an O–P bond of 1.50 Å and induces a small deformation of the underlying phosphorene lattice. The O–P bond tilts away around 44.5° from the surface normal direction. This oxidation process is exothermic with a net gaining energy of forming P–O bond. Electronic structural analysis shows that dangling oxygen is an electronically neutral defect which induces no defective states within the gap of phosphorene. The second lowest energy configuration of oxygen dopant has an interstitial geometry with the O atom occupying a middle position of a P–P bond. This oxygen adsorbate is also electronically neutral. Despite an inert electronic character, these adsorbed oxygen species form the P–O bonds which are highly polar. This will greatly modify the chemical activity of the surface with a more hydrophilic character to adsorb water or other polar molecules. Indeed, the adsorption energy of a water molecule around the pre-adsorbed oxygen atom is −0.22 eV, which is enhanced compared with adsorption above pristine surface (−0.09 eV). Another popular oxidizing group is the hydroxyl species. It is found that there is a partially filled defective level associated with the adsorbed hydroxyl group.

Formation of these oxygen species in phosphorene from a direct dissociation of oxygen molecules is estimated to require an

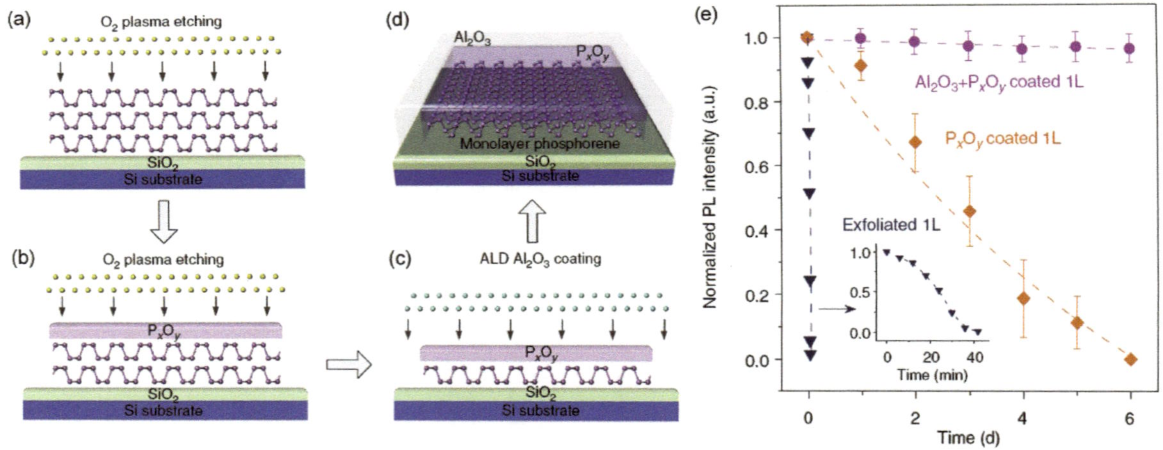

Figure 5.12 Schematics showing the process of the production of phosphorene oxide based on phosphorene. (a) O_2 plasma etching of the exfoliated phosphorene layers supported by the SiO_2/Si substrate. (b) Penetration of O_2 through the P_xO_y layers to the underlying phosphorene layers. (c) Obtaining a designated number of layers of phosphorene due to a dynamic equilibrium formed between oxidation of the phosphorene and physical removal of the P_xO_y layer. (d) Coating of an Al_2O_3 protective layer by ALD above the P_xO_y layer, which separates the underlying phosphorene from the precursor gases used in the ALD process. (e) Variation of the PL intensity with time for three different phosphorene samples: exfoliated 1L phosphorene (blue), P_xO_y coveraged phosphorene via O_2 plasma etching (orange), and ALD Al_2O_3 coveraged P_xO_y/phosphorene layers (purple). Reprinted with permission from Pei et al.[30] Copyright (2016) Nature Publishing Group.

energy barrier of around 0.81 eV, which is impossible through solely thermal activation.[29] It is likely that the photoexcitation can greatly decrease the barrier and enhance the oxidation. Another possible reason for lowering the barrier is related to the presence of intrinsic defects of phosphorene. The first-principles calculations[29] showed that the energy barrier can be reduced to 0.59 eV at a single phosphorus vacancy compared with the dissociation at the perfect lattice of 0.81 eV. This will lead to around 5000 faster oxidizing rate at the vacancy site than at the perfect site at room temperature. Therefore, in principle, synthesis and storage of phosphorene in vacuum and dark environment can suppress the oxidation of phosphorene. Passivation of atomic defects and edges of phosphorene can also help to enhance the chemical stability of phosphorene.

Experiments on the oxide of phosphorene were thoroughly investigated by Pei et al.[30] through oxygen plasma etching of phosphorene. In their experiments, a mechanically exfoliated thick phosphorene flake was supported by a SiO_2/Si substrate and treated with O_2 plasma etching (Fig. 5.12a). It is expected the etching process initiates at the top surface layers which are transformed into P_xO_y. These oxide layers act as protecting layers and suppress the degradation of the underlying phosphorene. The continuing etching process thins down the phosphorene layer and increases the thickness of the oxide layers, while the O_2 plasma also sputters away some of the oxide layers via nuclei collisions. A dynamic equilibrium and a constant thickness of the top P_xO_y layer are achieved after the plasma pre-treatment (Fig. 5.12b). The layer numbers of the phosphorene underneath can be controlled due to a constant etching rate (Fig. 5.12c). The XPS analysis revealed that the P_xO_y layer is dominated by P_2O_5. While the P_xO_y layer prolongs the lifetime of the phosphorene, a further coating of an Al_2O_3 layer above the P_xO_y layer promotes the durability. As shown in Fig. 5.12d, the additional Al_2O_3 layers were deposited via ALD method, where the P_xO_y layer prevents the reaction between the underlying phosphorene and the precursor gases used in the ALD.

The time-dependent behaviors of the structure for three phosphorene samples fabricated by three different methods are shown in Fig. 5.12e, which presents a varied PL intensity in ambient conditions. The exfoliated naked phosphorene layer only survived for around 30 min, while the phosphorene layer with a top P_xO_y layer shows a much better lifetime due to the suppressing of the environmental oxygen molecules and water. However, the phosphorus oxide layer can degrade slowly due to the moisture and disappear in a week's time. Adding a Al_2O_3 layer is found to greatly prolong the lifetime of the P_xO_y and phosphorene layers up to two months in air.

Cu: The doping of Cu adatoms onto few-layer black phosphorus has been theoretically and experimentally investigated.[31] It was found that the addition of the Cu species can induce a n-type conduction in phosphorene, which is well-known to be a p-type semiconductor. A scalable, completely 2D material complementary inverter with both n- and p-type conducting channels was fabricated with the Cu-doped phosphorene, a BN gate dielectric, and a graphite gate. First-principles calculations were performed to reveal the electronic structures of four-layer phosphorene with Cu being doped above the surface or within the interlayer. For both systems, the Cu adatom forms strong chemical bonds with the host sheets and donates electrons to phosphorene, thus changing the polarity of phosphorene from p-type to n-type without breaking the sheets. However, the theoretical calculations showed that the adsorbed and interstitial Cu dopants can lead to a different trend of the bandgap: the gap is reduced to 0.09 eV for the surface adsorbing Cu, but increased to 0.32 eV for the intercalated Cu dopant compared with the 0.18 eV in the undoped phosphorene. The underlying mechanism is due to the different internal electric field built between the Cu dopant and the sheet. The adding of Cu dopants causes a nearly rigid upward shift of the Fermi level to the conduction band without any large changes in the conduction and valence band dispersions near the band edges. This is consistent with the threshold voltage for electron conduction, as revealed in experiment. The coexistence of n-type and p-type conducting channels within a single 2D material allows the fabrication of a CMOS-like inverter without the need of normal local electrostatic gating.

K: Doping potassium into phosphorene has been experimentally investigated.[32] In situ surface doping of K atoms into few-layer black phosphorus shows an effective modulation of the bandgap through angle-resolved photoemission spectroscopy (ARPES). The mechanism can be related to the vertical electric field built at the dopant–phosphorene interface due to a large charge transfer. Owing to a low electron negativity, the K atoms adsorbed above the phosphorene surface donate electrons to phosphorene, leading to a formation of 2D electron gas near the surface. The large charge transfer and the giant Stark effect of phosphorene leads to a semiconductor–semimetal transition after K doping. Band inversion and formation of Dirac-like electronic properties can occur upon the electric field exceeding a certain value.

At low K doping density, the phosphorene layers become slightly n-doped with the bandgap larger than 0.6 eV. The bandgap reduces gradually with increasing K intensity, which can be attributed to the enhanced electric field. The electric field gradually reduces the conduction band edge and increases the valence band edge. The doping-sensitive bandgap allows a wide and continuous tuning of the electronic properties of phosphorene from a moderate bandgap to a band-inverted anisotropic Dirac semimetal, where the bands are linear along the armchair direction and quadratic in the zigzag direction. A linear fit of the ARPES measured momentum–energy curves yields a velocity of carriers around 5.1×10^5 m/s, half of that of graphene.

Al: Phosphorene layers with intercalation of Al were thoroughly investigated for fabricating transistors and photovoltaics.[33,34] Compared to the addition of Cu via transient one-second sputtering,[31] the doping of Al atom was shown to have a better controllability of the concentration of the dopant. With a smaller atomic radius, Al intrinsically diffuses easier into phosphorene sheets than Cu. Effectively transforming intrinsic p-type conduction in naked phosphorene to n-type transport, Al-doped phosphorene was used for fabricating complementary n-FETs with a high electron mobility up to 1495 $cm^2V^{-1}S^{-1}$ at 260 K. It was shown that the transport mechanism is Mott variable range hopping (VRH) transport through the Al localized state at low temperature. The effect of Al doping on the structure and electronic structures of phosphorene was investigated by first-principles calculations (Fig. 5.13).

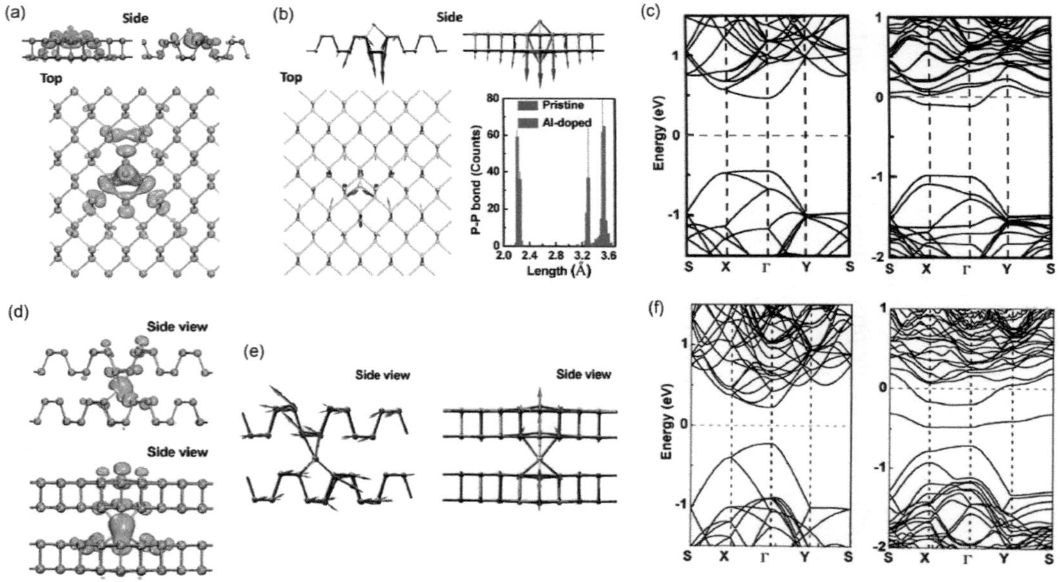

Figure 5.13 (a–c) Al dopant above the single phosphorene sheet: (a) the differential charge density; (b) atomic displacement of phosphorus atoms upon Al doping and distribution of the P–P bond length; (c) comparison of the band structure for pristine (left) and Al-doped (right) phosphorene; (d–f) Al dopant between the sheets in a bilayer phosphorene; (d) the differential charge density; (e) atomic displacement of phosphorus atoms upon Al doping; (f) comparison of the band structure for pristine (left) and Al-doped (right) phosphorene. Reprinted with permission from Prakash et al.[33] and Liu et al.[34] Copyright (2017) John Wiley and Sons.

For the monolayer phosphorene sheet, each Al dopant forms three covalent Al–P bonds with around 1.60 Å above phosphorene and a binding energy of −2.04 eV. The differential charge density plot (Fig. 5.13a) shows that the Al atom acts as a donor with donating significant electrons (around 1.75 e per Al atom based on Bader charge analysis) to phosphorene. The phosphorus atoms around the Al adatom are displaced away from their original sites (Fig. 5.13b). There is a clear strain field developed around the Al dopant with anisotropic tensile (compressive) strains up to 2.11% (−1.35%). Interestingly, the strain fields attenuate slower in the zigzag direction than in the armchair direction. Comparison of the distribution of the P–P bond length is presented in Fig. 5.13b, where a broadening of the distribution of the P–P bond length occurs upon Al doping. Figure 5.13c presents the band structures of pristine and Al-doped phosphorene. The Al introduction shifts the Fermi level into the conduction band of phosphorene with a clear distortion of the valence and conduction band edges.

For adsorption in multilayer phosphorene, the Al adatom at both surface site adsorption and the intercalated site donates electrons. Figure 5.13d shows the charge transfer between Al and a bilayer phosphorene with the Al atom intercalated and chemically bonded with the phosphorene layers. The stablest adsorption adopts a six Al–P bonded structure with bond length ranging from 2.39 to 2.61 Å and binding energy of −2.21 eV. The Al adsorption induces large outward and in-plane atomic displacements of the neighboring P atoms and a large broadening of the distribution of the P–P bond length (Fig. 5.13e). Similarly, the Fermi level moves into the conduction band of bilayer phosphorene upon Al doping (Fig. 5.13f).

Based on the Al-doped phosphorene, a p–n homojunction in a single phosphorene sheet was fabricated for high-performance photovoltaic application.[34] A rectification behavior with a near unity ideality factor of 1.001 and an ON/OFF ratio of around 1.6×10^3 at 2 V were observed. It was found that the dark current can be greatly suppressed in the photovoltaic applications. The temporal short-circuit current and open-circuit voltage in the absence of external bias allows for self-powered photovoltaic

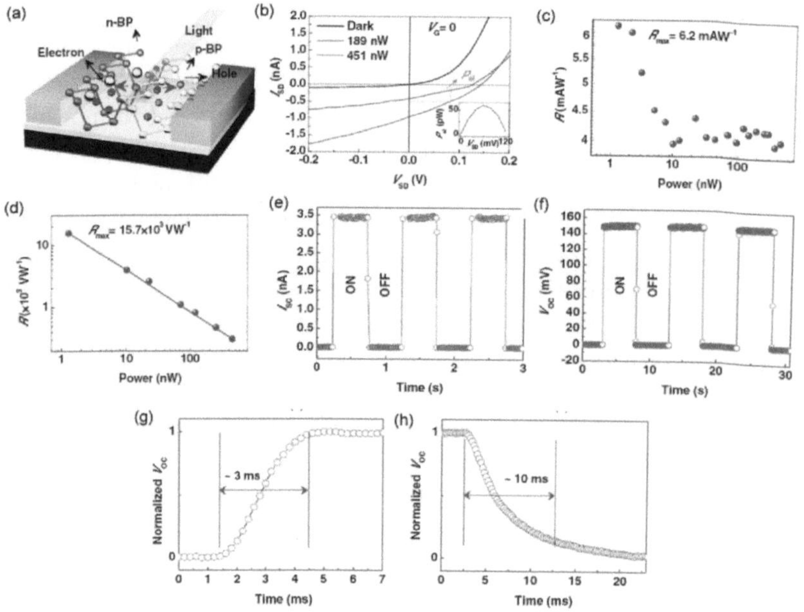

Figure 5.14 Photovoltaic effect of Al-doped phosphorene. (a) Schematic showing the production of light-induced electron and hole which are separated due to the built-in electric field at the interface between pristine and Al-doped part of phosphorene. (b) I_{SD}–V_{SD} curves measured under light illumination with different intensities without gate modulation. The relationship of the electrical power and V_{SD} under 451 nW light illumination is shown in the inset. (c, d) Dependence of photoresponsivity on the incident light power. (e, f) Switching ON/OFF of the electrical signals (short-circuit current I_{SC} and open-circuit voltage V_{OC}) with laser light illumination. (g, h) The transient rise and fall of V_{OC} as a function of time. Reprinted with permission from Liu et al.[34] Copyright (2017) John Wiley and Sons.

and photodetection applications. The production and transport of photoexcited carriers in the p–n diode of phosphorene was measured based on a schematic device structure as shown in Fig. 5.14a. The current–voltage profiles across the source and drain electrodes (I_{SD}–V_{SD}) under light illumination are shown in Fig. 5.14b. Under light illumination with a 1550 nm laser, a finite voltage and current were produced even when the applied voltage was zero, suggesting a nonzero open-circuit voltage and short-circuit current and a photovoltaic behavior. The electrical power (P_{el}–V_{SD}) relationship (inset of Fig. 5.14b) shows that a maximum output power of 54.3 pW can be obtained. The maximum photoresponsivity is measured to be around 6.2 mA W^{-1} and 15.7 $\times$ 10^3 VW^{-1} (Fig. 5.14c,d). This is around 20 times of the value of a gate-defined p–n diode made of a 30 nm thick phosphorene. The underlying mechanism for the strong photovoltaic should be related to the built-in field in the Al-doped phosphorene p–n junction, which can separate the electron–hole pairs effectively. Most importantly, the phosphorene diode has a zero dark current and power consumption which is favored for low-noise spectroscopy and optical communications.

By taking advantage of the light-triggered current and voltage in the device, it is possible to make a dynamic photovoltaic switch. Figure 5.14e,f plots the ON/OFF short-circuit current and open-circuit voltage with an incident power of 860 nW with a good reversibility after hundreds of ON/OFF cycles. Such behavior was only observed after introduction of the Al dopant. The temporal time-dependent changes of the open-circuit voltage V_{OC} are shown in Fig. 5.14g,h. The rise and fall time of V_{OC} are 3 and 10 ms, respectively, which is comparable to the gate-defined p–n junction.[35] The advantage of the current homojunction is related to its facile and efficient operation without any external electrical supply. It is expected that the additional application of a source–drain voltage can further increase the response speed, favoring high-speed photodetecting applications.

5.3.3 Molecular Doping and Sensing

Owing to the high surface–volume ratio, atomically thin 2D materials are very sensitive to the external physisorbed or chemisorbed adsorbates. This allows the modulation of the properties of 2D materials through doping and functionalization with external molecules. For example, hydrogenation of MoS_2[36] was found to lead to a conducting surface. On the other hand, physisorbed small molecules can lead to marked effects on the electronic properties on modifying carrier density and shifting the Fermi level. Therefore, 2D materials can serve as effective gas sensors owing to the intrinsic weak electronic screening effect compared to bulk materials. Since small molecules like O_2, H_2O, NH_3, NO_2, CO, H_2, and NO are ubiquitous in environment and cannot be fully removed from surfaces, a clear understanding of their effect on the 2D materials is always a prerequisite for their potential applications.

Density functional theory–based calculations have been performed to study the effects of small molecules on phosphorene with respect to energetics and charge transfer.[37] The calculation showed that the H_2O, CO, NH_3, and H_2 molecules are weak donors, while O_2 and NO_2 are strong acceptors. Interestingly, while an NO molecule is a donor when adsorbed above graphene, it receives electrons after adsorption above phosphorene. Phosphorene becomes electronically polarized upon coming in contact with some of these small gas molecules. Due to its atomically thin thickness, a tiny amount of charge transfer can trigger marked changes in the carrier density in phosphorene. The open-shell molecules like O_2, NO_2, and NO generally show a stronger binding and interaction with phosphorene than those of the closed-shell molecules. Phosphorene normally shows in-gap states involved with these adsorbed open-shell molecules, which may induce new recombination centers for excitons, thus affecting the optical properties.

In addition, the moderate binding energy between H_2 and phosphorene suggests that phosphorene is a promising hydrogen storage material at room temperature. These distinct adsorption behavior and charge transfer occurring between these environmental molecules and phosphorene allow phosphorene to be used as a gas sensor. In addition, the environmental molecules can tune the polarity and density of conducting carriers in phosphorene. Figure 5.15 presents the charge transfer of phosphorene adsorbed with these molecules. A detailed discussion on the adsorption energetics and the donor/acceptor nature of the molecules is given below for the case of each molecular adsorbate (Table 5.2).

CO adsorption: The lowest energy configuration for CO molecules adsorbed above phosphorene is a tilt geometry with the CO sitting above the ridge with the C–O bond slightly deviated from the normal direction to the surface. The binding energy for single CO adsorption is −0.31 eV. A large charge redistribution occurs between phosphorene and CO, which greatly broadens the

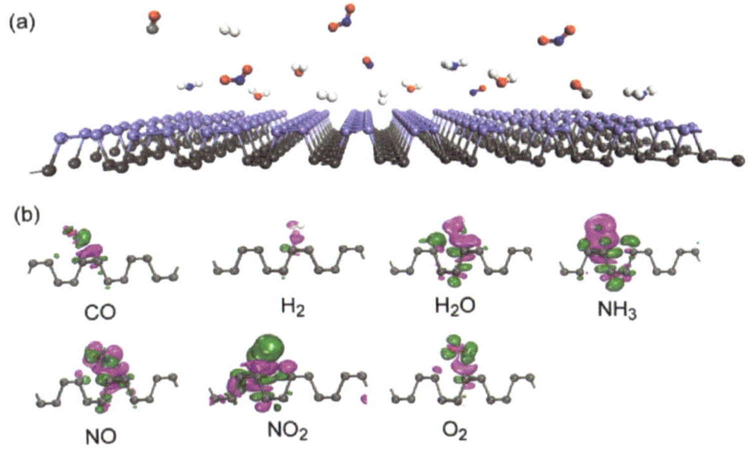

Figure 5.15 (a) Schematic of gas adsorption above phosphorene. (b) Isosurface plot with a value of 0.005 Å^{-3} of the differential charge density of small molecules adsorbed phosphorene. The violet (green) color denotes the loss (gain) of electrons. Reprinted with permission from Cai et al.[37] Copyright (2015) American Chemical Society.

Table 5.2 The binding energy (E_b), amount of transferred charges (q) from the molecules to phosphorene, and the length of bonds (B_{X-P}) formed between the molecules and the nearest P atoms, where X is the closest atom of the molecule to the surface. The negative value of q indicates a loss of electrons in phosphorene. The donor/acceptor character of the molecule is listed and compared with adsorptions above graphene. Data adapted from Cai et al.[37]

Molecules	E_b (eV)	q	$B_{X\text{-}P}$ (Å)	Molecule on phosphorene	Molecule on graphene[45]
CO	−0.31	0.007	3.06−3.15	Donor	Donor
H_2	−0.13	0.013	2.46	Donor	
H_2O	−0.14	0.035	2.71	Donor	Acceptor
NH_3	−0.18	0.050	2.59	Donor	Donor
NO	−0.32	−0.074	2.75−2.88	Acceptor	Donor
NO_2	−0.50	−0.185	2.27	Acceptor	Acceptor
O_2	−0.27	−0.064	2.76	Acceptor	

carbon-centered 5σ HOMO state, while the CO 2π*(LUMO), 1π (HOMO-1), and 4σ (HOMO-2) states are almost unchanged. The net charge transfer from CO to phosphorene is found to be 0.007 electron with the CO serving as a weak donor.

H_2 adsorption: The H_2 molecules prefer to align in a parallel configuration with one of the H atoms located above the P atom and the H–H bond along the armchair direction. The adsorption energy is −0.13 eV, which satisfies the criteria for hydrogen storage with the binding energy between −0.15 and −0.3 eV per H_2 molecule.[38,39] One advantage of phosphorene for hydrogen storage is related to its buckled structure, allowing a large uptake of hydrogen molecules related to the convex areas of the surface. The presence of curvature in carbon fullerenes, nanotubes, and graphene is found to lead to a promoted gas adsorption.[40–43] Charge transfer analysis shows that H_2 is a donor, transferring around 0.013 e per molecule to phosphorene.

H_2O adsorption: The most stable adsorbing configuration for H_2O can be found with one of the O–H bonds being normal to the surface and the other being parallel to the surface along the armchair direction with the binding energy of −0.14 eV. The charge difference plot shows that the H_2O induces a great charge redistribution of the phosphorene layer due to the relatively large dipole moment of the molecule, suggesting that water can trigger great effects on the performance of the phosphorene device, including the carriers' mobility and durability. This strong interaction is rooted in the relatively large state hybridization between H_2O and phosphorene. The 3a_1 (HOMO-1) orbital has a clear orbital mixing with the nearest phosphorus atom, which leads to a large broadening of this level. It should be noted that the lowest energy configuration of H_2O above phosphorene is quite similar to the lowest energy geometry of H_2O adsorbed above graphite.[44] However, the oxidation state of adsorbed H_2O above phosphorene strongly differs from that of adsorption above graphene: H_2O is a strong donor above phosphorene, but a charge acceptor on graphene surface. Such difference can be rooted in the much larger work function of phosphorene than graphene, which favors the electron transport from water to phosphorene. As the work function of phosphorene layers is strongly dependent on the number of layers according to previous section, the effects and charge transfer due to water molecule can be highly sensitive to the thickness of phosphorene, with weaker ability to donate electrons to the thicker phosphorene layers compared with the adsorption above the monolayer phosphorene.

NH_3 adsorption: The strongest binding configuration for NH_3 molecule shows the H atoms points away from the surface and the N atom is directly above the phosphorus atom with the N–P bond length of 2.59 Å and the binding energy of −0.18 eV. There exists a strong charge redistribution between NH_3 and phosphorene with coupling between the lone-electron pair and the N atom–centered $3a_1$ (HOMO) state of NH_3. This nonbonding HOMO level of NH_3 is strongly affected by the P atoms, as indicated by a large broadening of the related peak in LDOS. Charge transfer analysis shows that the NH_3 molecule is a donor, with phosphorene receiving about 0.05 e.

NO adsorption: The ground state for the NO adsorption is quite similar to that of the CO adsorption above phosphorene. The binding energy is −0.32 eV. This relatively large adsorption is rooted in the half-filled, doubly degenerated NO 2π (HOMO) level, which favors back and forward charge transfer through orbital hybridization with phosphorene. According to the isosurface plot of differential charge density, there is an electron depletion in phosphorene, while an accumulation in the NO molecule with a total amount of transferred electron around 0.074 e. After contacting with phosphorene, the singly occupied HOMO level of NO shifts above the Fermi level and becomes empty. The net gain of electrons of NO results from the strong state hybridization of lower-lying HOMO-1 and HOMO-2 states with the 3s orbital of phosphorus atoms, which leads to a larger amount of electrons being transferred from phosphorene to NO than that induced by the depopulation of the HOMO orbital. Interestingly, this acceptor character of NO above phosphorene is in sharp contrast to that of NO above graphene, where NO is an effective donor.

NO_2adsorption: The NO_2 molecule was shown to have the strongest adsorption among all the investigated molecules with the binding energy up to −0.50 eV, according to Cai et al.[37] The ground adsorbing state is that the two N–O bonds are parallel to the zigzag direction, with the N atom sitting above the phosphorus atom and a N–P bond length of 2.27 Å. This configuration differs from that of

NO_2 above graphene, where the oxygen atoms are pointing toward the carbon atoms. The NO_2 is an acceptor, receiving around 0.19 e from phosphorene with a similar mechanism as the NO case.

O_2 adsorption: The oxygen molecule is the main source of structural degradation of phosphorene. Understanding the interaction of physisorbed O_2 with phosphorene is highly important since oxygen is ubiquitous in air. The lowest energy configuration for oxygen adsorption is found with one of the oxygen atoms locating above the P atom and the O–O bond aligned nearly parallel to the armchair direction and tilted around 30° away from the surface normal direction. The binding energy is −0.27 eV. The oxygen molecule is a charge acceptor, with obtaining around 0.06 e from phosphorene. The bond length of the O–O bond becomes 1.24 Å after adsorbing above phosphorene; this value is slightly larger than 1.22 Å for the free gas molecule.

Owing to their ultrathin character, atomically thin 2D materials can be easily bended with the formation of ripples and bubbles during material synthesis and device fabrication. According to studies on graphene and carbon nanotube,[46–49] the curved surface shows greatly different chemical activity toward gas adsorption. For phosphorene, formation of ripples and wrinkles is more likely to occur due to its highly flexible and softer structure compared with other 2D materials. Therefore, understanding the curvature effect on the energetics and kinetics of gas interaction is highly important.

The effect of structural rippling on the physisorption of the NO molecule was studied using first-principles calculations.[50] The rippled phosphorene was simulated by constructing waved phosphorene structures along the armchair direction (Fig. 5.16). The adsorptions of the NO molecule in the convex and concave parts of the ripples with different degrees of curvature were both examined and the lowest energy configurations of adsorbed NO are shown in Fig. 5.16. By comparing with the adsorption above the planar phosphorene sheet, the strength of interaction between the molecule and the waved phosphorene is significantly enhanced. The binding energy decreases as the degree of rippling increase for both convex and concave adsorption. Bader charge analysis shows the NO molecule receiving more electrons for adsorption at wrinkle

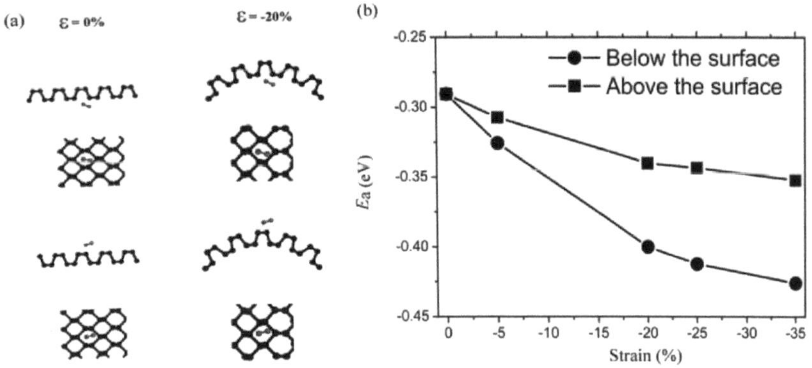

Figure 5.16 (a) Adsorption of the NO gas molecule at the planar phosphorene (left) and concave (below the surface) and convex (above the surface) regions of phosphorene. The red, blue, and black balls represent the oxygen, nitrogen, and phosphorus atoms, respectively. (b) The adsorption energy of NO above the rippled phosphorene surface as a function of the curvature induced by different strain. Reprinted with permission from Kistanov et al.[50] Copyright (2016) American Chemical Society.

sites compared with the adsorption at the planar part. The gaining of electrons of NO is 0.206 e for adsorption at −30% strained wrinkles, three times larger than the 0.084 e for adsorption at structures without rippling. The underlying reason for the enhanced adsorption and charge transfer can be due to the curvature effect. The convex and concave regions in the ripples suffer from strong tensile and compressive strains, respectively, which can modify the orbital hybridization and carrier density. Intentionally controlling the rippled structures in phosphorene can be promising for gas-sensing applications. On the other hand, the curvature effect may pose some effects on the structural stability of phosphorene due to an enhanced adsorption of gas molecules.

Experimental measurements have been made to investigate the sensitivity of 2D phosphorene nanosheets.[51] It was found that the phosphorene FET sensor shows a very high sensitivity to NO_2 in dry air environment. The 4.8 nm thick phosphorene sensor shows the maximum response with a sensitivity up to 190% at 20 parts per billion (p.p.b.) at room temperature. By recording the electrical conductance at a constant source–drain bias of 0.6 V, the time-dependent dynamic-sensing response ($\Delta G/G_0$) for a 4.8 nm thick phosphorene layer is shown in Fig. 5.17a with concentrations of NO_2 varying from 20 to 1000 p.p.b. Since NO_2 is an acceptor molecule that withdraws electrons from the phosphorene layer, as shown in the previous section, the conductivity of the sensor increases upon exposure to NO_2 as the hole concentration in the phosphorene conducting channel increases accordingly. This scenario is similar to other p-type semiconducting sensors like graphene. There is a positive correlation between the response amplitude and the gas concentrations.

The 4.8 nm thick phosphorene sensor shows an excellent response of 190% at a low concentration of 20 p.p.b. and 1600% for a concentration of 100 p.p.b. This performance of the phosphorene sensor is competitive to other highly sensitive sensors. For instance, the lower detection limits of reduced graphene oxide,[52] chemically modified graphene,[53] semiconduting single-walled carbon nanotubes,[54] and 2D MoS_2 nanosheets decorated with Pt nanoparticles[55] are 50, 70, and 44, and 2 p.p.b., respectively. The phosphorene sensor was found to have a high stability in dry air upon examining the behavior of the 40 nm thick sensor with different aging durations.

Rooted in its thickness-dependent band alignments and work function,[4] the phosphorene sensor shows a thickness-dependent sensitivity. The sensing performance of the phosphorene sensor as a function of thickness against 500 p.p.b. NO_2 is shown in Fig. 5.17b. Generally, a thinner phosphorene sensor was found to have a higher sensitivity. However, for a phosphorene flake thinner than 4.8 nm, the sensitivity drops dramatically as the thickness decreases further. The line profile of the thickness response is presented in Fig. 5.17d, showing that the sensitivity degrades with thicker layers with the best performance at around 5 nm. The response as a function of the concentration of NO_2 for the 4.8 nm thick sensor is shown in Fig. 5.17c, where the sensitivity generally increases linearly with increasing concentration. However, two different slopes can be found from 20 to 200 p.p.b. with high sensitivity and from 200 to 1000 p.p.b. with low sensitivity. This can be ascribed to the different conducting performance of phosphorene: at high concentration, poor conductivity due to the scattering of the

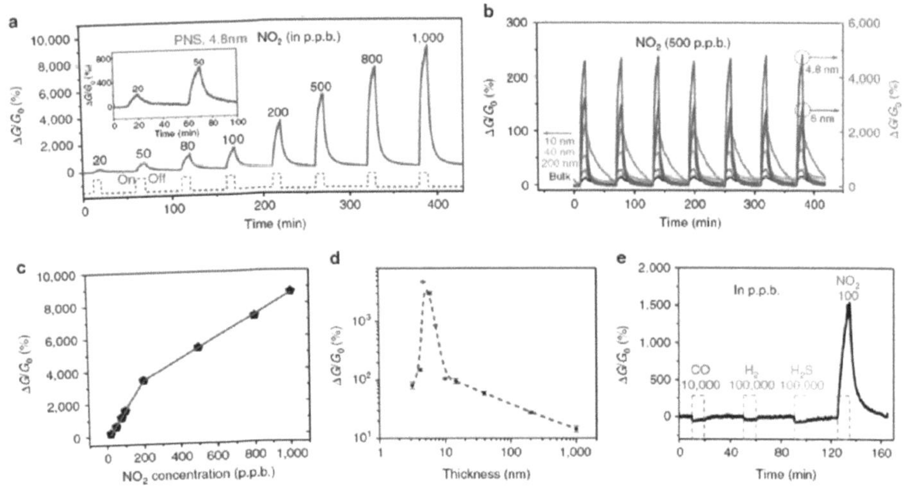

Figure 5.17 (a) The dependence of the relative conductance change with time for 4.8 nm thick phosphorene upon exposure to NO_2 from 20 to 1000 p.p.b. (balanced in dry air). The dashed lines show the switching on/off of NO_2 gas. (b) Thickness-dependent performance of the phosphorene sensor to 500 p.p.b. NO_2. The relative conductance change as a function of the NO_2 concentration (c) and the thickness (d) of the phosphorene. (e) Comparison of the sensing response curve of the 4.8 nm thick phosphorene to 10,000 p.p.b. CO, 100,000 p.p.b. H_2, 10,000 p.p.b. H_2S, and 100 p.p.b. NO_2. Reprinted with permission from Cui et al.[51] Copyright (2015) Nature Publishing Group.

physisorbed NO_2 molecules, and at low concentration, good mobility of phosphorene with limited effect of NO_2 molecules. The sensing ability of phosphorene for other gases, including 10,000 p.p.b. CO, 10,000 p.p.b. H_2S, and 100,000 p.p.b. H_2, was also studied in a same setting as for the NO_2 measurement.[51] The dynamic response to the other molecules is shown in Fig. 5.17e. It can be seen that the phosphorene sensor shows a weaker response to other gases despite a higher gas concentration than NO_2, which can be due to the strongest interaction of NO_2 according to a first-principles study.[37] This reveals that the phosphorene sensor has a strong specific adsorbing ability for NO_2. The superior performance of the phosphorene-based FET sensor is highly promising for a wide range of applications in environmental monitoring.

5.3.4 Heterostructure

Hybrid 2D materials consisting of at least two different types of 2D materials such as graphene/MoS_2, BN/MoS_2, graphene/BN, and MoS_2/WS_2 have been extensively explored owing to their intriguing properties, which are absent in each isolated 2D component. The different electronic band structure and work function of each single component allows the modulation of the barrier height for improving efficiency of carrier injection. For phosphorene, another benefit related with the integration with other 2D materials is improved structural stability of phosphorene layers in devices. The structural degradation of phosphorene can be suppressed by a passivation of oxide layers or more inert graphene and BN layers. This strategy of combining chemically stable 2D layers like graphene and BN with chemically sensitive 2D structures like silicene has already been demonstrated.[56] Promoted stability was also found in the BN or graphene-passivated TMD layers.[57]

Cai et al. investigated the energetics, band alignment, and interfacial charge transfer in graphene/phosphorene (G/BP) and BN/phosphorene heterostructures (BN/BP) by first-principles calculations.[58] They found that there is a similar stacking structure, energetics, and charge transfer between G/BP and BN/BP bilayers. In both cases, the interlayer interaction between the phosphorene and graphene or BN layer is weak due to the noncovalent bonding. The phosphorene becomes slightly n-doped in both heterostructures. The direct bandgap and the linear dichroism are preserved after capping the graphene or BN layer. However, it was shown that there is a great redistribution of electrostatic potential at the interface, which can affect the recombination of the electron–hole pair. Therefore, in addition to the protection of the phosphorene layer, they suggested that the graphene and BN can act as an active layer to tailor the contacting potential and the Schottkey barrier by tuning the number of the graphene or BN atomic layers.

To maintain lattice commensuration, the bilayer G/BP and BN/BP structures are modeled by combining an orthogonal supercell ($\sqrt{3} \times 1$) of graphene and a 3×1 supercell of phosphorene to minimize the lattice mismatch (smaller than 3%).[58] The calculations with the optB88 method reveal an interlayer spacing of around 3.49 Å and the binding energy of −71 meV per phosphorus atom. The orbital-decomposed band structure of G/BP bilayer (Fig. 5.18a) shows that the direct bandgap of phosphorene and Dirac-cone characteristics of graphene are maintained in the hybrid layer. Owing to the asymmetric potential distribution normal to the graphene layer in G/BP, there is a small gap opening around 1.3 meV in the Dirac cone associated with graphene, and the Fermi level is located inside the small Dirac gap. The bandgap associated with the phosphorene layer in G/BP is unchanged compared with the isolated layer. Therefore, the graphene can be a perfect layer to protect or support the phosphorene layer for improving the contact performance while not degrading the electronic properties of phosphorene.

As shown in Fig. 5.18a, there is a clear orbital hybridization of states of graphene and phosphorene along the S–X and X–Γ directions, while such interaction is less apparent along the Y–S and Γ–Y directions. This indicates that there is an angular dependent interlayer coupling between the two layers with a stronger electronic interaction along the zigzag direction than along the armchair direction. The intrinsic linear dichroism of phosphorene tends to be unaffected by the contacting graphene layer due to the transparency of the latter to the linear x- and y-polarized lights.

For the BN/BP case, the binding energy between the two layers is around −79 meV per P atom and the interlayer distance is 3.46 Å, almost the same as that of the G/BP case. This similar adsorbing energetics and geometry between the two cases is surprising as the polar BN is supposed to have larger electrostatic interaction with phosphorene than the graphene case. One possible reason is that the dispersive force with phosphorene in the graphene case is larger than that of the BN/BP case, which compensates the favorable electrostatic interaction in the BN/BP case. Figure 5.18b shows the orbital decomposed band structure of BN/BP, which shows that the energetic positions of the frontier orbitals related with phosphorene are sandwiched between those of the large gap of BN. The BN/BP has a type I (a stradding gap) band alignment between phosphorene and BN based on the calculated work function and

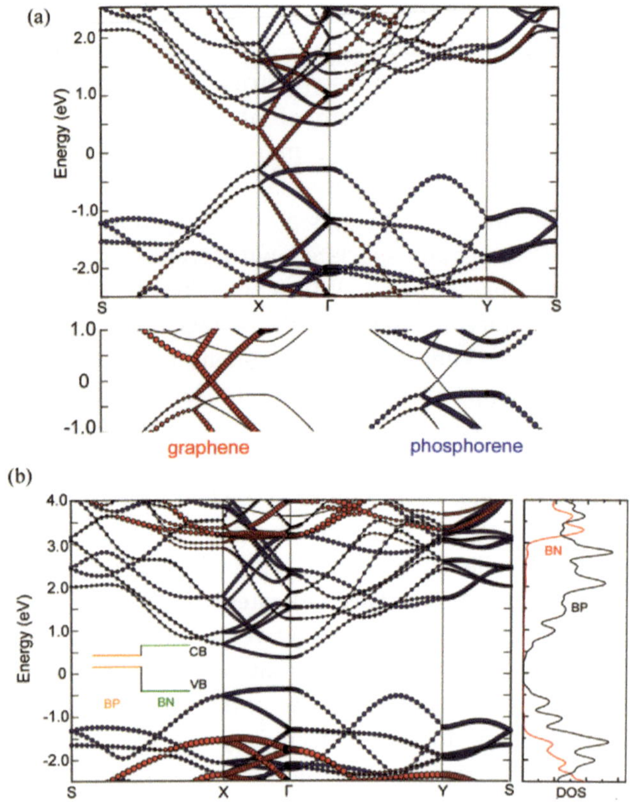

Figure 5.18 (a) Projected band structure of the G/BP bilayer calculated by HSE06. The components of graphene (phosphorene) are proportional to circles of the red (blue) circles. Bottom insets are the enlarged part around the Fermi level. (b) Projected band structure and DOS of the BN/BP bilayer. The components of BN (phosphorene) are proportional to circles of the red (blue) circles. The inset shows type I band alignment of the heterostructure. Reprinted with permission from Cai et al.[58] Copyright (2015) American Chemical Society.

bandgap of BN and phosphorene.[58] This very large bandgap of BN, together with the fully buried electronic states of phosphorene, suggests that BN is transparent and a promising protecting layer for phosphorene.

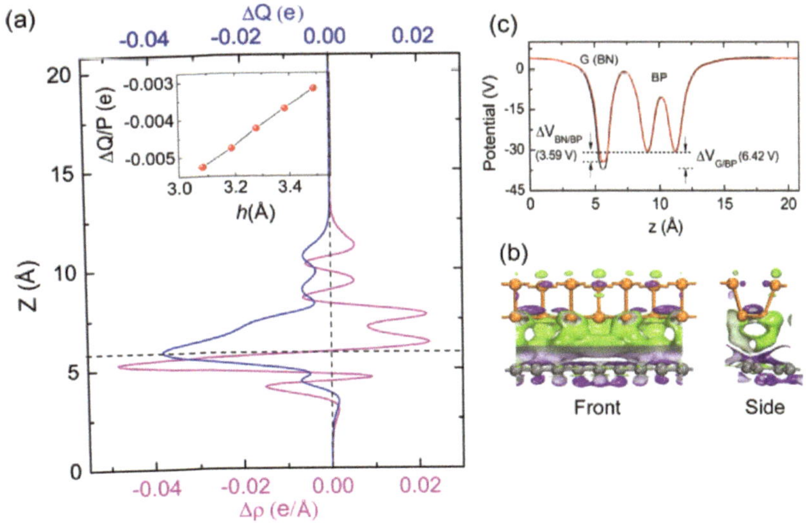

Figure 5.19 (a) Charge transfer in the G/BP bilayer. The plane-averaged differential charge density $\Delta\rho(z)$ (red line) and the amount of transferred electron at z point $\Delta Q(z)$ (blue line). Inset shows the variation of transferred electrons per phosphorus atom with interlayer distance h. The negative value of ΔQ shows a gain of electrons in phosphorene. (b) Isosurface plot of the differential charge density. The violet and green colors represent the loss and gain of electrons, respectively. (c) Distance–potential profile of the G/BP (black) and BN/BP (red) along the normal direction to the interface. Reprinted with permission from Cai et al.[58] Copyright (2015) American Chemical Society.

The charge transfer between the phosphorene and graphene/BN layer[58] was investigated by using the planar integration method, which is widely adopted in layered and surface systems. Starting from the planar averaged differential charge density along the direction normal to the surface $\Delta\rho(z)$, the amount of transferred charge up to the z point $\Delta Q(z)$ can be determined by integrating $\Delta\rho(z)$ from the bottom infinity to the z point, which is shown in Fig. 5.19a for the G/BP bilayer. It is shown that the phosphorene layer receives electrons from the graphene layer, resulting in a p-doping (n-doping) in the graphene (phosphorene) sheet. This trend of charge transfer across the graphene–phosphorene interface is vividly shown in Fig. 5.19b. The total amount of transferred electrons is around 0.003 e per P atom. The tiny charge transfer is consistent with the weak binding energy between graphene and phosphorene, indicating that the p_z state of phosphorus atoms weakly couples that of the π state of graphene. It is shown that the amount of charge transfer is proportional to the compressive strain along the normal direction (inset of Fig. 5.19a). The planar averaged potential across the interface of the G/BP heterostructure is plotted in Fig. 5.19c, where a large potential drop of 6.42 eV is found across the interface. This suggests that the charge transport and electron–hole recombination rates can be different from the behaviors of isolated graphene or phosphorene layer due to the newly built-in electric field across the interface. For the BN/BP bilayer structure, a similar charge doping ability of BN as graphene in the G/BP structure. However, a much smaller potential drop with only 3.59 eV is found at the BN/BP interface than the G/BP bilayer.

Since phosphorene is very sensitive to environmental molecules, the major role played by the capping graphene or BN layer is to protect phosphorene from these molecules. In addition, the presence of lone-pair electrons makes phosphorene prone to corrugation upon contact with the metal electrodes, resulting in the degraded performance of phosphorene. The insertion of a relatively more stable graphene or BN layer between the electrodes and phosphorene sheets can improve the structural integrity and allow the modulation of the contacting resistance at the interface.

Experiments have successfully demonstrated the encapsulation of ultrathin black phosphorene with boron nitrides under inert gas condition.[59] A field-effect transistor was fabricated by further utilizing graphene as source/drain electrodes. A linear current–voltage behavior together with negligible temperature dependence was revealed, suggesting that graphene electrodes can have a small contacting resistance and a low Schottky barrier. The atomically thin graphene electrodes and BN layer can also improve the stability of black phosphorus by sealing the exposed edges and surfaces. The graphene–phosphorene–BN architecture with complete 2D components could also be suitable for other 2D channel and metallic materials.

5.4 Dielectric and Optical Properties

5.4.1 Excitons

Exciton, one of the elemental quasiparticles, is closely related to the excited properties of semiconductors through the many-body effect. A neutral exciton, similar to the neutral hydrogen atom, is a bound quasiparticle state of a coupled electron and hole through Coulomb interaction. Through incorporating the many-body self-energy calculations, the excitonic effects in phosphorene were theoretically studied by first-principles GW–Bethe–Salpeter equation (BSE) method.[6] By comparing the quasiparticle bandgap and the optical spectral, the effects of electronic screening in few-layer phosphorene and bulk black phosphorus can be understood. The many-body corrections significantly increase the electronic bandgap from 0.8 eV based on normal DFT calculation to 2 eV based on GW calculation, thereby decreasing the lowest-energy optical absorption peak down to 1.2 eV due to the strong excitonic effect associated with the quasi-one-dimensional electronic structure of phosphorene. This unique feature allows a strong anisotropic response to polarized light excitations. In addition, the exciton binding energy and the optical spectra are highly sensitive to thickness, which allows the engineering of the photon absorption and emission through controlling the layers number.

Figure 5.20 presents the optical absorption spectra of the multilayer phosphorene due to the shining of the polarized light along the armchair (Fig. 5.20a–e) and zigzag (Fig. 5.20f–j) directions. It was found that the excitonic interactions play critical roles in the light–matter interaction in phosphorene. For instance, the binding energy of exciton in monolayer phosphorene can reach up to 800 meV with the first absorption peak at 1.2 eV. Such a strong excitonic effect, comparable to those in other 1D and 2D structures,[60–63] can arise from the suppressed electronic screening due to the low dimensionality. However, for bulk phosphorus, the inclusion of the electron–hole interactions slightly tailors the absorption spectrum, and there is a weak excitonic interaction with the exciton binding energy around 30 meV (Fig. 5.20d,e), which is dramatically weaker than that of the bulk layered BN with an exciton binding energy

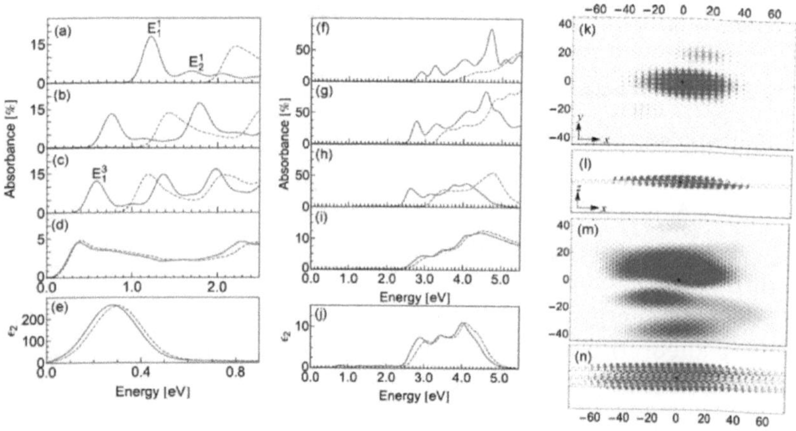

Figure 5.20 Optical absorption spectra of monolayer phosphorene with light polarized along the armchair (a–e) and zigzag (f–j) directions. The single-particle optical absorption spectra (dashed lines) and spectra with electron–hole interaction (solid lines) are compared for monolayer (a,f), bilayer (b,g), trilayer (c, h), and bulk (d, e, i, and j) phosphorene. (d) and (i) are spectra per layer in bulk black phosphorus. Reprinted with permission from Tran et al.[6] Copyright (2014) American Physical Society.

of 600 eV.[64] The underlying reason would be related to a stronger interlayer interaction in black phosphorus, its behavior being more similar to the traditional three-dimensional material.

By comparing the spectra between the armchair and zigzag polarized light, it was found that there exist different energy ranges of light absorption. For instance, monolayer phosphorene is transparent to the zigzag polarized light from 1.1 to 2.8 eV. A strong light absorption of armchair polarized light occurs in the above energy window. Rooted in the optical selection rules for the polarized light coupled with the frontier orbitals, phosphorene can be a good linear polarizer, which is potentially useful in displays and optical quantum computers. Importantly, this polarization energy window can be continually tuned from 0.3 to 2.8 eV by controlling the thickness of the phosphorene layer.

Figure 5.20k–n shows the spatial distribution of the excitons associated with the E_1^1 and E_3^1 peak in Fig. 5.20a,c of monolayer and trilayer phosphorene. The excitons, forming striped patterns, are highly anisotropic with more delocalization along the armchair direction. The more extending behavior in the armchair direction would be due to the more mobile carriers along this direction. Interestingly, these more extending excitons along the armchair direction can only be triggered by polarized light along the same direction. As shown by trilayer phosphorene (Fig. 5.20m,n), the excited electrons can well extend in top and bottom sheets despite the hole being fixed in the middle layer, suggesting that the energetics and distribution of excitons can be modulated by the number of layers.

The above theoretical predictions of anisotropic and strong excitonic effect were corroborated by the experiment by using polarization-resolved photoluminescence measurements at room temperature.[65] It was found that the excitons in monolayer phosphroene were strongly bound and highly anisotropic. By shining differently polarized light, the emitted light is always linearly polarized along the light effective mass direction and centered at around 1.3 eV. By recording the intensity of photoluminescence, the quasiparticle bandgap of 2.2 eV was determined, where the strongest absorption increase rate was observed around the quasiparticle band edge. The exciton banding energy of around 0.9 eV was obtained. It was also shown that the dielectric screening with surface passivation of phosphorene can slightly reduce the quasiparticle bandgap and exciton binding energy.

In contrast to a neutral exciton consisting of an electron–hole pair, a trion is a charged exciton which consists of a neutral exciton coupled with an additional electron or hole similar to H^- or H_2^+. Therefore, the density of trions can be controlled by applying the gate voltage, allowing tunable optoelectronic properties. Moreover, due to this additional charge with nonzero spin, characteristics of trions can be highly important for spin manipulation. For potential applications, a large binding energy of trions is highly desired to overcome thermal fluctuations. The dynamics of excitons and trions in monolayer phosphorene were studied by controlling the photocarrier injection.[66] It was found that the exciton binding energy is only 0.3 eV for phosphorene supported by the SiO_2/Si substrate, greatly smaller than that of the free-standing layer, as shown before. The lifetime of the triggered exciton was found to be around 220 ps, comparable to that of MoS_2. The binding energy of trion was shown to be 0.1 eV, which is huge and around 5 times larger than that of MoS_2.

The peak energy of the photoluminescence spectrum corresponds to the optical gap (E_{opt}), which is the difference between the electronic quasiparticle bandgap (E_g) and the binding energy of excitons (E_b). Based on the measured E_{opt} of 1.75 eV and the reported value of E_g of 2.05 eV in a previous study[67,68] of monolayer phosphorene, the exciton binding energy of the SiO_2/Si-supported phosphorene amounted to 0.3 eV. The exact peak energies are 1.75 eV (1L), 1.29 eV (2L), 0.97 eV (3L), 0.84 eV (4L), and 0.80 eV (5L). This rapid decrease in E_{opt} with increasing thickness can be due to the strong quantum confinement effect. The thickness-dependent behavior of E_{opt} can be fitted by a power law: $E_{opt} = 1.486/N^{0.686} + 0.295$, where N is the number of layers.

By using back-gated metal–oxide–semiconductor (MOS) architecture, similar findings of high binding energy of trions were reported in few-layer phosphorene, up to 0.16 eV at room temperature,[68] much larger than that of transition metal dichalcogenide ($20 - 30$ meV) and quasi-2D quantum wells ($\sim 1 - 5$ meV). Clearly, such strongly bound trions in phosphorene, which have also been observed in 1D CNTs, are rooted in their quasi-1D structure along the ridge direction. The advantage of phosphorene over traditional 1D materials like CNTs is related to its intrinsically large optical cross sections owing to its large 2D area.

In principle, the population and dynamics of trions can be modulated by tuning the carrier doping concentration through electrostatic gating and chemical doping. By monitoring the evolution of photoluminescence with gate voltage, the reversible electrostatic tunability of the neutral and charged excitons were investigated in 3L phosphorene MOS device.[68] There are two peaks located at around 1100 and 1300 nm, which can be assigned to the exciton emission and positively charged trion emission, respectively. Upon applying a gate voltage (V_g), the intensities and position of both peaks evolve gradually. For a negative V_g, phosphorene can be effectively hole-doped, inducing an increase in the content of the trions. When V_g reaches -50 V, almost all the excitons become trions and the peak at 1100 nm corresponding to neutral excitons nearly disappears. On the contrary, applying a positive V_g can suppress positive trions while enhancing the population of neutral excitons.

5.4.2 Linear Dichroism

Two-dimensional materials like graphene and MoS_2 have been found to show promising applications in optoelectronics as photodetectors. However, the ability to detect polarized light is elusive. Linear dichroism is a phenomenon related to different absorptions of parallel or perpendicular polarized light to an orientation axis. Normally, it was realized in an extrinsic way by creating a proper conformation and orientation of material structures to be oriented in anisotropic patterns[69,70] or in an intrinsic way relying on an anisotropic crystal structure.[71,72] Black phosphorus, owing to its highly anisotropic structure along the armchair and zigzag directions, allows a highly anisotropic interaction of electrons and photons within the layers. The presence of mirror reflection symmetry only in the armchair direction allows photodetection of linearly polarized lights.

The optical absorption spectra and linear dichroism of phosphorene were initially investigated by first-principles calculations.[73] The thickness-dependent absorption spectra of incident light along the normal direction and linearly polarized in the armchair (x) and zigzag (y) directions are shown in Fig. 5.21a. The linear dichroism can be seen from the different band edges determined from the absorption spectra between the light polarized in zigzag and armchair directions. By measuring the linear dichroism, the orientation of the phosphorene sample can be determined by optical spectroscopy (Fig. 5.21b). By adopting an incident light source normal to the surface and linearly polarized, the zigzag or armchair direction can be determined by rotating the sample.

The underlying mechanism accounting for the linear dichroism is related to the symmetries of the electronic wavefunctions. Since phosphorene, both bulk and few-layer, possesses mirror reflection symmetry (M) and inversion symmetry (parity) only in the y (zigzag) direction, the electronic states at the Γ point can be labeled as odd or even state according to polarity and mirror symmetries of their wavefunctions. The optical selection rules allow the electronic

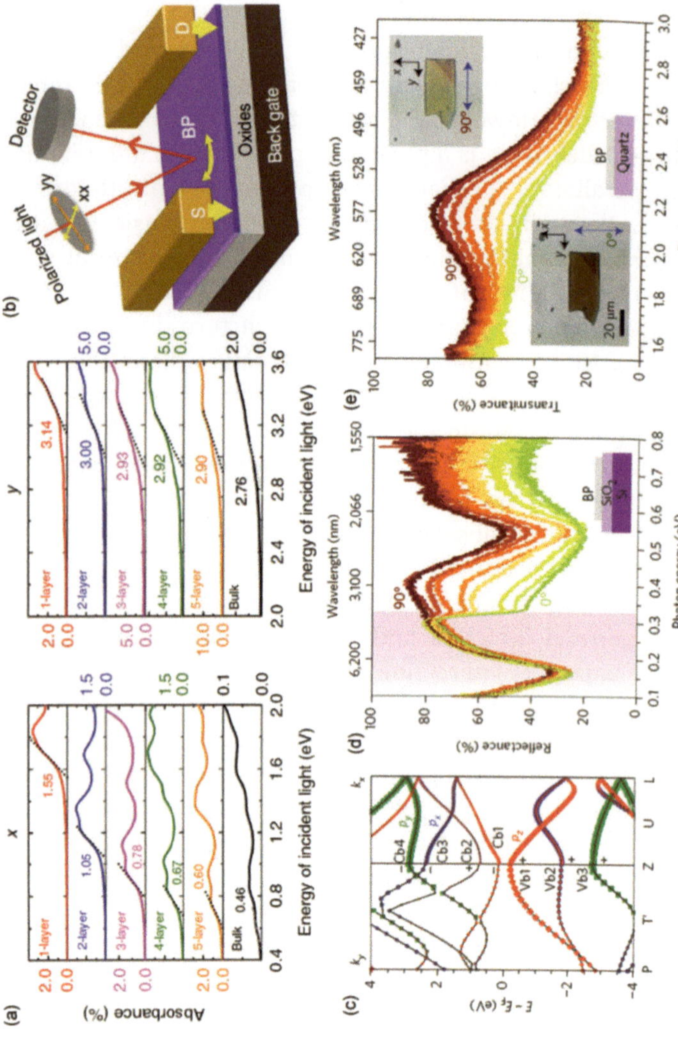
(a)
Absorbance (%)
Energy of incident light (eV)
(b)
Polarized light
Detector
BP
Oxides
Back gate
(c)
(d)
Reflectance (%)
(e)
Transmittance (%)
Wavelength (nm)
Photon energy (eV)
Quartz

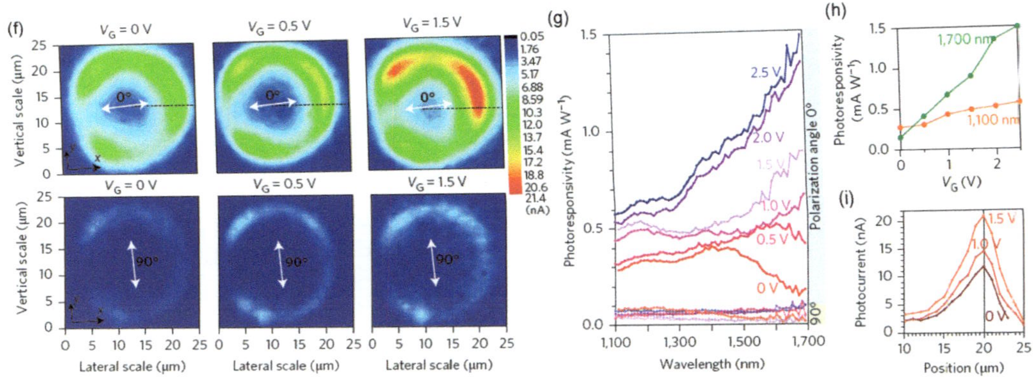

Figure 5.21 (a) Thickness dependence of the optical absorption spectra of phosphorene with normal incident light polarized in x and y directions. (b) Schematic experimental setup for determining the orientation of phosphorene samples via optical spectroscopy. Reprinted with permission from Qiao et al.[73] Copyright (2014) Nature Publishing Group. (c) Parity and orbital composition of band structure of bulk black phosphorus. The parity of the electronic state at the zone center is represented by "+" and "−". The p_z, p_x, and p_y orbitals in the projected band dispersion are represented by the red, blue, and green colors, respectively. (d) Reflection spectrum of infrared light as a function of the polarized angle. (e) Transmission spectrum of visible light as a function of the polarized angle. In both (d) and (e), the incident light is linearly polarized in an angular step of 15° from the armchair (green/0°) to zigzag (dark red/90°) directions. (f) Images of the photocurrent of the gate-controlled black phosphorus p–n junction on illuminating with 1500 nm linearly polarized light under different gate voltages. The white arrows indicate the armchair (0°) and zigzag (90°) directions of black phosphorus. (g) Photoresponsivity as a function of wavenumber and polarized angle of the incident light. (h) Photoresponsivity as a function of gate voltage for two different lights polarized along the armchair direction. (i) Site-dependent photocurrent curves from the centre of the ring to the metal edge along the black dashed line in (f) under different gate voltages for 1500 nm light polarized along the armchair direction. Reprinted with permission from Yuan et al.[74] Copyright (2015) Nature Publishing Group.

transitions from odd to even electronic states due to the odd nature of the dipole operator. Therefore, the parity and polarization selection rules determine that optical absorption can only occur between two electronic states with opposite parities and a proper polarized electric state under linearly polarized light. In particular, the polarization selection rules directly lead to linear dichroism. Absorption of *x* (*y*)-polarized photons can only be allowed with even (odd) electronic states of mirror symmetry.

The parity (represented by "+" and "−") and the polarized nature (p_x, p_y, or p_z) of the electronic states of bulk phosphorene are shown in Fig. 5.21c. According to the optical selection rules discussed above, the lowest-energy optical transition at the Z point between Vb1 and Cb1 bands is allowed due to their opposite parities. The Vb1 and Cb1 at the Z point and $k_y = 0$ plane (Z–L line) are even under the mirror symmetry in the *y* direction. Therefore, the absorption of *y*-polarized light is forbidden. In contrast, the states with zero *x* moment (Z–P line) have no mirror symmetry, thus allowing absorption of both *x*- and *y*-polarized light. Since the transition between Vb1 and Cb1 states, permitting only *y*-polarized light, involves photons with energies from 0.33 to 4 eV, a linear polarized absorption can occur in a wide range of energy in the absorption spectrum. Despite the polarization sensitivity being weakened in high-energy photons above 1.2 eV, the polarization-sensitive absorption can be quite apparent from 0.33 eV to 3.0 eV, which is enough for potential applications.

By taking advantage of linear dichroism and in-plane optical anisotropy of black phosphorus, a broadband photodetector based on transistor geometry with light-polarized sensitivity from ~400 to 3750 nm was fabricated by using black phosphorus.[74] The gate-inducted perpendicular built-in electric field can suppress the recombination of electrons and holes and promote the performance of photodetection. Figure 5.21d,e shows the reflection and transmission spectra of black phosphorus. The linear dichroism can be seen from the asymmetric infrared spectra for photons polarized along armchair and zigzag directions: a sudden drop of reflecting intensity is found for incident light polarized along the armchair direction, while it is absent for light polarized along the zigzag direction. There is a gradual evolution of the transmission spectra with the polarization angles for the visible light. For light polarized along the armchair direction, the transmittance is low (42% at 2.3 eV) due to a stronger absorption. In contrast, for the light polarized along the zigzag direction, the transmittance is high (80% at 2.3 eV) and the sample appears transparent due to a lower absorption.

The 2D photocurrent images of a ring-shaped black phosphorene sample were produced under different gate voltages (Fig. 5.21f). The photoresponsivity increases one order of magnitude by adopting a positive gate voltage (Fig. 5.21g,h). The voltage-dependent characteristics of the photocurrent differ from differently polarized light. For instance, upon increasing the gate voltage from 0 to 1.5 V, the photocurrent under light polarized along armchair direction increases significantly, while no significant changes occur for light polarized along the zigzag direction (Fig. 5.21f). The difference between the photoresponsivities measured under different polarizations increases with increasing gate bias. Upon applying the built-in electric field, the downward band bending can effectively separate the light-excited carriers, thus reducing the recombination rate of electrons and holes. This vertical p–n junction setup can make the effect and performance of linear dichroism of black phosphorus more significant.

5.4.3 Effects of Rippling on the Electronic and Photonic Properties

Formation of the ripples and bubbles is very common in the atomically thin materials due to substrate effect, external molecules, and stress. Modulating the bandgap through such local-strain-related deformation of 2D materials is particularly suitable and an interesting avenue for tunable optotelectronic applications as they can sustain unprecedented nonhomogeneous deformations before rupture. One advantage of phosphorene is its extraordinarily bendable flexibility related to its intrinsic puckered structure and soft P–P bonds. Owing to the absence of surface dangling bonds, phosphorene nanomaterials can withstand a large localized deformation such as ripples and bubbles. Kistanov et al. systematically investigated the effect of ripple formation on the electronic properties of phosphorene by using DFT calculations.[50]

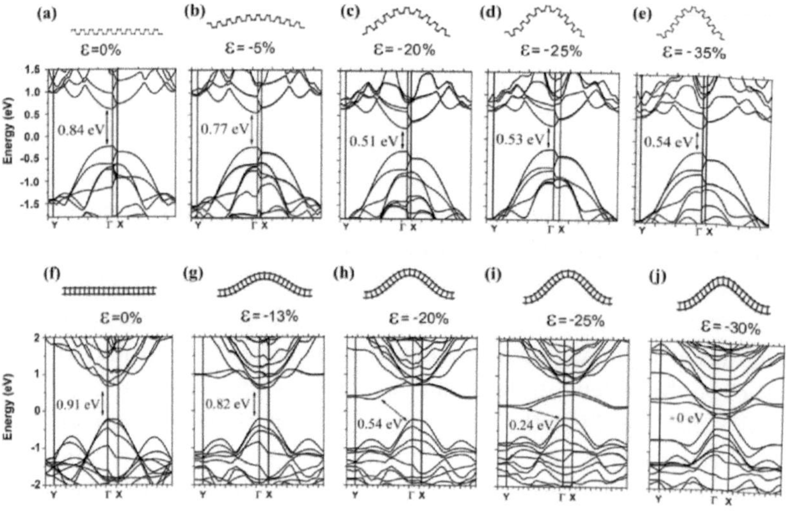

Figure 5.22 Variation of the atomic structure and band structure of monolayer phosphorene with structural rippling quantified by different strains along the armchair (a–e) and zigzag (f–j) directions. Reprinted with permission from Kistanov et al.[50] Copyright (2016) American Chemical Society.

Rippled phosphorene structures are modeled by creating waved structures based on phosphorene supercells (40 phosphorus atoms) by applying strain along the zigzag or armchair direction (Fig. 5.22). Depending on the directions of the periodicity of the ripples, rippled phosphorene shows different electronic properties compared with those of the perfect flat phosphorene. For instance, the bandgap of the initial flat phosphorene decreases from 0.84 eV to 0.51 eV for rippled structure along the armchair direction with a −20% strain, and the bandgap is changes less as the strain is further increased up to −30%. The direct bandgap is maintained for ripples along the armchair direction for the considered strain up to −35% (Fig. 5.22a–e). In contrast, the structural rippling along the zigzag direction triggers a direct-to-indirect transition of the bandgap at −20% strain with the bottom of the conduction band shifting from Ã to a point between the Ã and Y points. With further compression, a semiconductor–metal transition initiates at around −30% (Fig. 5.22f–j).

Structural rippling can also change the orbital composition near the band edges of phosphorene. For the perfect phosphorene without ripples, the s, p_x, p_y, and p_z states are almost equally distributed within the valence band and the frontier orbital mainly involves the p_z orbital. However, the components of p_y (p_x) state at the top of the valence band and the bottom of the conduction band increases upon increasing the degree of rippling along the armchair (zigzag) direction. One important difference between rippled and perfect flat phosphorene lies in the different composition of atomic orbitals at the valence top. For the phosphorene without curvature effect, there is only p_y state at the valence top, the p_x state being absent, which is the underlying reason for the linear dichroism. Upon the phosphorene becoming rippling along the zigzag direction, the p_x and p_y states coexist at the valence top, which may lead to the quenching and loss of the linear dichroism.

In-depth examination of the local electronic properties of atoms at different parts of the ripple shows spatially varying local bandgap and alignment. To compare the shift and alignment of states for atoms along the line profiles of ripples, the local DOS (LDOS) of each pair of atoms at different parts of the ripples is given in Fig. 5.23. For the phosphorene with ripples along the armchair direction, there is no shift in the LDOS for the atom pair and the gaps are almost overlapped. However, for the ripples along the zigzag direction, there is a great shift in the LDOS with the position of atoms in the ripples. A redshift of the states is shown for atoms located at the peaks of the ripples, while blueshift of states is predicted for atoms away from the peaks of ripples. These highly spatially dependent electronic properties are useful for fabricating devices by controlling the optical properties and efficiency of carrier transport based on rippled phosphorene. Such a strong spatially dependent alignment of the electronic states, producing type II band alignment of heterostructures, allows an efficient separation of electrons and holes, favoring photovoltaic applications. In addition, the relative shift of the valence and conduction bands along the periodic direction of ripples can induce the funnel effect, which can enhance the efficiency of photovoltaic energy. The electronic properties are more robust for ripples formed along the armchair direction than those formed along the zigzag direction. The strong

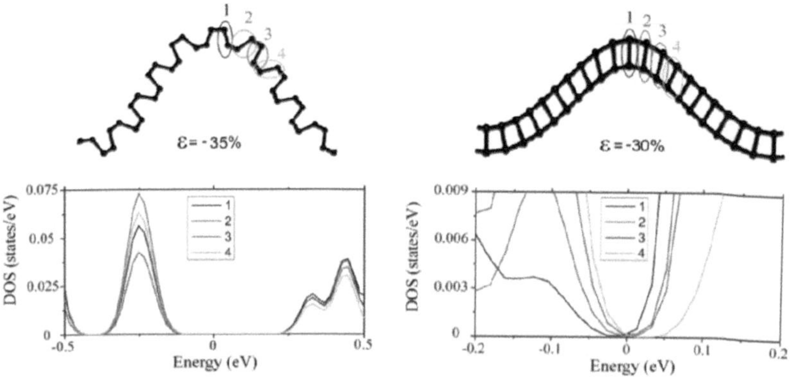

Figure 5.23 The LDOS for atomic pairs of rippled phosphorene along the periodic along the armchair (a) and zigzag (b) direction. Reprinted with permission from Kistanov et al.[50] Copyright (2016) American Chemical Society.

site-dependent behavior of the local electronic properties for ripples along the zigzag direction shown above allows the modulation of the optical properties of phosphorene.

Experimental measurement of the effect of ripples on the optoelectronic properties of black phosphorus was performed by applying periodic stress in the samples.[75] To intentionally produce periodic ripples, the exfoliated black phosphorus flakes were transferred onto a prestretched elastomeric gel-film substrate followed with a sudden release of strain. A moderate initial strain (<30%) avoids the formation of the delamination of the flake with the substrate and induces a periodic rippling in the black phosphorus flake through trade-off between the adhesion force, the compressive stress, and the bending rigidity. The initiated ripples in black phosphorus in this way show alternating tensile (compressive) strain on the summits (valleys) at the surface of the flake.

Figure 5.24a shows the transmission mode optical microscopy image of the ripples in a 10 nm thick (~18 layers) black phosphorus flake. It was found that the rippling period is proportional to the flake's thickness, and for 10 nm thick black phosphorus flake the rippling period is around 1 μm. The height of the ripples is determined by atomic force microscopy and shown in the bottom

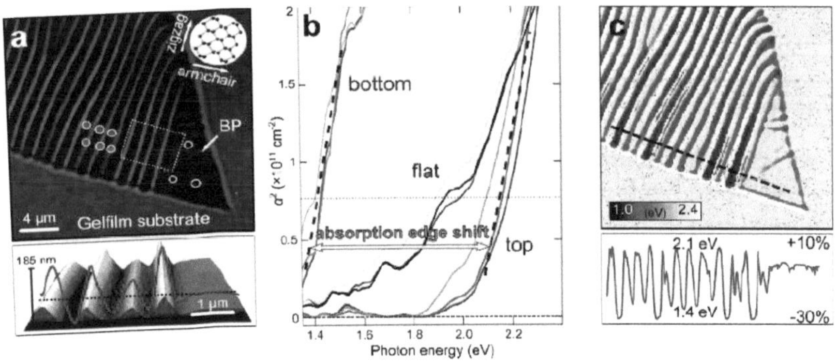

Figure 5.24 (a) Optical image of the ripples of a 10 nm thick black phosphorus above a gel-film substrate. The inset shows the lattice orientation of black phosphorus. The bottom inset shows the atomic force microscopy topography image of the region highlighted with the dashed rectangle. (b) Comparison of the optical absorption spectra measured at the summits, valleys, and flat regions each for three times with colored circles in (a). (c) Iso-absorption map with spatially dependent absorption spectra. The bottom inset shows the linecut obtained along the dashed line in the iso-absorption map. Reprinted with permission from Quereda et al.[75] Copyright (2016) American Chemical Society.

of Fig. 5.24a. On the basis of the strong angular-dependent linear dichroism of phosphorene, the orientation of the ripples relative to the orthorhombic lattice is determined and found to be almost aligned parallel to the zigzag direction. This is reasonable as the bending stiffness along the armchair direction is much smaller than along the zigzag direction.

The optical absorption spectra measured at different positions of the rippled flake, the summits, valleys, and flat regions, are shown in Fig. 5.24b. It can be seen that the absorption edge is blueshifted by around 0.7 eV on the summits relative to the valleys. Compared to the flat flake, the absorption edge at the valleys is more shifted than at the summits. The shifts of the optical absorption edge could be induced by the tensile and compressive strains in the valleys and summits of the ripples. The spatial variation of the energy of the absorption edge was measured and shown in Fig. 5.24c. This iso-absorption map, showing the same rippling pattern, reflects the onset of the absorption edge across the rippled black phosphorus and a strong correlation of the optical properties with localized tensile and compressive strain fields. The line-cut plot shows that a blueshift of around 10% on the summits and a redshift of around 30% on the valleys compared to the flat region. The shifts induced by the ripples in black phosphorus are much stronger than that in MoS_2 subjected with comparable strain, where only 5% change in bandgap is observed. The modulated local electronic properties of black phosphorene with periodic compressive and tensile stress imply its promising applications in exciton funnels which occur when a spatially dependent bandgap is present. The large shift of band edges at the summits and the valleys of the ripples leads to a realignment valence and conduction bands which effectively confine the carriers. This makes the excitons to funnel into the confined regions, thus enhancing the efficiency of photovoltaic devices.

5.5 Summary

Phosphorene materials are well suited to act as the 2D semiconducting channel for nanoelectronic applications owing to their high mobility and suitable bandgap. The soft structure and remarkably high mobility of phosphorene, even larger than those of MoS_2 and silicon, suggest its promising applications in flexible electronics. Its wide tunable optical and quasiparticle bandgap is highly promising for optical applications in a very broad wavelength range, from visible to mid-infrared light. In addition, few-layer phosphorene can be applied as an on-chip optical interconnect. Its unique anisotropy in electronics and optical properties enable on-chip light polarization manipulation. Different from the out-of-plane light polarization in graphene optical devices, the in-plane light polarization in phosphorene is highly desirable for integration with silicon photonics. In addition, the presence of unique quasi-1D channels along the ridge direction leads to the extraordinarily bound quasi-1D trions in phosphorene, which is absent in other common 2D materials. The one order larger trion binding energy than TMDs, together with a large optical cross section inherent in a 2D material system, allows extensive optoelectronic applications like photodetectors and tunable light sources.

References

1. Churchill HOH, Jarillo-Herrero P, (2014). Phosphorus joins the family. *Nat. Nanotechnol.*, **9**:330–331.
2. Ameen TA, Ilatikhameneh H, Klimeck G, Rahman R, (2016). Few-layer phosphorene: an ideal 2D material for tunnel transistors. *Sci. Rep.*, **6**:28515.
3. Batmunkh M, Bat-Erdene M, Shapter JG, (2016). Phosphorene and phosphorene-based materials-prospects for future applications. *Adv. Mater.*, **28**:8586–8617.
4. Cai Y, Zhang G, Zhang Y-W, (2014). Layer-dependent band alignment and work function of few-layer phosphorene. *Sci. Rep.*, **4**:6677.
5. Das S, Zhang W, Demarteau M, Hoffmann A, Dubey M, Roelofs A, (2014). Tunable transport gap in phosphorene. *Nano Lett.*, **14**:5733–5739.
6. Tran R, Soklaski R, Liang Y, Yang L, (2014). Layer-controlled band gap and anisotropic excitons in few-layer black phosphorus. *Phys. Rev. B*, **89**:235319.
7. Rudenko N, Katsnelson MI, (2014). Quasiparticle band structure and tight-binding model for single- and bilayer black phosphorus. *Phys. Rev. B*, **89**:201408(R).
8. Buscema M, Groenendijk DJ, Blanter SI, Steele GA, van der Zant HSJ, Castellanos-Gomez A, (2014). Fast and broadband photoresponse of few-layer black phosphorus field-effect transistors. *Nano Lett.*, **14**:3347–3352.
9. Liu H, Neal AT, Zhu Z, Luo Z, Xu X, Tománek D, Ye PD, (2014). Phosphorene: an unexplored 2D semiconductor with a high hole mobility. *ACS Nano*, **8**:4033–4041.
10. Chakraborty B, Gupta SN, Singh A, Kuiri M, Kumar C, Muthu DVS, Das A, Waghmare UV, Sood AK, (2016). Electron-hole asymmetry in the electron-phonon coupling in top-gated phosphorene transistor. *2D Mater.*, **3**:015008.
11. Jin Z, Mullen JT, Kim KW, (2016). Highly anisotropic electronic transport properties of monolayer and bilayer phosphorene from first principles. *Appl. Phys. Lett.*, **109**:053108.
12. Liao B, Zhou J, Qiu B, Dresselhaus MS, Chen G, (2015). Ab initio study of electron-phonon interaction in phosphorene. *Phys. Rev. B*, **91**:235419.
13. Li P, Appelbaum I, (2014). Electrons and holes in phosphorene. *Phys. Rev. B*, **90**:115439.

14. Low T, Rodin AS, Carvalho A, Jiang Y, Wang H, Xia F, Castro Neto AH, (2014). Tunable optical properties of multilayer black phosphorus thin films. *Phys. Rev. B*, **90**:075434.
15. Rudenko N, Yuan S, Katsnelson MI, (2015). Toward a realistic description of multilayer black phosphorus: From GW approximation to large-scale tight-binding simulations. *Phys. Rev. B*, **92**:085419.
16. Lado JL, Fernández-Rossier J, (2016). Landau levels in 2D materials using Wannier Hamiltonians obtained by first principles. *2D Mater.*, **3**:035023.
17. Ehlen N, Senkovskiy BV, Fedorov AV, Perucchi A, Di Pietro P, Sanna A, Profeta G, Petaccia L, Grüneis A, (2016). Evolution of electronic structure of few-layer phosphorene from angle-resolved photoemission spectroscopy of black phosphorous. *Phys. Rev. B*, **94**: 245410.
18. Peng X, Wei Q, Copple A, (2014). Strain-engineered direct-indirect band gap transition and its mechanism in two-dimensional phosphorene. *Phys. Rev. B*, **90**:085402.
19. Çakır D, Sahin H, Peeters FM, (2014). Tuning of the electronic and optical properties of single-layer black phosphorus by strain. *Phys. Rev. B*, **90**:205421.
20. Fei R, Yang L, (2014). Strain-engineering the anisotropic electrical conductance of few-layer black phosphorus. *Nano Lett.*, **14**:2884–2889.
21. Rodin AS, Carvalho A, Castro Neto AH, (2014). Strain-induced gap modification in black phosphorus. *Phys. Rev. Lett.*, **112**:176801.
22. Xiang ZJ, Ye GJ, Shang C, Lei B, Wang NZ, Yang KS, Liu DY, Meng FB, Luo XG, Zou LJ, Sun Z, Zhang Y, Chen XH, (2015). Pressure-induced electronic transition in black phosphorus. *Phys. Rev. Lett.*, **115**:186403.
23. Gong P-L, Liu D-Y, Yang K-S, Xiang Z-J, Chen X-H, Zeng Z, Shen S-Q, Zou L-J, (2016). Hydrostatic pressure induced three-dimensional Dirac semimetal in black phosphorus, *Phys. Rev. B*, **93**:195434.
24. Kikegawa T, Iwasaki H, (1983). Electrical investigation of phase transition in black phosphorus under high pressure. *Acta Crystallogr. B*, **39**:158.
25. Cartz L, Srinivasa SR, Riedner RJ, Jorgensen JD, Worlton TG, (1979). Electric field effect in ultrathin black phosphorus. *J. Chem. Phys.*, **71**:1718.
26. Brown A, Rundqvist S, (1965). Refinement of the crystal structure of black phosphorus. *Acta Crystallogr.*, **19**:684.

27. Wang V, Kawazoe Y, Geng WT, (2015). Native point defects in few-layer phosphorene. *Phys. Rev. B*, **91**:045433.

28. Ziletti A, Carvalho A, Campbell DK, Coker DF, Castro Neto AH, (2015). Oxygen defects in phosphorene. *Phys. Rev. Lett.*, **114**:046801.

29. Kistanov AA, Cai Y, Zhou K, Dmitriev SV, Zhang Y-W, (2017). The role of H_2O and O_2 molecules and phosphorus vacancies in the structure instability of phosphorene. *2D Mater.*, **4**:015010.

30. Pei J, Gai X, Yang J, Wang X, Yu Z, Choi D-Y, Luther-Davies B, Lu Y, (2016). Producing air-stable monolayers of phosphorene, and their defect engineering. *Nat. Commun.*, **7**:10450.

31. Koenig SP, Doganov RA, Seixas L, Carvalho A, Tan JY, Watanabe K, Taniguchi T, Yakovlev N, Castro Neto AH, Özyilmaz B, (2016). Electron doping of ultrathin black phosphorus with Cu adatoms. *Nano Lett.*, **16**:2145–2151.

32. Kim J, Baik SS, Ryu SH, Sohn Y, Park S, Park B-G, Denlinger J, Yi Y, Choi HJ, Kim KS, (2015). Observation of tunable band gap and anisotropic Dirac semimetal state in black phosphorus. *Science*, **349**:723.

33. Prakash A, Cai Y, Zhang G, Zhang Y-W, Ang K-W, (2017). Black phosphorus N-type field-effect transistor with ultrahigh electron mobility via aluminum adatoms doping. *Small*, **13**:1602909.

34. Liu Y, Cai Y, Zhang G, Zhang Y-W, Ang K-W, (2017). Al-Doped black phosphorus p–n homojunction diode for high performance photovoltaic. *Adv. Funct. Mater.*, **27**:1604638.

35. Buscema M, Groenendijk DJ, Steele GA, van der Zant HS, Castellanos-Gomez A, (2014). Photovoltaic effect in few-layer black phosphorus PN junctions defined by local electrostatic gating. *Nat. Commun.*, **5**:4651.

36. Cai Y, Bai Z, Pan H, Feng YP, Yakobson BI, Zhang Y-W, (2014). Constructing metallic nanoroads on a MoS_2 monolayer via hydrogenation. *Nanoscale*, **6**:1691.

37. Cai Y, Ke Q, Zhang G, Zhang Y-W, (2015). Energetics, charge transfer, and magnetism of small molecules physisorbed on phosphorene. *J. Phys. Chem. C*, **119**:3102–3110.

38. Ding F, Yakobson BI, (2011). Challenges in hydrogen adsorptions: From physisorption to chemisorptions. *Front. Phys.*, **6**:142–150.

39. Fair KM, Cui XY, Li L, Shieh CC, Zheng RK, Liu ZW, Delley B, Ford MJ, Ringer SP, Stampfl C, (2013). Hydrogen adsorption capacity of adatoms on double carbon vacancies of graphene: a trend study from first principles. *Phys. Rev. B*, **87**:014102.

40. Goler S, Coletti C, Tozzini V, Piazza V, Mashoff T, Beltram F, Pellegrini V, Heun S, (2013). Influence of graphene curvature on hydrogen adsorption: toward hydrogen storage devices. *J. Phys. Chem. C*, **117**:11506–11513.

41. Tozzini V, Pellegrini V, (2011). Reversible hydrogen storage by controlled buckling of graphene layers. *J. Phys. Chem. C*, **115**:25523–25528.

42. Park S, Srivastava D, Cho K. (2003). Generalized chemical reactivity of curved surfaces: carbon nanotubes. *Nano Lett.*, **3**:1273–1277.

43. Boukhvalov DW, Katsnelson MI, (2009). Enhancement of chemical activity in corrugated graphene. *J. Phys. Chem. C*, **113**:14176–14178.

44. González BS, Hernández-Rojas J, Bretón J, Gomez Llorente JM, (2007). Global potential energy minima of $(H_2O)_n$ clusters on graphite. *J. Phys. Chem. C*, **111**:14862.

45. Leenaerts O, Partoens B, Peeters FM, (2008). Adsorption of H_2O, NH_3, CO, NO_2, and NO on graphene: a first-principles study. *Phys. Rev. B*, **77**:125416.

46. Okamoto Y, Miyamoto YJ, (2001). Ab initio investigation of physisorption of molecular hydrogen on planar and curved graphenes. *Phys. Chem. B*, **105**:3470–3474.

47. Vaźquez de Parga AL, Calleja F, Borca B, Passeggi MCG, Hinarejos Jr JJ, Guinea F, Miranda R, (2008). Periodically rippled graphene: growth and spatially resolved electronic structure. *Phys. Rev. Lett.*, **100**:056807.

48. Atanasov V, Saxena A. (2010). Tuning the electronic properties of corrugated graphene: confinement, curvature, and band-gap opening. *Phys. Rev. B*, **81**:205409.

49. Sun L, Li Q, Ren H, Su H, Shi QW, Yang J, (2008). Strain effect on electronic structures of graphene nanoribbons: a first-principles study. *J. Chem. Phys.*, **129**:074704.

50. Kistanov AA, Cai Y, Zhou K, Dmitriev SV, Zhang Y-W, (2016). Large electronic anisotropy and enhanced chemical activity of highly rippled phosphorene. *J. Phys. Chem. C*, **120**:6876–6884.

51. Cui S, Pu H, Wells SA, Wen Z, Mao S, Chang J, Hersam MC, Chen J, (2015). Ultrahigh sensitivity and layer-dependent sensing performance of phosphorene-based gas sensors. *Nat. Commun.*, **6**:8632.

52. Cui S, Pu H, Mattson EC, Wen Z, Chang J, Hou Y, Hirschmugl CJ, Chen J, (2014). Ultrasensitive chemical sensing through facile tuning defects and functional groups in reduced graphene oxide. *Anal. Chem.*, **86**:7516–7522.

53. Yuan WJ, Liu AR, Huang L, Li C, Shi GQ, (2013). High-performance NO_2 sensors based on chemically modified graphene. *Adv. Mater.*, **25**:766-771.

54. Li J, Lu Y, Ye Q, Cinke M, Han J, Meyyappan M, (2003). Carbon nanotube sensors for gas and organic vapor detection. *Nano Lett.*, **3**:929–933.

55. He Q, Zeng Z, Yin Z, Li H, Wu S, Huang X, Zhang H, (2012). Fabrication of flexible MoS_2 thin-film transistor arrays for practical gas-sensing applications. *Small*, **8**:2994–2999.

56. Berdiyorov GR, Neek-Amal M, Peeters FM, van Duin ACT, (2014). Stabilized silicene within bilayer graphene: a proposal based on molecular dynamics and density-functional tight-binding calculations. *Phys. Rev. B*, **89**:024107.

57. Shi Y, Zhou W, Lu A-Y, Fang W, Lee Y-H, Hsu AL, Kim SM, Kim KK, Yang HY, Li L-J, Idrobo J-C, Kong J, (2012). van der Waals epitaxy of MoS_2 layers using graphene as growth templates. *Nano Lett.*, **12**:2784–2791.

58. Cai Y, Zhang G, Zhang Y-W, (2015). Electronic properties of phosphorene/graphene and phosphorene/hexagonal boron nitride heterostructures. *J. Phys. Chem. C*, **119**:13929–13936.

59. Avsar A, Vera-Marun IJ, Tan JY, Watanabe K, Taniguchi T, Neto AHC, Özyilmaz B, (2015). Air-stable transport in graphene-contacted, fully encapsulated ultrathin black phosphorus-based field-effect transistors. *ACS Nano*, **4**:4138–4145.

60. Qiu DY, da Jornada FH, Louie SG, (2013). Optical spectrum of MoS_2: many-body effects and diversity of exciton states. *Phys. Rev. Lett.*, **111**:216805.

61. Yang L, Cohen ML, Louie SG, (2007). Excitonic effects in the optical spectra of graphene nanoribbons. *Nano Lett.*, **7**:3112.

62. Spataru CD, Ismail-Beigi S, Benedict LX, Louie SG, (2004). Excitonic effects and optical spectra of single-walled carbon nanotubes. *Phys. Rev. Lett.*, **92**:077402.

63. Wang F, Dukovic G, Brus LE, Heinz TF, (2005). The optical resonances in carbon nanotubes arise from excitons. *Science*, **308**:838.

64. Wirtz L, Marini A, Rubio A, (2006). Excitons in boron nitride nanotubes: dimensionality effects. *Phys. Rev. Lett.*, **96**:126104.

65. Wang X, Jones AM, Seyler KL, Tran V, Jia Y, Zhao H, Wang H, Yang L, Xu X, Xia F, (2015). Highly anisotropic and robust excitons in monolayer black phosphorus. *Nat. Nanotechnol.*, **10**:517.

66. Yang J, Xu R, Pei J, Myint YW, Wang F, Wang Z, Zhang S, Yu Z, Lu Y, (2015). Optical tuning of exciton and trion emissions in monolayer phosphorene. *Light Sci. Appl.*, **4**:e312.
67. Liang L, Wang J, Lin W, Sumpter BG, Meunier V, Pan M, (2014). Electronic bandgap and edge reconstruction in phosphorene materials. *Nano Lett.*, **14**:6400–6406.
68. Xu R, Zhang S, Wang F, Yang J, Wang Z, Pei J, Myint YW, Xing B, Yu Z, Fu L, Qin Q, Lu Y, (2016). Extraordinarily bound quasi-one-dimensional trions in two-dimensional phosphorene atomic semiconductors. *ACS Nano*, **10**:2046–2053.
69. Wang JF, Gudiksen MS, Duan XF, Cui Y, Lieber CM, (2001). Highly polarized photoluminescence and photodetection from single indium phosphidenanowires. *Science*, **293**:1455–1457.
70. Nanot S, Cummings AW, Pint CL, Ikeuchi A, Akiho T, Sueoka K, Hauge RH, Léonard F, Kono J, (2013). Broadband, polarization-sensitive photodetector based onoptically-thick films of macroscopically long, dense, and aligned carbon nanotubes. *Sci. Rep.*, **3**:1335.
71. Dresselhaus G, (1957). Optical absorption band edge in anisotropic crystals. *Phys. Rev.*, **105**:135–138.
72. Blakemore JS, Nomura KC, (1962). Intrinsic optical absorption in tellurium. *Phys. Rev.*, **127**:1024–1029.
73. Qiao J, Kong X, Hu Z-X, Yang F, Ji W, (2014). High-mobility transport anisotropy and linear dichroism in few-layer black phosphorus. *Nat. Commun.*, **5**:4475.
74. Yuan H, Liu X, Afshinmanesh F, Li W, Xu G, Sun J, Lian B, Curto AG, Ye G, Hikita Y, Shen Z, Zhang S-C, Chen X, Brongersma M, Hwang HY, Cui Y, (2015). Polarization-sensitive broadband photodetector using a black phosphorus vertical p–n junction. *Nat. Nanotechnol.*, **10**:707–714.
75. Quereda J, San-Jose P, Parente V, Vaquero-Garzon L, Molina-Mendoza AJ, Agraït N, Rubio-Bollinger G, Guinea F, Roldán R, Castellanos-Gomez A, (2016). Strong modulation of optical properties in black phosphorus through strain-engineered rippling. *Nano Lett.*, **16**:2931–2937.

Chapter 6

Black Phosphorus Transistors

Xiaomeng Ma and Zhiyong Zhang
Key Laboratory for the Physics and Chemistry of Nanodevices, Department of Electronics, Peking University, Beijing 100871, China
zyzhang@pku.edu.cn

Since high-performance field-effect transistors (FETs) were first fabricated based on few-layer black phosphorus in 2014,[1] the potential applications of black phosphorus in electronics began to attract extensive attention. It is well known that the preferred channel materials for constructing sub-10 nm transistors should present two essentials, ultrathin body and high carrier mobility. A thinner body of the conducting channel may lead to better gate electrostatics and controllability, which would be propitious to minimize the short-channel effect during the scaling down of FETs' dimensions. Higher carrier mobility promises higher speed operation of FETs with fixed dimension (channel length). However, compromise between thin body and high mobility is inevitable in conventional bulk semiconductors, in which the carrier mobility is severely degraded when the body is thinned to several nanometers. The emerging 2D materials, including graphene and transition metal dichalcogenides, have demonstrated their advantages for constructing sub-10 nm FETs[2–4] owing to their ultrathin body without obvious carrier degradation. Unfortunately, these famous

Phosphorene: Physical Properties, Synthesis, and Fabrication
Edited by Yongqing Cai, Gang Zhang, and Yong-Wei Zhang

ISBN 978-981-4774-64-2 (Hardcover), 978-0-203-71061-6 (eBook)
www.jennystanford.com

2D materials suffer from intrinsic deficiency and are therefore considered unsuitable for constructing high-performance FETs.

For example, graphene is famous for its ultrahigh mobility and atomic thickness, but the lack of bandgap results in no switching off or current saturation in graphene transistors, which limit its application in digital and logic circuits.[2,5] Transition metal dichalcogenides like MoS_2 and WSe_2 have a bandgap larger than 1 eV. However, transition metal dichalcogenides always suffer from low carrier mobility[6–9] and are not fit for high-performance applications in electronics. Black phosphorus (BP), presenting both a relatively large bandgap and high mobility, has drawn much attention recently as a new 2D semiconducting material. Layered like graphene, the phosphorus atom is covalently bonded with three neighboring atoms in each layer, with weak Van der Waals forces holding the layers together. The puckered layers in black phosphorus should lead to a bandgap of ~2 eV[10,11] for a monolayer and ~0.3 eV for the bulk.[12,13] Moreover, BP presents high mobility up to 200–3000 $cm^2/V{\cdot}s$, which is much larger than that of transition metal dichalcogenides. The combination of high mobility and layer-dependent bandgap makes BP one of the most attractive 2D materials for nanoelectronics and optoelectronics.[14–18]

6.1 Black Phosphorus Field-Effect Transistors

6.1.1 Characteristics and Performance Parameters of Field-Effect Transistors

The field-effect transistor (FET), as a typical voltage-controlled current source, is a transistor that uses an electric field to control the electrical conductivity of a channel consisting of semiconductor material. Generally, a FET contains three terminals: the source (S), through which the carriers enter the channel; the drain (D), through which the carriers leave the channel; and the gate (G), which modulates the channel conductivity. When a FET works in a common-source mode in which S is connected to the ground, voltages V_{ds} and V_{gs} are applied respectively to D and G. Therefore, there are two kinds of characteristic curves for FETs according to

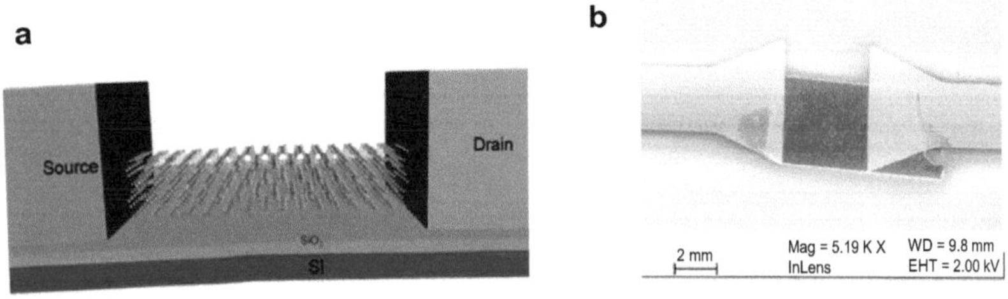

Figure 6.1 Structure of field-effect transistors with back gate. (a) Structural diagram of BP FET. (b) SEM image of a typical back-gated BP FET with a channel length of 4 μm and of 3.6 μm. Reproduced with permission from Ma et al.[22] Copyright 2015 AIP Publishing LLC.

the measuring conditions. One is transfer characteristic (V_{gs}–I_{ds}), in which the source–drain current I_{ds} is obtained by sweeping the gate voltage (V_{gs}) with a fixed drain voltage (V_{ds}). The other is output characteristic (V_{ds}–I_{ds}), in which I_{ds} is measured by sweeping V_{ds} under a fixed V_{gs}.

Figure 6.1 shows the structure diagram and scanning electron microscope (SEM) image of black phosphorus (BP) back gate FET, which uses a BP flake as the conducting channel and a doped silicon substrate as the back gate. Figure 6.2a demonstrates a typical measured transfer characteristic of a fabricated back-gated BP FET. Theoretically, we can get any I_{ds} at a certain bias condition (V_{gs}, V_{ds}) if the step of V_{ds} in transfer curves is small enough. Therefore, the transfer characteristic in principle can contain every feature of the BP FET. The typical output characteristic of the same BP FET in Fig. 6.2a is shown in Fig. 6.2b. Similarly, if the step of V_{gs} in transfer curves is small enough, I_{ds} at a certain bias condition (V_{gs}, V_{ds}) can also be obtained. Therefore, the output characteristic also can contain all the features of the BP FET. It is said that for a FET, the transfer characteristic and the output characteristic can transmute into each other if the measured data is dense enough.

To characterize the performance of a FET quantitatively, some key parameters have been defined and widely used for nanomaterial-based FETs, typically involving transconductance, subthreshold slope swing, and current on/off ratio. Transconductance is the key device parameter to reflect the amplifying ability

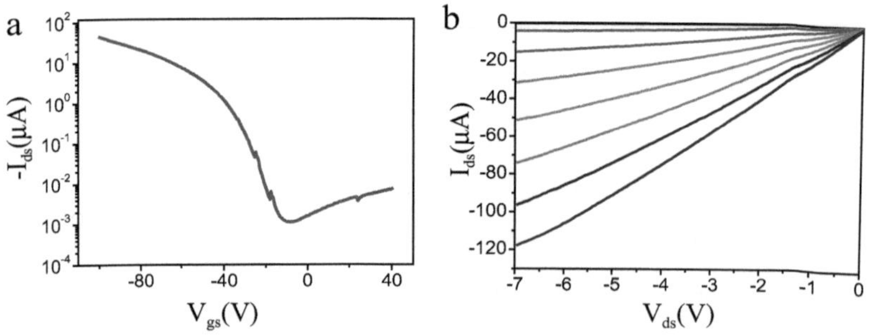

Figure 6.2 Electronic measurements of black phosphorus-based field-effect transistors with back gate. (a) Transfer characteristic of the FET at $V_{ds} = -1$ V, and its (b) output characteristics, in which back gate voltage ranges from 0 to -70 V with steps of -10 V from top to bottom. Reproduced with permission from Ma et al.[22] Copyright 2015 AIP Publishing LLC.

of a FET, and can be defined as

$$g_m = \left.\frac{\partial I_{ds}}{\partial V_{gs}}\right|_{V_{ds}=\text{const}} \tag{6.1}$$

For a FET with a long channel, the transconductance can be written as

$$g_m = \mu C_g \frac{W}{L_G} V_{ds} \tag{6.2}$$

where W and L_G are respectively the width and the length of the gate, C_G is the gate capacitance, and μ is the carrier mobility. It is obvious that the transconductance of a FET is not only dependent on its dimension and the source drain bias voltage, but also strongly dependent on the optimization degree of structure, contact, and gate efficiency of FETs. Compared with transconductance, carrier mobility is a preferred parameter to benchmark the potential of a FET which is still immature in structure and fabrication, and thus it is widely used as an important parameter to characterize the performance of a nanomaterial-based FET.

As the most significant parameter for black phosphorus FETs, mobility is crucial in evaluating the electrical conductivity performance of the semiconductor and determining the electrical conductivity together with carrier concentration, besides influencing the speed of the devices. The most common method to extract mobility of BP FET is the maximum transconductance method, that is

$$\mu = \frac{L_G \cdot g_m}{W \cdot V_{ds} \cdot C_g}. \tag{6.3}$$

Another widely used parameter for a nanomaterial-based FET is the current on/off ratio I_{on}/I_{off}, which is mainly dependent on the bandgap of the nanomaterial channel intrinsically. High current on/off ratio always means high on-state current and low off-state current, and is required for a FET. Higher on-state current leads to higher speed for a FET, and lower off-state leads to smaller power consumption. Therefore, in order for a device to be excellent, the on-state current should be as large as possible and the off-state current should be as small as possible. In other words, a high current on/off ratio is preferred in FETs.

6.1.2 Black Field-Effect Transistors

Black phosphorus is a p-type semiconductor material, and monolayer BP with perfect crystal is predicted to present ultrahigh carrier mobility ranging from 10,000 to ~26,000 cm^2/Vs through theoretical simulation.[19] Therefore, BP, especially the few-layer BP, has been considered as an ideal channel material to build ultrasmall FETs. Since the first demonstration of few layers in building FETs in 2014, the new two-dimensional semiconductor material has drawn extensive attention as a promising semiconducting channel material. BP FETs were first built based on exfoliated black phosphorus with multilayers, and demonstrated an excellent mobility of up to 1000 cm^2/Vs and a current on/off ratio of about 10^5 at room temperature.[1] It is worth mentioning that mobility and current on/off ratio in BP FETs were found to be strongly dependent on the thickness (or layer number) of the BP sample, since the energy band of BP is dependent on the layer number. The bandgap and carrier effective mass increases with decreasing layer number of BP. As a result, mobility increases with an increase in the thickness of BP, and reaches its maximum at a thickness of about 10 nm. On the contrary, the current on/off ratio of BP FETs demonstrated an absolutely opposite trend: it increases with a decrease in thickness. Although the on-state current will gently decrease with decreased thickness owing to the degradation of carrier mobility in BP, the off-state current will decrease exponentially with layer number because of an increase in the bandgap. Therefore, selecting BP samples with suitable thicknesses forms the basis for constructing FETs with high performance or low power dissipation. Current research on black phosphorus electronic applications usually chose samples with a thickness of about 10 nm owing to the following reasons. First of all, currently, black phosphorus samples are mainly fabricated based on the mechanical exfoliation method, and thus have non-uniform thickness and random location. BP samples with a thickness of 10 nm are easy to obtain and found in optical microscopy, and thinner ones, especially monolayer flakes, are very rare in exfoliated BP samples. Monolayer black phosphorus samples were fabricated by the plasma-assisted fabrication method,[20] but it is still difficult to build FETs based on monolayer samples. Second, black phosphorus is not stable in ambient conditions, since it will suffer irreversible performance degradation by reacting with water/oxygen molecules. Performance degradation is much more terrible for thinner samples, and longtime exposure in ambient conditions even etches away the thinner parts.[21,22] Passivation treatments can protect black phosphorus from water/oxygen and are necessary during device fabrication, but the fabricating process of BP FETs before passivation is exposed in air. Third, considering both high-performance and low-power applications and black phosphorus with thickness of about 10 nm can balance mobility and the current on/off ratio.

FETs based on BP with a thickness of about 10 nm are ambipolar owing to the small bandgap of BP. The interlayer spacing of black phosphorus is 0.53 nm, so that samples of about 10 nm usually used in experiments can be considered as bulk black phosphorus, so that the bandgap is about 0.3 eV. The transfer characteristics during the measurements all demonstrate ambipolar characteristics because of the narrow bandgap of the samples. This phenomenon can be generally seen in published works.[1,23,24]

Gate structure is a very important aspect for nanomaterial-based FETs. In the current stage, most BP FETs have adopted a simple back-gated structure, which is similar to the early stage of other nanomaterial-based FETs. The main advantage of the back-gated structure is the simple fabrication process, which is beneficial to maintain pristine performance of nanomaterials. However, back-gated BP FETs always suffer from low gate efficiency and the disability to control FETs individually, which will severely limit the performance and applications of BP FETs. Compared with the back-gated FET, the top-gated FET has a much more advanced structure, and can therefore achieve high performance and individual control of transistors. The fabrication of the top-gated BP FET must involve the growth of the insulator layer as gate dielectric on BP channel, and the growth is always executed through the atomic layer deposition (ALD) process, or other methods. Maintaining the performance of BP sample during the growth of gate insulator is the key for the fabrication of high-performance top-gated BP FETs.

6.1.3 Effect of Contact Metal for Black Phosphorus FETs

Contact resistance between metal and semiconductor is crucial for electronic devices, especially for FETs. Realizing low contact resistance is therefore the precondition for building high-performance FETs. For a FET, the source/drain contact resistance should be much smaller than the channel resistance to ensure that the semiconducting channel dominates the whole output resistance, and devices with higher mobility and shorter semiconducting channel require lower contact resistance. In conventional metal–oxide–semiconductor field-effect transistors (MOSFETs), the ohmic contacts are always formed by highly doping the source and drain region of the semiconductor to make the Schottky barrier thinner at contact. When a valid local doping method is absent in nanomaterial-based FETs, ohmic contact must be realized through other ways. It is theoretically predicted that there is no or weak Fermi pinning at the surface of a low-dimensional semiconductor, which means the Schottky barrier between low-dimensional semiconductor and metal is mainly determined by the difference of their work functions. The weak Fermi pinning in low-dimensional semiconductor then provides the possibility to adjust the height of the Schottky barrier at the metal–semiconductor interface.

For black phosphorus transistors, similar to the case of carbon nanotubes, the contact metal with a high work function makes its

Fermi level aligned closed to the valence band of black phosphorus. Under these circumstances, the barrier is small for holes and very large for electrons. Transistors turn on at negative gate voltage and turn off at positive gate voltage, showing p-type performance. On the contrary, for low work function metals, the Fermi level of the metal is close to the conduction band of black phosphorus, so that the barrier is small for electrons and large for holes. Transistors turn off at negative gate voltage and turn on at positive voltage, performing as n-type FETs.

Aiming to optimize the performance of p-type BP FETs, researchers chose metals with different work functions as the source/drain contacts. Some widely used metals such as Ti, Ni, Pd, and Al have been used as contacts for BP FETs to study the effect of the work function of contact metals on the performance of BP FETs.[23,24] As is well known, metals with different work functions lead to different heights of the Schottky barrier at the junction between metal and black phosphorus owing to the weak Fermi pinning at the surface of BP. Experiment results demonstrated that contact metals with high work function (like Pd) can reduce the contact resistance efficiently and further suppress the ambipolar characteristic (or n-type branch) of black phosphorus FETs. Besides, metals with low work function (like Al) make the n-type behavior much more obvious and even better than the p-type performance. By applying proper gate and drain voltages, even n-type FETs can be obtained based on black phosphorus, and then it is possible to realize ambipolar CMOS circuits based on BP FETs. These results provided guidance for researchers to choose appropriate contact metals according to experimental purpose.

6.1.4 Electrical Anisotropy Performance of Black Phosphorus

Black phosphorus is a layered material, and each phosphorus atom covalently bonded with three other phosphorus atoms forms a folded ring structure, the layers being stacked by Van der Waals forces. The unique structure of black phosphorus helps the adjacent atomic layers keep a certain sequence, which leads to the

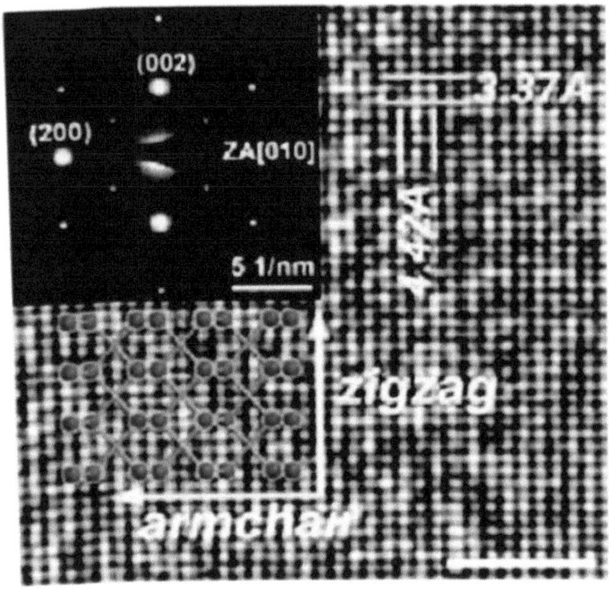

Figure 6.3 TEM image of black phosphorus. Scale bar: 2 nm. Reproduced with permission from Lu et al.[19] Copyright 2014 AIP Publishing LLC. Copyright 2007 Nature Publishing Group.

anisotropy of black phosphorus and makes it typically different from graphene.[5,25,26]

Researchers from Yale University, Peking University, and Zhejiang University all did further research on the anisotropy performance of black phosphorus.[19,27,28] To determine the relation between structure anisotropy and transport performance anisotropy of BP accurately, Chuanhong Jin et al. have designed an experiment systematically to clarify the relationship between black phosphorus's orientation and Raman spectrum/electrical conductance. The first part of the research is to study the corresponding relationship between orientation and the optical anisotropic properties. The crystal orientation is confirmed by transmission electron microscopy (TEM), shown in Fig. 6.3, and according to the crystal orientation along the direction from 0° to 180° at a step of 30° measured Raman spectrum. During the measurement, 0° is along the armchair direction and 90° is along the zigzag direction, and the

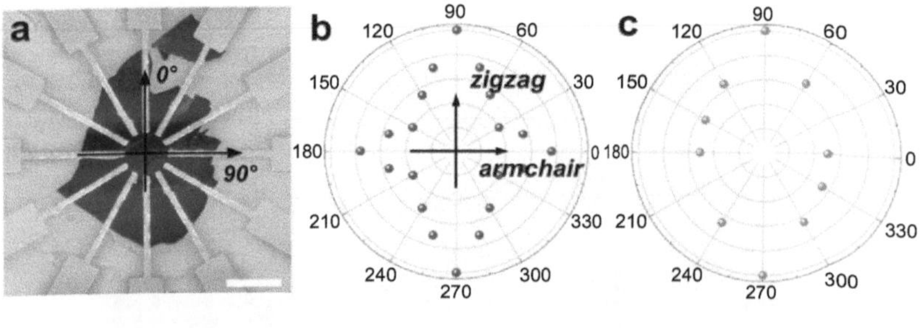

Figure 6.4 (a) SEM image of a device with six pairs of electrodes spaced 30° apart. Scale bar: 5 μm. (b) Normalized intensity of the A_g^2 mode. (c) Normalized conductance of BP channels between six pairs of electrodes. Reproduced with permission from Lu et al.[19] Copyright 2014 AIP Publishing LLC.

crystal orientation is confirmed by TEM. The normalized intensity of the A_g^2 mode reaches its maximum at 90°, and according to TEM images it is along the zigzag direction, whereas the normalized intensity of the A_g^2 mode obtained a local maximum value along the armchair direction. The second part is to clarify the corresponding relationship between optical anisotropy and electrical anisotropy. Based on the results of the first part, the crystal orientation was confirmed by the Raman spectrum, and they fabricated six pairs of electrodes from 0° (armchair direction) to 180° spaced by 30°. Standard electronic measurements were taken on six pairs of electrodes.

Figure 6.4 demonstrates that the maximum conductance was obtained at 90°, the direction at which A_g^2 reaches its maximum and, according to the results of the first part, which is the zigzag direction, whereas the minimum conductance was obtained along 0°, the armchair direction. In conclusion, along the zigzag direction, the normalized intensity of the A_g^2 mode and the electrical conductance reach their maximum. Six pairs of the electrode structure also helped to find out the ratio of the conductance along zigzag and armchair directions, which is about 2. In order to study the accurate ratio of the mobility, researchers fabricated two normal, terminal back-gate field effect transistors along the zigzag and armchair directions at the same continuous sample region. The transfer characteristics are shown in Fig. 6.5. The ratio for the mobility along

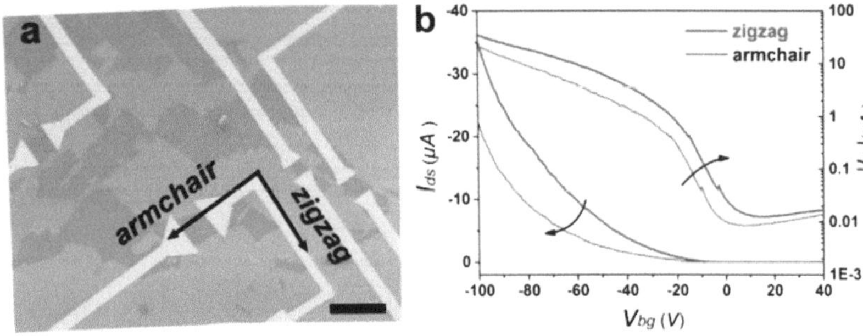

Figure 6.5 (a) SEM image of field-effect transistors with the channels fabricated along the zigzag or armchair direction. The scale bar is 20 μm. (b) Transfer characteristics along the zigzag and armchair direction with $V_{ds} = -1$ V at room temperature. Reproduced with permission from Lu et al.[19] Copyright 2014 AIP Publishing LLC.

the zigzag direction and that along the armchair direction is about 1.4. Compared to previous works,[17,27–29] Prof. Jin gives a completely different argument, which corrected the relationship between crystal orientation and electrical conductance. This conclusion must lead to new discussions on the anisotropic performance of black phosphorus. Meanwhile, Prof. Jin's work gives an easy access to distinguish the crystal orientation by Raman spectrum and puts forward the explicit relationship between crystal orientation and the electrical anisotropic property, thus facilitating the way to distinguish crystal orientation and helping a lot in fabricating high-performance black phosphorus field-effect transistors referring to crystal orientation.

The anisotropic properties of BP may trigger both positive and negative effects on applications. On the one hand, FETs fabricated on a BP channel with different crystal directions should present different carrier mobilities. As a result, BP samples must be oriented before device fabrication, and the performance uniformity of BP FETs is possibly realized by fabricating FETs on BP channels with the same direction. On the other hand, the anisotropy on structure and performance will lead to richer physics for BP, and then bring about more possibilities for designing devices based on BP. Designing novel devices using anisotropy provides a good opportunity for physicists and engineers on devices, and may be also a big challenge.

6.1.5 Stability of Black Phosphorus in Ambient Conditions

Black phosphorus has a broad prospect in electrical applications on account of its superior electrical properties. However, the published results demonstrate that long-time exposure in ambient conditions make black phosphorus easy to degrade.[15,21,30] Therefore, the stability of black phosphorus devices is the premise of the realization of different kinds of research. BP crystals are fabricated by white phosphorus under high temperature and pressure, and fabricated BP is the most stable allotrope of phosphorus at room temperature. However, the stability of BP samples exposed in air has been always under suspicion. Researchers observed the droplets on the surface of black phosphorus by optical microscope after devices deposited in ambient conditions. The same phenomenon can also be observed through an atomic force microscope (AFM) image, as shown in Fig. 6.6. The droplets grew up with time. Short-time exposure in ambient condition induces the adsorption of oxygen and water molecules above the surface. More seriously, long-time exposure destroyed samples. Joshua D. Wood et al. used X-ray photoelectron spectroscopy to assess the mechanism of degradation in ambient conditions. It turned out that the chemical composition of the droplets is PO_x, which confirmed that with long-time exposure in ambient conditions, black phosphorus suffers from irreversible chemical reaction with oxygen and water molecules.

A research group from Peking University systematically studied the effect of air conditions on black phosphorus devices.[22] They fabricated back-gate field-effect transistors of black phosphorus with standard micro- and nanofabrication processes. Transistors were deposited in ambient conditions and the electronic properties were measured in a fixed time interval. Figure 6.7a shows the transfer characteristics of a typical device at different times. The source–drain voltage is -1 V, the channel length is about 5.6 μm, and the width is about 1.4 μm. The transfer characteristics indicated that current decreases sharply by two orders of magnitudes, and therefore the threshold and the transfer curves change obviously. The mobility of as-fabricated black phosphorus field-effect transis-

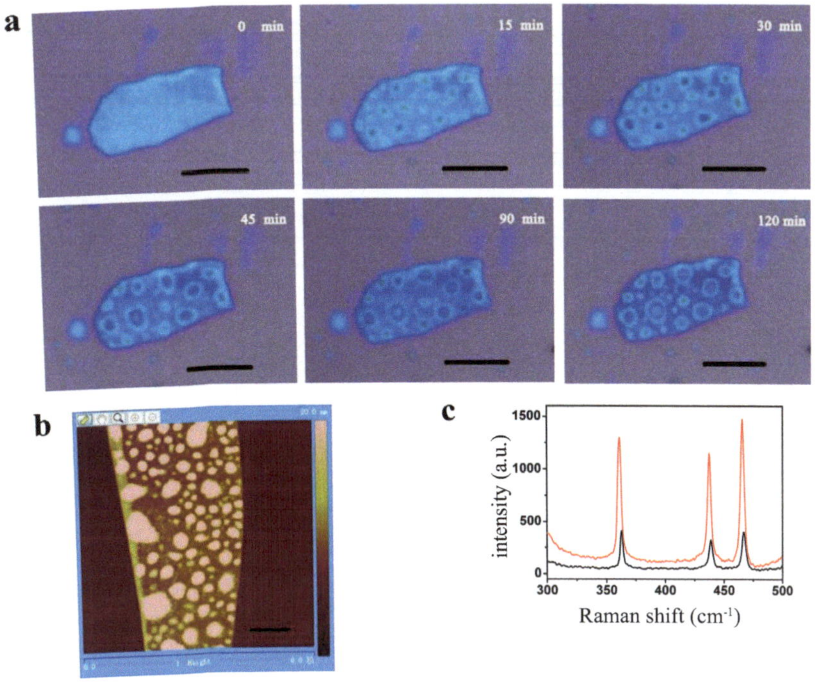

Figure 6.6 Effect of water effect on black phosphorus. (a) A sequence of optical images observed every 15 minutes in air at a humidity of 85%. Scale bar: 5 μm. (b) AFM image acquired during a two-day interval (humidity 35%). Scale bar: 1 μm. (c) Raman spectrum obtained from the black phosphorus at 0 h (red) and 30 h (black) after fabrication in air. The wavelength of the excitation laser is 633 nm. Reproduced with permission from Ma et al.[22] Copyright 2015 AIP Publishing LLC.

tors retrieved from the transfer curve is about 132 cm^2/Vs and degraded to 1.8 cm^2/Vs after exposure to air for 216 h. Figures 6.7b and 6.7c show the output characteristics measured at 0 h and 216 h after fabrication. The output curve measured immediately after fabrication indicated excellent performance, including the linear I–V characteristics at low voltage bias and the outstanding current saturation at high voltage bias. Nevertheless, 216 h after fabrication, the output characteristics exhibit poor performance of nonlinear I–V characteristics at low bias and severe decay of current.

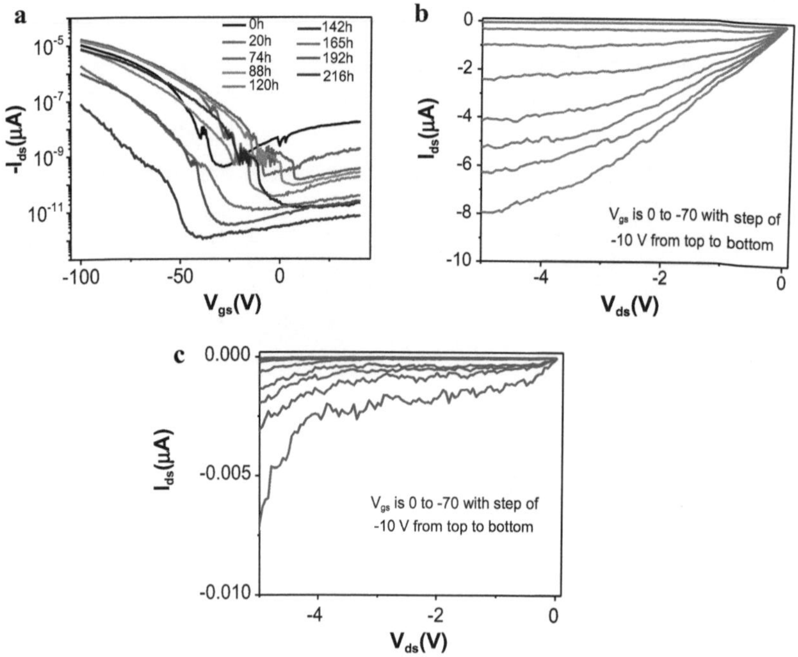

Figure 6.7 (a) Time-dependent transfer characteristics of a typical BP FET (with channel length and width of 5.6 μm and 1.4 μm, respectively) measured at different times under ambient conditions. (b) Output properties of the as-fabricated BP FET at 0 h and (c) after exposure in air for 216 h. Reproduced with permission from Ma et al.[22] Copyright 2015 AIP Publishing LLC.

Furthermore, in order to demonstrate the effect of the degradation on the performance of FETs in ambient conditions quantitatively, two of the important parameters, namely carrier mobility and the current on–off ratio, were extracted from a group of devices in ambient conditions. The specific variation tendency of mobility and the current on–off ratio measured at different times is presented in Fig. 6.8, which indicates that at the first 20 hours of exposure in air there is no detectable degradation in the mobility. This conclusion provides researchers a reference for device fabrication—i.e., the first 20 hours after preparation of BP are the "golden 20 hours"

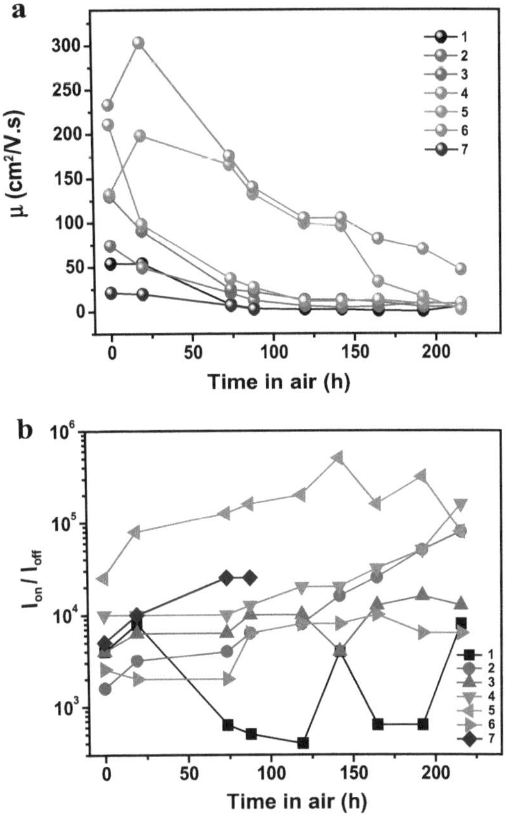

Figure 6.8 Time-dependent electronic characteristics of BP FETs in air. (a) Carrier mobility and (b) on–off current ratio of seven BP FETs versus ambient exposure time. Reproduced with permission from Ma et al.[22] Copyright 2015 AIP Publishing LLC.

for fabrication of devices. After that, obvious decline in mobility appears after about 120 hours' exposure in ambient conditions; the mobility of five devices drops to less than 10% of the original, and one of them even broke down. Six of the devices suffered from a decline in mobility to less than 10 cm^2/Vs in 216 h. The current on–off ratio exhibits a totally different tendency—a rising trend shown in Fig. 6.8b. Therefore, long-time exposure in ambient

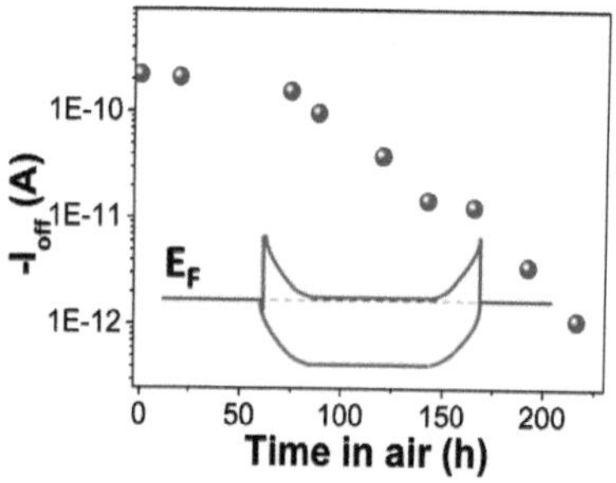

Figure 6.9 The ambient exposure time-dependent minimum current of the BP FET in Fig. 6.7a. Inset: band diagram of the BP FET at minimum current situation. Reproduced with permission from Ma et al.[22] Copyright 2015 AIP Publishing LLC.

conditions results in a severe degradation of black phosphorus and simultaneously in the performance, including a decrease in mobility and an increase in the current on–off ratio.

By analyzing the transfer characteristics measured at different times, researchers speculated that the degradation of the electric performance is attributed to the decline of the minimum current. Figure 6.9 demonstrates the time-dependent minimum current of black phosphorus field-effect transistors in ambient conditions for a typical transistor. After 216 hours' exposure to air, there is a two orders of magnitude decline in the minimum current, from 2.2×10^{-10} to 1.1×10^{-12}. For the Schottky barrier transistors, the off-state current leads to two kinds of leakage current, thermal emission and tunneling. When the Fermi level of black phosphorus tuned to a near-conduction band, tunneling happened between the channel and contact electrodes, as shown in the inset of Fig. 6.9. The potential distribution of Schottky barrier is influenced at the same time. The thickness of the Schottky barrier is dependent on gate efficiency: the higher the gate efficiency, the thinner the Schottky barrier. Tunneling of the current is strongly dependent on the thickness of the Schottky barrier, while it is approximately equal to the thickness of the gate insulator.[31] The thickness shown in the experiment is about 290 nm, and so is the Schottky barrier, which means the tunneling current can be ignored. Therefore, the off-state current originates from terminal emission. The gate electrodes modulate the energy band under the off-state, as shown in the inset of Fig. 6.9, which is a big potential barrier for both electron and hole injections. Carriers injected into the channel by terminal excitation and the current depend on $1/T$ exponentially,

$$I_{\text{off}} \propto \exp\left(-\frac{E_{\text{b}}}{kT}\right) \tag{6.4}$$

where k is the Boltzmann constant, and E_{b} is the height of the barrier. The off-state current I_{off} reaches its minimum when the barrier is the highest for both electrons and holes. The barrier is approximately equal to the bandgap E_{g}, and increasing the gate voltage further will lead to the tunneling of electrons between source/drain and conduction band and further lead to an increase in the current.

$$I_{\text{min}} \propto \exp\left(-\frac{E_{\text{g}}}{kT}\right) \tag{6.5}$$

This conclusion indicates the exponential relation between the minimum current I_{min} and the bandgap E_{g}.

$$\Delta E_{\text{g}} = -\Delta \ln I_{\text{min}} \tag{6.6}$$

Referring to Fig. 6.9 and the formula above, the bandgap calculated from the minimum current can be approximately increased by 137 meV. Previous results demonstrate that the bandgap decreases as the number of layers of black phosphorus increases, from which the effective layer number is obtained. From the measured changes in the electrical properties, researchers deduced that the degradation of black phosphorus in ambient conditions occurs layer by layer, which is consistent with the change in the surface morphology of black phosphorus in ambient conditions.

To further confirm the influence of oxygen and water, a contrast experiment was carried out with devices stored in vacuum conditions. The transfer characteristics and output characteristics for a typical device are plotted in Fig. 6.10 a–c, which shows a smart stability compared to devices in ambient conditions. Five of the black

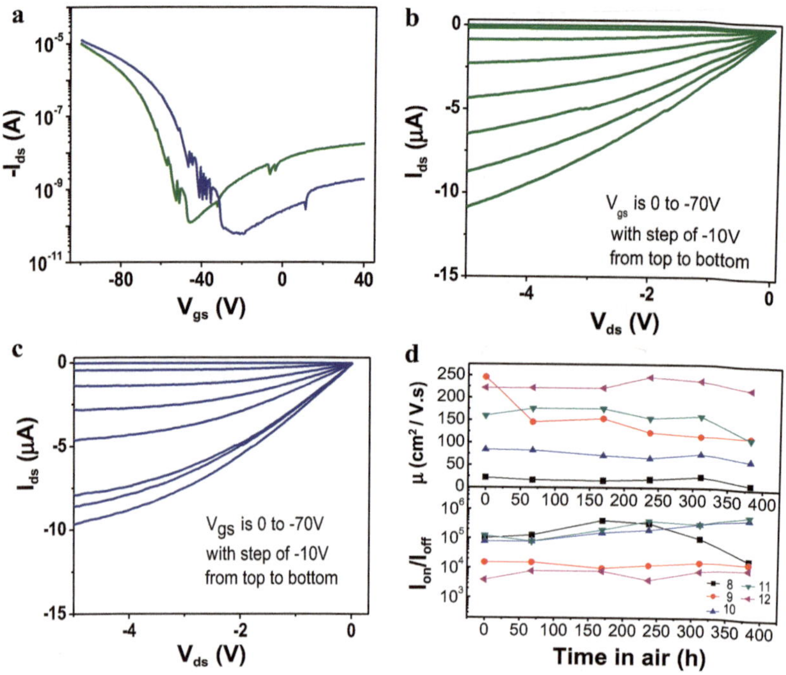

Figure 6.10 Electronic characteristics of BP FETs stored in vacuum. (a) Transfer characteristics measured at 0 h (olive) and 238 h (blue) after fabrication of a BP FET under $V_{ds} = -1$ V. The channel length and width of the FET is 6 μm and 3.5 μm, respectively. The output characteristics of the FET at 0 h (b) and 238 h (c) in vacuum. (d) Time-dependent mobility and on–off ratio for five devices. Reproduced with permission from Ma et al.[22] Copyright 2015 AIP Publishing LLC.

phosphorus transistors were fabricated with the same process and stored in vacuum. Similarly, transfer and output characteristics were focused on. The statistical results of mobility and current on-off ratio are presented in Fig. 6.10d. There is no obvious degradation in about 380 h, but for tiny fluctuations, which are attributed to the release of oxygen and water molecules after storing in vacuum for the first time. Therefore, vacuum deposition between each step of device fabrication should be performed within the golden 20 h after black phosphorus is obtained. In addition, passivation is crucial for the applications of black phosphorus.

6.2 Applications of Black Phosphorus Transistors

6.2.1 CMOS Circuits Based on Black Phosphorus

Constructing integrated circuits (ICs) is one of most important applications of BP FETs, and the demonstration of ICs can give researchers confidence for BP FETs. Modern ICs are always constructed by complimentary MOS (CMOS) FETs, which contain two types of FETs, n-type and p-type ones. Compared with pure pMOS or nMOS ICs, CMOS ICs can provide much lower static power dissipation and higher noise margin, and have been also considered as the preferred circuit style for nanomaterial-based FET circuits. As the most basic and simplest CMOS IC, CMOS inverter is always fabricated in nanoelectronics to demonstrate the possibility of the realization of integrated circuits. The CMOS inverter is constituted by a p-FET in series with an n-FET, and then the "inverting" function is realized through a push–pull work style of n- and p-FETs under appropriate voltages. However, the fabrication process for CMOS inverters is much more complicated than pure n-FET or p-FET ICs, since both p- and n-FETs must be fabricated compatibly on the same chip. Therefore the fabricating process of two types of FETs must be compatible, and performance matching is essential as well. Generally, a BP FET presents a typical p-type FET with high mobility, high on–off ratio, and excellent saturation performance, and has been considered to have huge potential in digital and analog circuit applications. At present there are two approaches to prepare CMOS inverters based on BP FETs. One is hybrid heterogeneous integration, i.e., using BP as the channel material for p-type transistors and MoS_2 as the channel material for n-type transistors.[17,32] The other is utilizing the ambipolar performance of BP to fabricate p-type transistors and n-type transistors, respectively, and further constitute the ambipolar CMOS inverter.[33]

Since BP is a p-type semiconductor, and MoS_2 is a typical n-type semiconductor, heterogeneous integration between BP and MoS_2 is the right way to realize CMOS ICs. Prof. Ye et al. from Purdue University have successfully fabricated p-FETs based on BP and n-FETs based on MoS_2 on the same substrate, and then constructed a CMOS inverter.[17,32] A complex procedure is required to design and fabricate CMOS FETs based on the combination of BP and MoS_2. In the aspect of device fabrication, black phosphorus and MoS_2 prepared by the mechanical exfoliated method are scattered randomly on the substrate; the thickness and the relative location of the samples are beyond control. In order to transfer exfoliated samples to the substrate with appropriate location during the experiment, researchers introduced the optical alignment stage, which demands high levels of accuracy and a complicated transfer process. In addition, protection of the two kinds of 2D semiconductors in the complicated fabrication process is also a big technique challenge. In the aspect of CMOS FET design, since BP presents a higher mobility than MoS_2, the p-FET should exhibit higher performance than the n-FET with the same dimension. Therefore, the device dimension, including gate length and width, should be designed elaborately to make the performance of p-type BP FETs and n-type MoS_2 FETs match. Besides, other problems should be considered during heterogeneous integration, including threshold voltage and contact resistance of two kinds of FETs. Since heterogeneous integration has significantly improved, researchers are exploring new methods based on black phosphorus itself to fabricate CMOS ICs with a simple technological process and superior performance.

Compared to heterostructures integrated with other n-type semiconducting materials, constructing ambipolar CMOS FETs based on BP films to realize CMOS ICs is more feasible. Taking advantage of the ambipolar performance of BP FETs, p-type and n-type FETs have been realized on the same transistor with different measurement conditions; inverters, too, have been constructed.[33,34] Generally, BP FETs exhibit typical p-type properties, and the n-branch is very weak. It should be noted that some valid methods must be adopted to enhance the n-branch of BP FETs. For example, utilizing Ti instead of Pd as S/D contacts can enhance the n-branch of BP FETs[33,34] since the lower function contacting metal is helpful for lowering the Schottky barrier for electrons. For a bottom-gated BP FET, the n-branch was proved to be enhanced by covering the BP channel by a Al_2O_3 film, since the Al_2O_3 film can provide n-doping or O_2/H_2O isolation from air.[34] The enhanced ambipolar

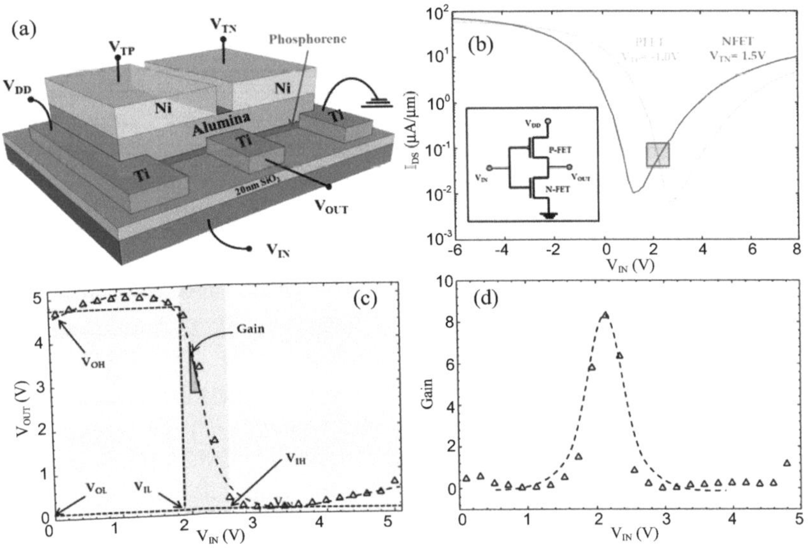

Figure 6.11 (a) Structure diagram of a double-gated black phosphorus FET used for a logic inverter. (b) Transfer characteristics of a p-type black phosphors FET and an n-type black phosphorus FET. (c) Transfer characteristics of a black phosphorus logic inverter. (d) Gain of the inverter as a function of the input voltage. Reprinted with permission from Das et al.[34] Copyright 2014 American Chemical Society.

BP FETs exhibit excellent n-type FET properties with a record peak transconductance of 66 μS/μm, and presented linear I_{ds}–V_{ds} at low voltage and perfect saturation at high voltage in output characteristics.[33,35] Saptarshi Das et al. have explored the use of ambipolar BP FETs to construct CMOS inverters, as shown in Fig. 6.11.[34] A pair of BP FETs has been fabricated by using Ti as contacts and a Si substrate as back gate. An Al_2O_3 film was then grown by ALD to cover both the FETs, and a top-gate electrode was fabricated on each FET. The top-gate was used to dope the BP channel beneath and shift the threshold voltage of BP FETs. To obtain a suitable threshold for p-FETs and n-FETs in ambipolar CMOS inverters, the top gates of V_{TP} and V_{TN} were biased with a negative and positive voltage, respectively. CMOS inverters with good performance have been realized, as shown in Fig. 6.11c,d, including a large noise margin of 2.25 V for high input and 1.85 V for low input at a supply voltage of 5 V.[34]

The main advantage of ambipolar CMOS is the simple process since only one kind of FET is required. However, the realization of ambipolar CMOS in the work of Das et al.[34] is much complex in fact. What is most surprising is that two additional supply voltages (V_{TP} and V_{TN}) are required to operate the inverter. The additional requirement of two supply voltages will significantly increase the burden of IC design and fabrication. Therefore, instead of applying voltages, some solid strategies are necessary to characterize the n-FET and p-FET, respectively, in ambipolar BP CMOS FETs. For example, metal with different work functions can be used as top-gate electrode to adjust the threshold voltage of n-FET or p-FET to a suitable value. The technology to fabricate CMOS ICs based on BP is still immature, and further explorations and developments are required and expected in the future.

6.2.2 Black Phosphorus Radio-Frequency Transistors

Benefitted with an ultrathin body, 2D semiconducting materials present excellent gate electrostatics than conventional bulk semiconductors and were predicted to possess better potential for applications in FETs scaled down to sub-10 nm technology. One of main purposes for scaling FETs is to pursue a higher speed, since the FETs with a shorter channel should exhibit a higher speed in theory. Therefore, 2D semiconductor–based FETs were always expected to present high speed, which is important not only for digital IC applications but also for radio-frequency (RF) applications. Possessing the thinnest body and an ultrahigh carrier mobility/saturation velocity, graphene was considered the most promising of materials for constructing RF devices and ICs. Parameters for benchmarking performances of RF transistors include cut-off frequency and maximum oscillation frequency. The cut-off frequency (f_T) reflects the upper limit of the highest frequency for an RF transistor as amplifier; generally, it means the cut-off frequency of the current amplifier factor. The maximum oscillation frequency is another crucial frequency parameter, which gives the highest frequency for an RF transistor as oscillator. Transistors with a frequency more than the maximum oscillation frequency cannot be active devices anymore. Intensive explorations have been carried out into graphene RF FETs, and the record of f_T was continuously refreshed and then pushed to about 450 GHz.[36–40] However, the intrinsic shortcoming of graphene, i.e., the absence of bandgap and the induced absence of current saturation in the output property, limits the improvement of the maximum oscillation frequency (f_{max}) and thus the development of graphene RF FETs. There are also some works studying the RF application of TMD materials, including MoS_2 and WSe_2, but their low mobility limits the RF performance of these devices.[41,42]

Black phosphorus was expected to act as an excellent active material to build high-performance RF FETs because it can provide higher mobility than TMDs and a larger bandgap than graphene as a 2D semiconductor. BP-based RF FETs were successfully demonstrated by F. Xia et al. To fabricate BP RF FETs, three points must be noted. Firstly, a BP sample with a thickness of 8.5 nm is preferred since it can balance mobility, current on–off ratio, and stability well. Secondly, the width of the channel should be large enough to reduce the output resistance of the FET down to near 50 Ω, which matches the standard resistance of an RF measurement system. Thirdly, a high-performance top-gated FET should be fabricated with a small gate length and large transconductance. It is a big challenge to grow a high-quality insulator on a BP channel without an obvious degradation of the performance of BP. Atomic-layer deposition (ALD)-grown HfO_2 under 150° was used here as gate dielectric with a thickness of 21 nm and dielectric constant of 13. The top-gated BP FETs with a gate length of 300 nm demonstrated here show respectable current saturation with an on–off ratio larger than 2×10^3, and exhibit a peak transconductance of 180 mS/mm and peak on-state current of 270 mA/mm at a bias of $V_{ds} = -2$ V.[43] Using standard RF characterization techniques, short-circuit current gains, cutoff frequency f_T, and the maximum oscillation frequency f_{max} of BP FETs are measured, and the corresponding values are 12 GHz and 20 GHz, respectively, after de-embedding, as shown in Fig. 6.12. Compared with graphene-based FETs, BP FETs may offer advantages for high-frequency applications in terms of voltage and power gain due to the good current saturation

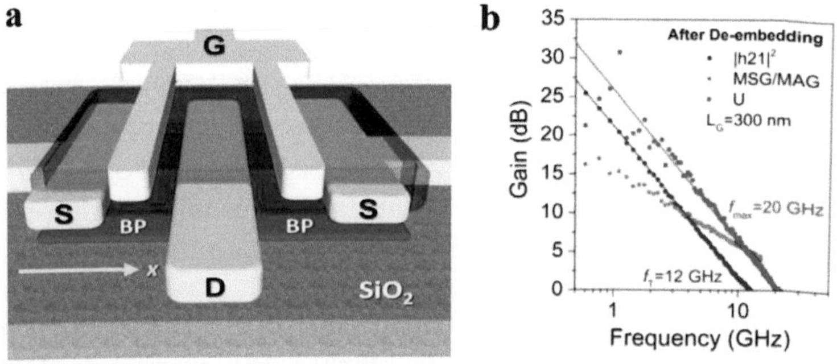

Figure 6.12 (a) Schematic of the BP transistor device structure. (b) The short-circuit current gain h_{21}, MSG/MAG, and unilateral power gain U of the 300 nm channel length device after de-embedding. $f_T = 12$ GHz and $f_{max} = 20$ GHz Reprinted with permission from Wang et al.[43] Copyright 2014 American Chemical Society.

properties arising from their finite bandgap. Therefore, BP FETs can be considered a promising candidate for future high-performance thin-film electronics technology for operation in multi-gigahertz frequency and beyond.[43]

6.2.3 Black Phosphorus Gas Sensors

Nanomaterials have been considered the most promising candidate in chemical sensing because of their ultrahigh inherent surface–volume ratio. The molecules absorbed on the surface of samples will change the charge density and the electrical conductivity of nanomaterials. Nanomaterials such as carbon nanotubes.[44,45] graphene,[46,47] and MoS_2[48,49] have been explored as active channels in gas-sensing applications. As a noxious gas in many fields, NO_2 can injure the respiratory system and damage humans. The published conclusion is that exposure to NO_2 with a concentration larger than 53 ppb leads to health problems. Meanwhile, NO_2 also contaminates water and soil. Therefore, it is very important to detect the gas concentration of NO_2. In the past experiments, researchers studied gas sensing for NO_2 by using graphene and MoS_2 and found that the detection sensitivity of MoS_2 is a few parts per million for the exfoliated samples to a few parts per billion for samples grown by CVD.[48,49] Recently, black phosphorus has been considered as

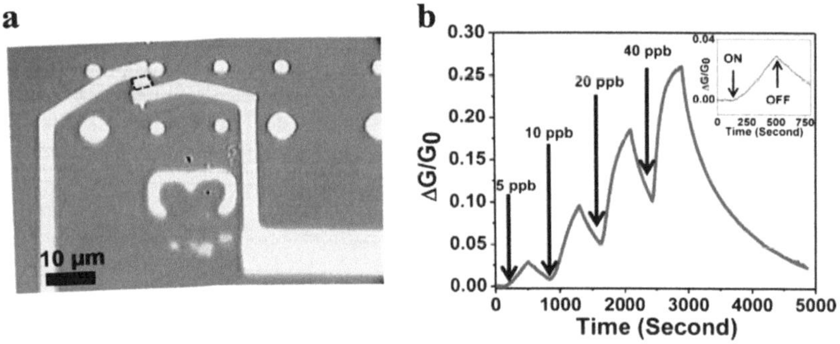

Figure 6.13 (a) An optical image of the multilayer BP flake between two Ti/Au electrodes. (b) Relative conductance change ($\Delta G/G_0$) vs. time in seconds for a multilayer BP sensor showing sensitivity to NO_2 concentrations (5–40 ppb). Inset shows a zoomed-in image of a 5 ppb NO_2 exposure response with identification of points in time where the NO_2 gas is switched on and off. Reprinted with permission from Abbas et al.[51] Copyright 2015 American Chemical Society.

another promising material for gas sensing because it has larger adsorption energy of gas molecules than graphene or MoS_2 and demonstrates excellent sensitivity in gas detecting.[50] Prof. Chongwu Zhou and his research group have researched the application of BP-based back-gate FETs in NO_2 gas detection.[51] Fabricated BP FETs were placed in a chamber with Ar protection during the measurement and the concentration of NO_2 was regulated by changing its quantity. The conductivity of devices demonstrated an increasing trend when NO_2 passed through the chamber and gradually declined when the gas changed from NO_2 to pure Ar. This phenomenon proved the hole doping of NO_2 into black phosphorus. It acquired different conductive responses by pumping in NO_2 with different concentrations, including 5 ppb, 10 ppb, 20 ppb, and 40 ppb; detailed results are shown in Fig. 6.13. The main mechanism for detecting NO_2 is proposed as the charge transfer between BP channel and absorbed NO_2, and the relationship between the conductivity/on-state current and the concentration of NO_2 fitted well with the Langmuir isotherm. BP FET–based NO_2 sensors demonstrated high sensitivity; i.e., the minimum detectable concentration for NO_2 is about 5 ppb,[51] which makes BP FET–

based gas sensors comparable with gas sensors based on other two-dimensional materials. The detecting sensitivity can be further improved by thinning the thickness of black phosphorus. In addition, there is no detectable change for the Raman spectrum of black phosphorus before and after NO_2 detecting, which means that BP FET sensors are stable and can be used repeatedly. These results lay a solid foundation for the applications of black phosphorus in chemical and biological sensors.

6.2.4 Black Phosphorus Based Heterojunction Structures

As a p-type 2D semiconductor, black phosphorus can be used to construct a heterojunction structure with other 2D materials, e.g., forming a PN junction with other n-type 2D semiconductors such as TMD materials, or forming an MIS structure with BN.[52] Black phosphorus–based heterojunction structures will lead to new physics and thus to the derivation of some new devices.

The representative work on BP-based heterojunction structures involved two famous 2D semiconductors, the p-type black phosphorus and n-type MoS_2. It is generally known that black phosphorus is a p-type semiconductor material and fabricated by the mechanical exfoliation method to get thinner samples. On interaction with van der Waals forces, the formation of a P–N junction with MoS_2 permits the mismatch of crystal lattice for different materials effectively. Meanwhile, ultrathin MoS_2 is conducive to the electrical modulation of band structure in the vertical direction. Figure 6.14a demonstrates a typical schematics of the P–N junction formed by black phosphorus and MoS_2. The P–N diodes fabricated by black phosphorus and MoS_2 exhibit strong gate-tunable current-rectifying characteristics, as shown in Fig. 6.14b. The maximum photodetection responsivity as a photodetector reaches up to 418 mA/W,[53] which is 100 times higher than that of the phototransistors fabricated by black phosphorus itself and 26 times higher than WSe_2 P-N diodes.[54–56] Black phosphorus is well suited to fabricate photodetectors with high sensitivities on account of its relatively small bandgap and high mobility. The P–N diodes fabricated with black phosphorus and MoS_2 possessed the peak EQE of 0.3%,[53] and

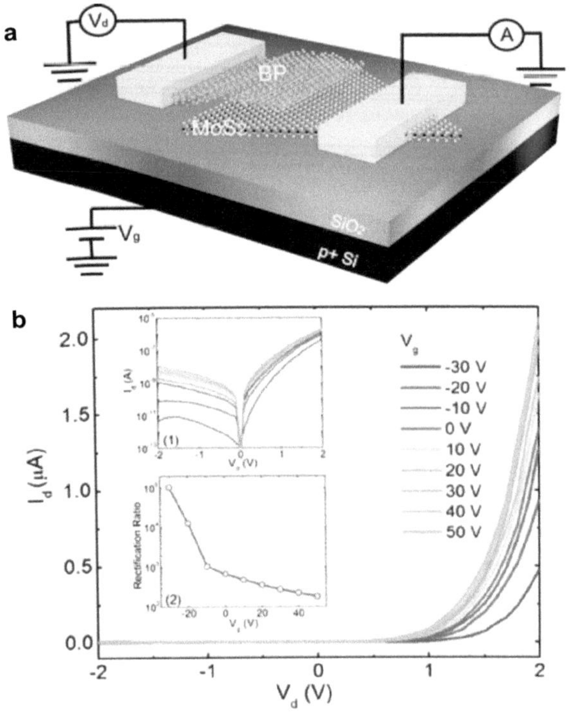

Figure 6.14 (a) Schematics of the device structure. (b) Gate-tunable I–V characteristics of the 2D p–n diode. The current increases as the back-gate voltage increases. Inset 1 shows the I–V characteristics under the semilog scale. Inset 2 shows the rectification ratio as a function of back-gate voltage V_g. Reprinted with permission from Deng et al.[53] Copyright 2014 American Chemical Society.

the properties of P–N diodes can be further improved by reducing the thickness of black phosphorus.

Degradation of properties under atmospheric conditions and high-density charge traps in black phosphorus have largely limited its actual mobility, thus hindering its future applications. Prof. Ning Wang from Hong Kong University of Science and Technology took advantage of the heterojunction structure with black phosphorus and BN to solve the problems.[57] The structure of BN–black phosphorus–BN provides ultraclean interfaces, and BN black phosphorus exhibits smart stability in air and can withstand high temperature annealing. In the Ar atmosphere with 500° annealing BN–black phosphorus–BN devices reduced the charge trap density efficiently and further improve the properties. Mobility in room temperature reached 1350 cm^2/Vs and the current on–off ratio is about 10^5. Furthermore, no hysteresis is observed in the heterojunction structure. BN–black phosphorus–BN performs well and offers a new idea in fabricating high-performance black phosphorus devices.

6.2.5 Passivation for Black Phosphorus Devices

Although BP demonstrates excellent electrical and optical properties, absence of stability in ambient conditions is the main obstacle for its application in devices. Black phosphorus is damaged in ambient conditions, as mentioned before, and that is the principal problem researchers need to solve. Results have already proved the irreversible chemical reactions of oxygen/water molecules with black phosphorus, and passivation of BP transistors is becoming crucial.[15,21,22,30,58]

Researchers tried a variety of methods to protect black phosphorus from ambient conditions and guarantee the stability of devices. Richard Martel et al. coated black phosphorus with 20 nm of parylene C to isolate black phosphorus from oxygen/water molecules. The Raman spectrum of black phosphorus with parylene C shows no detectable change in the observed 80 minutes.[30] Another way to protect black phosphorus is to cover it with an ALD-grown Al_2O_3 film.[21] The fabricated BP FETs with the protection of Al_2O_3 show excellent performance stability even after 335 hours of exposure in air.[21] The passivation of Al_2O_3 was widely used in the fabrication of black phosphorus devices and showed great achievements.[21,33–35] Besides, the BN film has also been demonstrated as an excellent insulator to protect black phosphorus. The BN film not only prevents the BP channel from air conditions, but also provides ultraclean contact interfaces.[57] Performance change and passivation are gradually attracting attention from researchers, who are aware of the fact that the isolation of oxygen and water molecules during the fabrication of BP devices is becoming much more necessary, and high-performance BP devices with extraordinary stability are going to be used in broader domains.

6.3 Conclusion and Perspectives

As an emerging 2D semiconductor, black phosphorus has drawn a lot of attention in nanoelectronic applications owing to its high carrier mobility and to the existence of bandgap. Intensive research has focused on BP-based devices, including field-effect transistors, radio frequency transistors, gas sensors, heterojunction-based diodes, and even simple integrated circuits. The superiority of black phosphorus over graphene is due to the direct bandgap of about 0.3–2 eV. Furthermore, the bandgap increases as the number of layers decrease. Compared with other transition metal dichalcogenides, black phosphorus possesses higher mobility, of up to 1350 cm^2/V. The optical and electrical anisotropic properties of black phosphorus are unique, and results show that along the zigzag direction both the normalized intensity of the A_g^2 mode and the conductivity reach their maximum. This conclusion provides a fast method to estimate the crystal orientation, which gives guidance needed to fabricate high-performance transistors. Referring to the relationship between crystal orientation and electrical conductivity, one can fabricate electrodes along the zigzag direction in order to obtain devices with a larger mobility. However, the applications of the anisotropic property have not been developed thoroughly, and BP-based devices are still immature. At the same time, there are some aspects researchers need to pay more attention to. Long-time exposure in ambient conditions does damage not only to the surface morphology but also to the electrical properties. There are a golden 20 hours for black phosphorus device fabrication. During the first 20 hours after black phosphorus is exfoliated from the bulk, there is no detectable degradation of mobility in devices. Therefore, devices should be fabricated as soon as possible, and samples need to be deposited in vacuum conditions between the adjacent two steps of fabrication to reduce the influence of oxygen/water molecules. Considering the practical use of devices, researchers should pay more attention to the passivation of devices. Proper passivation is needed to ensure the stability and the working life of devices. At present, the most common method to obtain layered black phosphorus is the mechanical exfoliation method. Layer number is hard to control accurately, and the distribution of locations is random. In light of these observations, researchers need to consider the batch preparation methods of black phosphorus while improving the performance of and realizing multifunctional devices.

It should be mentioned finally that the research on BP devices is still at a primary stage, and further explorations are required to push the applications of BP in electronics. There is big room for improving device structure, performance, stability, and repeatability of BP FETs and other kinds of devices. The explorations and applications of BP-based devices are still on the way.

References

1. Li L, Yu Y, Ye GJ, Ge Q, Ou X, Wu, H, et al. (2014). Black phosphorus field-effect transistors. *Nat. Nanotechnol.*, **9**:372–377.
2. Castro Neto AH, Guinea F, Peres NMR, Novoselov KS, Geim AK, (2009). The electronic properties of graphene. *Rev. Mod. Phys.*, **81**:109.
3. Liu H, Neal TA, Ye PD, (2012). Channel length scaling of MoS_2 MOSFETs. *ACS Nano*, **6**:8563.
4. Radisavljevic B, Radenovic A, Brivio J, Giacometti V, Kis A, (2011). Single-layer MoS_2 transistors. *Nat. Nanotechnol.*, **6**:147–150.
5. Du X, Skachko I, Barker A, Andrei EY, (2008). Approaching ballistic transport in suspended graphene. *Nat. Nanotechnol.*, **3**:491–495.
6. Radisavljevic B, Kis A, (2013). Mobility engineering and a metal-insulator transition in monolayer MoS_2. *Nat. Mater.*, **12**:815–820.
7. Fang H, Chuang S, Chang TC, Takei K, Takahashi T, Javey A, (2012). High-performance single layered WSe_2 p-FETs with chemically doped contacts. *Nano Lett.*, **12**:3788–3792.
8. Kim IS, Sangwan VK, Wood DJ, et al. (2014). Influence of stoichiometry on the optical and electrical properties of chemical vapor deposition derived MoS_2. *ACS Nano*, **8**:10551–10558.

9. Liu KK, Zhang W, Lee YH, Lin YC, Chang MT, Su CY, et al. (2012). Growth of large-area and highly crystalline MoS_2 thin layers on insulating substrates. *Nano Lett.*, **12**:1538–1544.

10. Morita A, (1986). Semiconducting black phosphorus. *Appl. Phys. A*, **39**:227–242.

11. Yukihiro T, Hideo A, Akira M, (1981). Electronic structure of black phosphorus in tight binding approach. *J. Phys. Soc. Jpn.*, **50**:3362.

12. Warschauer D, (1963). Electrical and optical properties of crystalline black phosphorus. *J. Appl. Phys.*, **34**:1853.

13. Keyes RW, (1953). The electrical properties of black phosphorus. *Phys. Rev.*, **92**:580–584.

14. Tran V, Soklaski R, Liang Y, Yang L, (2014). Layer controlled bandgap and anisotropic excitons in phosphorene and few layer black phosphorus. *Phys. Rev. B*, **89**:817–824.

15. Castellanos-Gomez A, Vicarelli L, Prada E, Island JO, Narasimha-Acharya KL, Blanter SI, Groenendijk DJ, Buscema M, Steele GA, Alvarez JV, Zandbergen HW, Palacios JJ, van der Zant HSJ, (2014). Isolation and characterization of few-layer black phosphorus. *2D Mater.*, **1**:23–29.

16. Qiao J, Kong X, Hu Z-X, Yang F, Ji W, (2014). Few-layer black phosphorus: emerging 2D semiconductor with high anisotropic carrier mobility and linear dichroism. *Nat. Commun.*, **5**:4475.

17. Liu H, Neal AT, Zhu Z, Luo Z, Xu X, Tománek D, Ye PD, (2014). Phosphorene: an unexplored 2D semiconductor with a high hole mobility. *ACS Nano*, **8**:4033–4041.

18. Dai J, Zeng XC, (2014). Bilayer phosphorene:effect of stacking order on bandgap and its potential applications in thin film solar cells. *J. Phys. Chem. Lett.*, **5**:1289–1293.

19. Lu W, Ma X, Fei Z, Zhou J, et al. (2015). Probing the anisotropic behaviors of black phosphorus by transmission electron microscopy, angular-dependent Raman spectra, and electronic transport measurements. *Appl. Phys. Lett.*, **107**:372.

20. Lu W, Nan H, Hong J, Chen Y, Zhu C, Liang Z, et al. (2014). Plasma-assisted fabrication of monolayer phosphorene and its Raman characterization. *Nano Res.*, **7**:853–859.

21. Wood JD, Wells SA, Jariwala D, Chen KS, Cho E, Sangwan VK, et al. (2014). Effective passivation of exfoliated black phosphorus transistors against ambient degradation. *Nano Lett.*, **14**:6964–6970.

22. Ma X, Lu W, Chen B, Zhong D, Huang L, Dong L, et al. (2015). Performance change of few layer black phosphorus transistors in ambient. *AIP Adv.*, **5**:107112.

23. Deng Y, Conrad NJ, Luo Z, et al. (2015). Towards high performance two dimensional black phosphorus optoelectronic devices the role of metal contacts. *Technical Digest International Electron Devices Meeting, IEDM*, 5.2.1–5.2.4.

24. Du Y, Liu H, Deng Y, Ye PD, (2014). Device perspective for black phosphorus field-effect transistors: Contact resistance, ambipolar and scaling. *ACS Nano*, **8**:10035–10042.

25. Pimenta MA, Dresselhaus G, Dresselhaus MS, Cancado LG, Jorio A, Saito R, (2007). Studying disorder in graphite-based systems by Raman spectroscopy. *Phys. Chem. Chem. Phys.*, **9**:1276–1291.

26. Novoselov KS, Geim AK, Morozov SV, Jiang D, Zhang Y, Dubonos SV, Grigorieva IV, Firsov AA, (2004). Electric field effect in atomically thin carbon films. *Science*, **306**:666–669.

27. Wu J, Mao N, Xie L, Xu H, Zhang J, (2015). Identifying the crystalline orientation of black phosphorus using angle-resolved polarized Raman spectroscopy. *Angew. Chem.*, **54**:2366–2369.

28. Xia F, Wang H, Jia Y, (2014). Rediscovering black phosphorus as an anisotropic layered material for optoelectronics and electronics. *Nat. Commun.*, **5**:4458.

29. Qiao J, Kong X, Hu ZX, Yang F, Ji W, (2014). High-mobility transport anisotropy and linear dichroism in few-layer black phosphorus. *Nat. Commun.*, **5**:4475.

30. Favron A, Gaufrès E, Fossard F, Lévesque PL, et al. (2014). Exfoliating black phosphorus down to the monolayer photo-induced oxidation and electronic confinement effects. *Int. J. Agric. Innovations Res.*, **145**:708–712.

31. Kim W, Javey A, Tu R, Cao J, Wang, Q, Dai H, (2005). Electrical contacts to carbon nanotubes down to 1 nm in diameter. *Appl. Phys. Lett.*, **87**:173101.

32. Su Y, Kshirsagar CU, Robbins MC, Haratipour N, Koester SJ, (2016). High-performance complementary logic inverter using integrated black phosphorus and MoS_2 transistors. *2D Mater.*, **3**:011006.

33. Haratipour N, Koester SJ, (2016). Ambipolar black phosphorus MOS-FETs with record n-channel transconductance. *IEEE Electron Device Lett.*, **37**:103–106.

34. Das S, Demarteau M, Roelofs A, (2014). Ambipolar phosphorene field effect transistor. *ACS Nano*, **8**:11730–11738.

35. Haratipour N, Robbins MC, et al. (2015). Black phosphorus n-MOSFETs with record transconductance. *73rd Annual Device Research Conference (DRC)*, 243–244.

36. Han SJ, Garcia AV, Oida S, Jenkins KA, Haensch W, (2014). Graphene radio frequency receiver integrated circuit. *Nat. Commun.*, **5**:3086.

37. Lin Y-M, Dimitrakopoulos C, Jenkins KA, Chiu H-Y, Grill A, Avouris P, (2010). 100-GHz transistors from wafer-scale epitaxial graphene. *Science*, **327**:662.

38. Wang H, Nezich D, Kong J, et al. (2009). Graphene frequency multipliers. *IEEE Electron Device Lett.*, **30**:547–549.

39. Liao L, Lin YC, Bao M, Cheng R, Bai J, Liu Y, et al. (2010). High-speed graphene transistors with a self-aligned nanowire gate. *Nature*, 467:305–308.

40. Wu Y, Lin YM, Bol AA, Jenkins KA, Xia F, Farmer DB, et al. (2011). High-frequency, scaled graphene transistors on diamond-like carbon. *Nature*, **472**:74–78.

41. Huang JK, Pu J, Hsu C-L, Chiu M-H, Juang Z-Y, et al. (2014). Large-area synthesis of highly crystalline WSe_2 monolayers and device applications. *ACS Nano*, **8**:923–930.

42. Zhu W, Low T, Lee YH, Wang H, Farmer DB, Kong J, et al. (2014). Electronic transport and device prospects of monolayer molybdenum disulphide grown by chemical vapour deposition. *Nat. Commun.*, **5**:3087.

43. Wang H, Wang X, Xia F, Wang L, Jiang H, Xia Q, et al. (2014). Black phosphorus radio-frequency transistors. *Nano Lett.*, **14**:6424–6429.

44. Qi P, Vermesh Q, Grecu M, Javey A, et al. (2003). Toward large arrays of multiplex functionalized carbon nanotube sensors for highly sensitive and selective molecular detection. *Nano Lett.*, **3**:347–351.

45. Kong J, Franklin NR, Zhou C, Chapline MG, Peng S, Cho K, Dai H, (2000). Nanotube molecular wires as chemical sensors. *Science*, **287**:622–625.

46. Abbas AN, Liu G, Liu B, Zhang L, Liu H, Ohlberg D, Wu W, Zhou C, (2014). Patterning, characterization, and chemical sensing applications of graphene nanoribbon arrays down to 5 nm using helium ion beam lithography. *ACS Nano*, **8**:1538–1546.

47. Huang L, Zhang Z, Li Z, Chen B, Ma X, Dong L, et al. (2015). Multifunctional graphene sensors for magnetic and hydrogen detection. *ACS Appl. Mater. Interfaces*, **7**:9581–9588.
48. Late DJ, Huang YK, Liu B, et al. (2013). Sensing behavior of atomically thin-layered MoS_2 transistors. *ACS Nano*, **7**:4879–4891.
49. Liu B, Chen L, Liu G, Abbas AN, Fathi M, Zhou C, (2014). High-performance chemical sensing using schottky-contacted chemical vapor deposition grown monolayer MoS_2 transistors. *ACS Nano*, **8**:5304–5314.
50. Kou L, Frauenheim T, Chen C, (2014). Phosphorene as a superior gas sensor: selective adsorption and distinct I-V response. *J. Phys. Chem. Lett.*, **5**:2675–2681.
51. Abbas AN, Liu B, Chen L, Ma Y, Cong S, et al. (2015). Black phosphorus gas sensors. *ACS Nano*, **9**:5618–5624.
52. Geim AK, Grigorieva IV, (2013). Van der Waals heterostructures. *Nature*, **499**:419–425.
53. Deng Y, Luo Z, Conrad NJ, et al. (2014). Black phosphorus–monolayer MoS_2 van der Waals heterojunction p–n diode. *ACS Nano*, **8**:8292–8299.
54. Buscema M, Groenendijk DJ, Blanter SI, Steele GA, et al. (2014). Fast and broadband photoresponse of few layer black phosphorus field effect transistors. *Nano Lett.*, **14**:3347–3352.
55. Baugher BW, Churchill HO, Yang Y, Jarillo-Herrero P, (2014). Optoelectronic devices based on electrically tunable p-n diodes in a monolayer dichalcogenide. *Nat. Nanotechnol.*, **9**:262–267.
56. Pospischil A, Furchi MM, Mueller T, (2014). Solar-energy conversion and light emission in an atomic monolayer p-n diode. *Nat. Nanotechnol.*, **9**:257–261.
57. Chen X, Wu Y, Wu Z, Han Y, Xu S, Wang L, et al. (2014). High quality sandwiched black phosphorus heterostructure and its quantum oscillations. *Nat. Commun.*, **6**:7315.
58. Koenig SP, Doganov RA, Schmidt H, Castro Neto AH, Özyilmaz B, (2014). Electric field effect in ultrathin black phosphorus. *Appl. Phys. Lett.*, **104**:103106.

Chapter 7

Electrochemical Applications

Qingqing Ke
Department of Materials Science and Engineering,
National University of Singapore, Singapore 117574, Singapore
msekqq@nus.edu.sg

Phosphorene, mono- or few-layer-thick form of black phosphorus (BP), exhibits a puckered honeycomb structure, in which each phosphorus atom binds with three neighboring atoms. It has emerged as a promising candidate for electrochemical applications due to its large surface area and highly desirable electrical properties. However, BP flakes are reported to chemically degrade rapidly upon exposure to ambient conditions, which greatly hinders its potential applications. This chapter first lays out the structural degradation problems of BP and then summarizes up-to-date strategies for BP protection. After overviewing these challenges and protection methods, the promising applications of BP in energy storage and oxygen production are reviewed.

Phosphorene: Physical Properties, Synthesis, and Fabrication
Edited by Yongqing Cai, Gang Zhang, and Yong-Wei Zhang

ISBN 978-981-4774-64-2 (Hardcover), 978-0-203-71061-6 (eBook)
www.jennystanford.com

7.1 Structural Degradation and Protection of BP Flakes

The fast degradation of phosphorene in ambient environments is an unresolved issue that may limit future applications. The presence of oxygen and humidity is supposed to be the possible cause of the degradation.[1–3] It was found that the pits and bubble-like structures formed on the surface of BP are revealed by atomic force and optical microscopy when BP is exposed to air.[4,5] Castellanos-Gomez et al. characterized the stability of exfoliated BP flakes by conducting experiments and doing calculations (see Fig. 7.1).[6] Their results show that droplets become visible on the flakes surface within one hour. On being exposed to air for a long time, the droplets keep

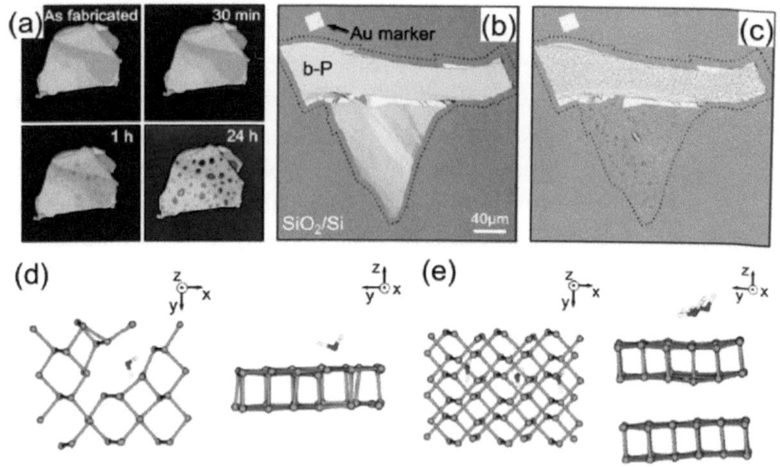

Figure 7.1 Evolution of black phosphorus flakes under ambient conditions. (a) Optical images of the exfoliated black phosphorus flakes acquired at different times. Droplet-like structures gradually form and grow on the surface of the flakes within several hours. (b, c) Morphologies of samples right after the transfer (b) and after two weeks (c) in air, respectively. (d) Atomic modelling of the atomic structure of a monolayer of black phosphorus adsorbing a water molecule simulated from DFT calculations. (e) Atomic models for a bilayer phosphorene with a higher concentration of water molecules. Reprinted with permission from Castellanos-Gomez et al.[6] Copyright (2014) by the Institute of Physics.

grow and deteriorate black phosphorous, even etching away the thinner parts of the BP flakes (Fig. 7.1b,c). The droplet structure was ascribed to the hydrophilic nature of the BP surface, which was further demonstrated by DFT calculations (Fig. 7.1d,e). Their theoretical results show that the BP lattice was greatly distorted with shrinkage by around 25% when one water molecule was deposited on it, the shrinkage resulting from the expected strong dipole–dipole interactions between water molecule and BP lattice.[7] Similarly, Island et al. also studied the environmental instability of mechanically exfoliated BP flakes in ambient conditions and characterized the water absorption at the surface of thin and isolated BP flakes by AFM.[8] Their results show that BP has strong affinity for water and over 200% increase in volume was observed as a result of water absorption over several days.

Although these investigations demonstrate water as a kind of primary species responsible for BP degradation, this conclusion indeed is not supported by the fact that devices consisting of BP used as humidity sensors can be stable in water for an extended exposure time without substantial degradations.[9] Moreover, Kang et al. further show that BP flakes dispersed in an aqueous solution rather than the commonly used anhydrous organic solvents deliver good optical and electronical performance.[10] This evidence questions the assumption that water is the possible reason for the degradation of BP flakes. Compared to humidity, oxygen atoms were recently suggested to be responsible for the degradation of BP flakes. Using first-principles calculations, Ziletti et al.[3] show that the degradation of BP is caused by surface reactions with oxygen atoms. In this work, three types of oxygen atoms, including dangling-, interstitial-, and bridging-type oxygen, are examined. Defects in dangling oxygen can change the surface hydrophilicity of BP, which is a starting point for the formation of more complex defects in water or other polar molecules.[6] Huang et al. experimentally examine the contributions of oxygen and water towards the degradation of BP together with ab initio calculations.[11] Their work evidently showed that BP rapidly degrades whenever oxygen is present, but is unaffected by deaerated (i.e., O_2 depleted) water. They further found that H_2O molecules are weakly physisorbed and do not dissociate on the BP surface, while BP oxidation is easily achieved through a facile dissociative chemisorption of O_2, thereby leading to rapid BP degradation.

For atomically thin 2D materials, lattice imperfections are very popular and can be introduced during the fabrication procedure or artificially produced using electron-beam or other high-energy particle excitations.[12–14] In the case of BP flakes, phosphorene vacancies are predicted to be more easily formed and abundant due to a low formation energy (1.65 eV) as well as an ultralow diffusion barrier of 0.3 eV.[15] Such defects in BF flakes may also play important roles in the instability of phosphorene in air through the interaction with environmental molecules, such as O_2 and H_2O. Considering the presence of the high population of vacancies in the BF flakes, Kistanov, et al. studied the interaction of H_2O and O_2 molecules with vacancy-contained phosphorene using first-principles calculations.[16] Their work shows that H_2O presents nearly the same adsorption energy at the vacancy site as that at the perfect one, which contrasts the common conception that physisorbed molecules are prone to perform a stronger adsorption at vacancy sites. For oxygen atoms, the story is quite different. It shows that the barrier for the O–O bond splitting to form O–P bonds is reduced from 0.81 eV at the perfect site to 0.59 eV at the site of defects, resulting in about 5000 times faster oxidizing rate at the site of defect than at the perfect site at room temperature (Fig. 7.2). It was concluded that the vacancy in phosphorene shows a stronger oxygen affinity than the perfect phosphorene lattice site. Their work further infers that structural degradation of phosphorene due to oxidization may occur rapidly at edges and grain boundaries, where vacancies tend to agglomerate.

Early experiments demonstrated that fast oxidation[1] and laser-induced degradation[17] occur in ambient conditions. However, a deeper understanding of this oxidation chemistry, which is critical to the applications of phosphorene-based electronics, has not been revealed. Recently, Favron et al. investigated thickness-dependent degradation of BP using in situ Raman and transmission electron spectroscopies.[18] They built a phenomenological model involving electron transfer and quantum confinement to clarify this chemistry. They prepared mono-, bi-, and multilayer 2D phosphorene in a glove box to clarify the effect of layer thickness on Raman modes.

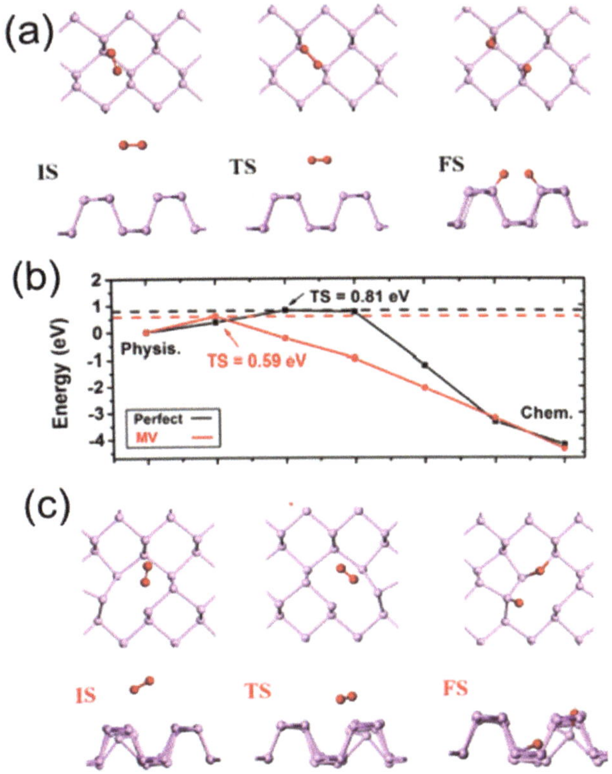

Figure 7.2 Formation of chemically adsorbed oxygen species from dissociation of the O_2 molecule on perfect (a) and single-vacancy-containing phosphorene (c). (b) The energy barriers for both processes. P and O atoms are represented with purple and red balls, respectively. Reprinted with permission from Kistanov et al.[16] Copyright (2017) by the Institute of Physics.

Photoinduced oxidation of *n*-layer 2D phosphorene in air was identified, which is greatly related to the photoassisted oxidation reaction with oxygen dissolved in adsorbed water. It shows that the rate of oxidation depends linearly on oxygen concentration and light intensity and exponentially on the square of the energy gap. In the review of the contribution from oxygen, isolating BP with water would be an efficient method to prevent the formation of more complex defects in BP, thereby avoiding sample oxidation.

Therefore, the degradation of phosphorene can be ascribed to the following factors: oxygen, water, light, and defects. Effective strategies for enhancing the stability of phosphorene can be developed from blocking the contacts with oxygen and water. To suppress light-induced photo-oxidation, phosphorene samples should be properly stored in a black environment. So far, various techniques have been effectively used to protect phosphorene, such as graphene/BN passivation[19,20] and Al_2O_3 coating with atomic layer deposition.[21,22] A high-quality, air-stable phosphorene film was fabricated by the use of oxygen plasma dry etching to form a thin oxide layer, followed by Al_2O_3 passivation.[23]

7.2 Black Phosphorus for Energy Storage Applications

7.2.1 Lithium-Ion Battery

With the rapid development of consumer electronics, including electric vehicles and mobile devices, energy storage techniques such as rechargeable battery with high capacity and rate capability have been gaining more and more scientific and technical attention. Lithium-ion battery (LIB) has been widely studied due to its apparent advantages in energy efficiency and portability.[24–26] In LIBs, the electrochemical reaction of lithium ions is achieved via varied chemical pathways, such as through alloying/de-alloying reaction, conversion reaction, and intercalation/de-intercalation reaction.[27] During these electrochemical conversion processes, the reversible shuttling of lithium ions between cathode and anode is critical and largely affects the overall performance of LIBs. Therefore, electrode materials are selected on the basis of their inherent properties, such as flat potential profile, high columbic efficiency, and good cycling performance.[28] Recent theoretical and experimental studies have shown that phosphorus-based materials can be potentially used as advanced battery materials, which can react with Li to form Li_3P compounds, giving a high specific capacity of 2596 mAh g^{-1} and a discharge potential window ranging from 0.4 to 1.2 V. Phosphorus has three main allotropes: white, red,

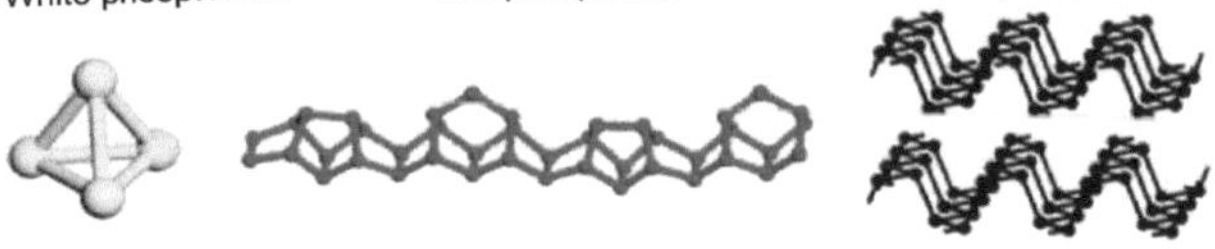

Figure 7.3 Schematics of white, red, and black phosphorus.

and black (Fig. 7.3).[29] Among these, white phosphorus (WP) is tetrahedral in structure, which is not stable due to the large bonding strain. It begins to burn at 30°C and is therefore unsuitable as electrode material for battery applications. Red phosphorus (RP) is amorphous and has low electron transfer capability, while black phosphorus (BP) is crystallized and orthorhombic with a layered structure. Compared to WP, BP and RP are thermodynamically stable allotropes and were studied as anode materials used at room temperature and atmosphere.[30]

RP is an abundant and eco-friendly material. However, due to its low conductivity and drastic volume change during Li alloying/de-alloying, the experimental capacity of RP is far below the theoretical value and greatly degraded after a few cycles.[31] To improve the conductivity of RP, construction of P–C composite was shown to be efficient.[32–34] In this composite, the conductive mesoporous carbon framework plays a role to improve the electrical conductivity of the supported materials.[35] Marino et al. fabricated an RP–carbon composite by vaporization and condensation, but the prepared composite delivered a reversible capacity of 900 mAh g^{-1} after 20 cycles together with poor Coulombic efficiency.[33] Ji et al. also reported combining nanostructuring sulfur composite materials with mesoporous carbon molecular sieves as the matrix material, which are beneficial in improving capacity and retention capability.[32] Gu et al. recently demonstrated porous carbon nanofibers (PCNFs) encapsulated with crystalline RP (P-PCNFs) to fabricate free-standing anodes for LIBs.[36] In this composite, porous carbon nanofibers were fabricated by electrospinning of polyacrylonitrile/poly(methyl methacrylate) (PAN/PMMA), followed by thermal carbonization. Afterwards, crystalline RP was grown onto PCNFs via a modified process of vaporization–adsorption–transformation (VAT) to form

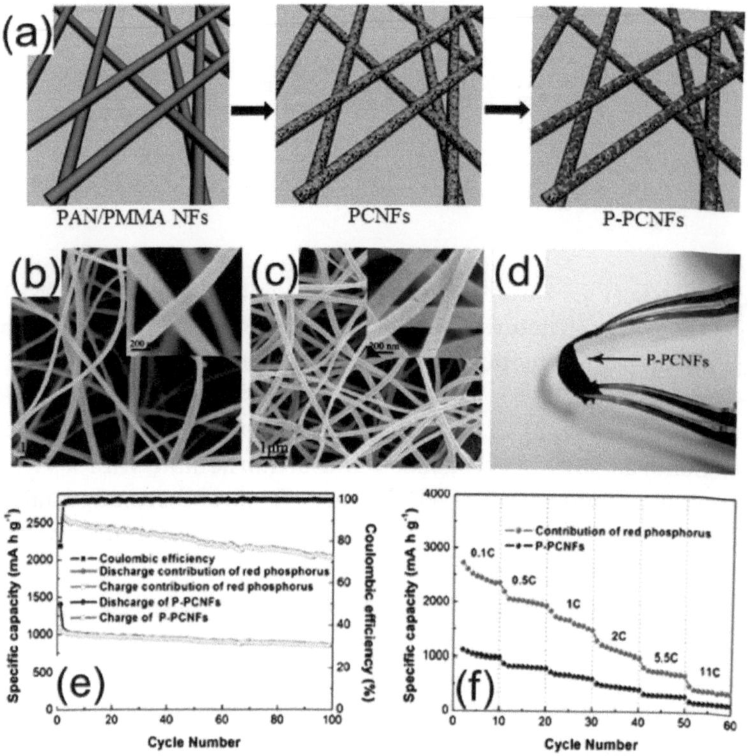

Figure 7.4 (a) Schematic plots showing the procedures of carbonizing electrospun PAN/PMMA bicomponent nanofibers to obtain PCNFs, and incorporating red phosphorus into the PCNF nanostructure. (b–d) The FESEM micrographs of PCNFs and P-PCNFs. The inset shows high-magnification images. (e) Capacity and Coulombic efficiency as a function of the cycle number for the P-PCNF electrode at a cycling rate of 0.1°C. (f) Discharge capacity of the P-PCNF electrode as a function of cycle number.[36] Reprinted with permission from Li et al.[36] Copyright (2014) Elsevier.

P-PCNF composites (Fig. 7.4a). In this composite, RP was shown to uniformly distribute inside the porous carbon matrix instead of being aggregated on the surface. The electrochemical performance shows that P-PCNFs delivered excellent cyclability (2030 mAh g^{-1} at 0.1°C rate after 100 cycles) and good rate capability of 380 mAh g^{-1} at 11°C (the capacity values were calculated based on the RP percentage). Even at a rate of 1°C, this composite exhibits a superior capacity of 1042 mAh g^{-1} based on the content of RP after 100 cycles. The improved electrochemical performance benefited from the pores of porous carbon nanofibers, which effectively accommodate the volume change of crystalline RP during cycling. Moreover, the carbon nanofiber matrix enables fast access of ions and provides RP with good conductivity and high mechanical flexibility.

To accommodate the volumetric expansion of the anode due to the increased mechanical stress during the Li^+ insertion-extraction process, amorphousizing[37–39] or dispersizing the active phases in an inactive buffering matrix[40–42] were shown to be effective methods.[43] Wang et al. also reported an amorphous P/carbon composite with highly reversible charge/discharge processes via a vaporization/adsorption strategy.[35] This composite shows a high lithium storage capacity of over 2413 mAh g^{-1}, as well as 87% retention after limited 50 cycles. Qian et al. prepared amorphous P/carbon composite via high energy ball-milling RP powders with amorphous carbon.[43] Their XRD and Raman results (Fig. 7.5) clearly demonstrated that the crystalline RP was transformed into amorphous P in the ball-milling process after 24 hours. The phase transformation was also revealed by TEM exanimation, where the ball-milled sample shows a disordered structure without any discernible lattice fringe. The electrochemical performance shows that the a-P/C composite electrode can deliver a high capacity of 2170 mAh g^{-1}, 1977 mAh g^{-1}, and 1750 mAh g^{-1} at high current densities of 1000 mA g^{-1}, 2000 mA g^{-1}, and 4000 mA g^{-1}, respectively. Even at a very high rate of 8000 mA g^{-1}, the reversible capacity can still reach 1457 mAh g^{-1}, 62% of its realizable capacity. Moreover, the a-P/C composite delivered an excellent reversible capacity over 100 cycles with capacity retention of 90% at very high rate of 2000 mA g^{-1}. This improved rate capability and cycling stability was ascribed to two reasons: (i) amorphousizing the active P phase may provide a stable and flexible host structure to alleviate high mechanical stress due to volumetric change during Li insertion and extraction; (ii) the P phase in a few nanometers encapsulated into the carbon matrix provided electrochemically favorable and

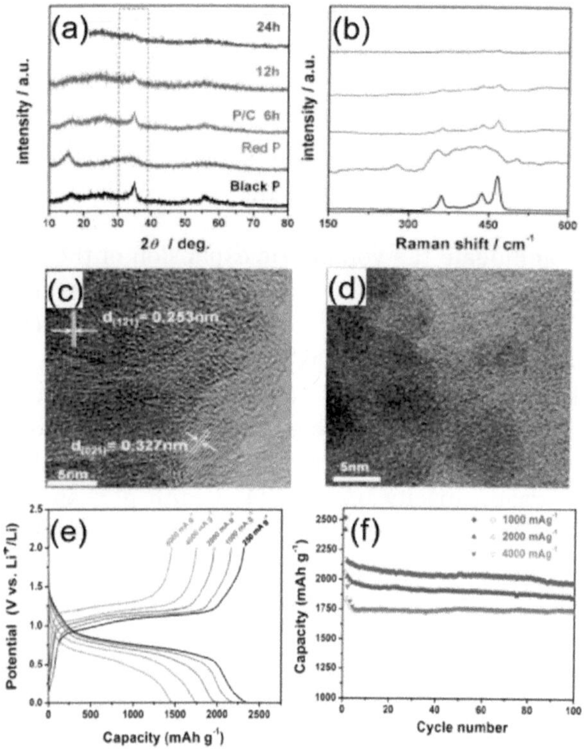

Figure 7.5 Comparison of the (a) XRD and (b) Raman spectra of the P/C composite ball-milled at different times with those of commercial red P and black P. HRTEM images of the P/C composite ball-milled for 12 hours (c) and 24 hours (d). (e) Charge–discharge curves at various current densities from 250 mA g^{-1} to 8000 mA g^{-1}. (f) Capacity as a function of cycle number at different rates of 1000, 2000, and 4000 mA g^{-1}. Reprinted with permission from Du et al.[43] Copyright (2012) by the Royal Society of Chemistry.

kinetically facile nanodomains for the Li insertion reaction, leading to good utilization of anode materials.[46]

Although RP is a kind of promising anode material due to its low cost and environmental benignness, there are still some challenges: (i) low electronic conductivity (~10–14 S cm^{-1})[30] and (ii) huge volume change during Li^+/Na^+ insertion/extraction.[44–46] These drawbacks would lead to mechanical fracture during cycling, fast degradation of capacity, and poor electrochemical performance at high current densities. In contrast, BP is thermodynamically the most stable, insoluble in most solvents, practically nonflammable, and chemically least reactive.[47–50] BP is very similar to graphite in terms of structure, appearance, and properties. Both of them are in layered structure with puckered sheets and exhibit good conductivity.[30] Moreover, one note that BP has high bulk electrical conductivity ($\sim 10^2$ S/m), low bandgap (0.34 eV), and reasonable density (2.69 g cm^{-3}), which are quite suitable for achieving both high power and energy densities.[51] Park et al. developed a simple method of transforming commercially available RP into BP using a high-energy mechanical milling (HEMM) technique at ambient temperature and pressure.[51] Their XRD pattern and TEM electron diffraction pattern confirm the amorphous nature of the RP, while orthorhombic structure is identified as BP by its lattice spacing (Fig. 7.6). The electrochemical performance of RP and BP is shown in Fig. 7.6c. It shows that the discharge and charge capacities of RP are 1692 and 67 mAh g^{-1}, respectively. BP shows an increased charge capacity of 1279 mAh g^{-1}. However, the first cycle efficiency is only 57%, which is due to the moderate electronic conductivity of BP inherited from its characteristic as a semiconductor. By mixing carbon with BP, the composite shows the first discharge and charge capacities of 2010 and 1814 mAh g^{-1}, respectively, together with first cycle efficiency of 90%. Through comparing the cycle performances made for the RP (voltage range: 0.0–2.0 V) and black P–carbon composite (voltage range: 0.0–2.0 V and 0.78–2.0 V) electrodes, they found that an improved cycle performance with a large capacity of more than 600 mAh g^{-1} (or 1600 mAh cm^{-3}) over 100 cycle is achieved in the case of the voltage range between 0.78 and 2.0 V.

It is known that the mechanical fracture induced by large volume change is able to cause the loss of electrical contact of active materials, thereby leading to the rapid fading of capacity. To break through this bottleneck, it is critical to maintain a good electric connection between BP and conductive carbon matrix. Sun et al. demonstrated a strategy for generating robust phosphorus–carbon (P–C) bonds between BP and a variety of carbon materials (e.g., graphite [G], graphite oxide [GO], carbon black [CB], and fullerene [C_{60}])) through the HEMM method.[30] The formation of P–C bonds

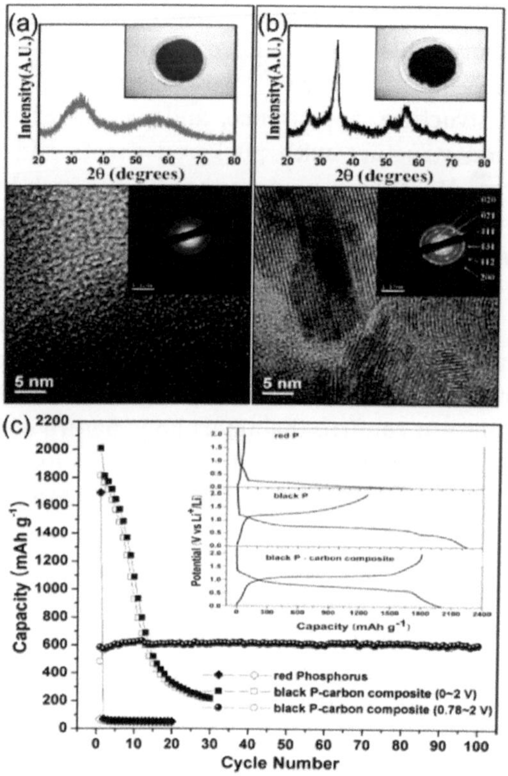

Figure 7.6 X-ray diffraction patterns and TEM images of red phosphorus (a) and black phosphorus (b). The color photo images and the corresponding lattice spacing are shown in the inset. (c) Comparison of cycle performance of capacities of red phosphorus and black phosphorus with black phosphorus–carbon composite. Inset shows the voltage profiles for first cycles. Reprinted with permission from Park et al.[51] Copyright (2007) by the John Wiley and Sons.

in BP–G is clarified by the lattice fringes of both BP and G in the HRTEM image, in which a coherent P–C bonding is found at the boundary of two different regions (Fig. 7.7a). Moreover, their Raman data collected at different stages of the electrochemical cycling demonstrated that the P–C bond is stable during lithium insertion/extraction (Fig. 7.7b–d), which is beneficial for a good electrical contact between BP and C. The cycling performance of the

BP linked with G (BP–G) was conducted under various high rates from 0.2 to 4.5 C (Fig. 7.7e). It shows that a high reversible capacity of 2270 mAh $\cdot$ g^{-1} was achieved at 0.2 C. Even at higher rates, such as 2, 3, and 4.5 C, the electrode retained a specific capacity of 1500, 1380, and 1240 mAh $\cdot$ g^{-1}, respectively. Compared to RP and mixture of BP and graphite (BP/G), BP–G shows a high initial charge capacity of 2786 mAh $\cdot$ g^{-1} and discharge capacity of 2382 mAh $\cdot$ g^{-1} and a stable cycle life of 100 cycles with 80% capacity retention (Fig. 7.7f). Their electrochemical impedance spectroscopy (EIS) in the work shows that BP-G electrode possesses a high electrical conductivity, a rapid charge-transfer process, and good lithium ion diffusivity for lithium uptake and extraction, which is responsible for the improved energy storage capability.[43]

These composites comprised of the unexfoliated BP–graphite display a good performance in LIB anodes. However, the BP used in these works are mainly bulk BP crystals. Recently new architectures using layered 2D materials have triggered wide research interest. In this regard, more Li atoms can be accommodated by the puckered honeycomb structure of few-layer phosphorene. Moreover, compared to the bulk BP crystal, the mono-layered phosphorene is rather robust and structure-stable during the lithiation process, which is expected for electrode materials in LIBs.[52] Therefore, an interesting question is raised: can the mono- or few-layer phosphorene be promising candidate materials for LIB anodes? To answer this question, both theoretical and experimental works have been performed. Li et al. have performed a theoretical study to investigate the binding and diffusion behavior of Li in phosphorene using density functional theory (DFT) calculations.[53] In their theoretical calculation, the Li atoms were shown to strongly bind with phosphorus atoms in the cationic state. The diffusion barrier is highly anisotropic. The value of the barrier along the zigzag direction is estimated to be 0.08 eV. This ultrahigh diffusivity is 100 times faster than that in MoS_2 and graphene at room temperature. In contrast, the diffusion barrier in a much higher value of 0.68 eV was calculated along the armchair direction, which greatly limits the diffusion in this direction. In this work, a remarkable large average voltage of 2.9 V is predicted in the phosphorene-based Li-ion battery (see Fig. 7.8). Moreover, they also predicted that a semiconducting to

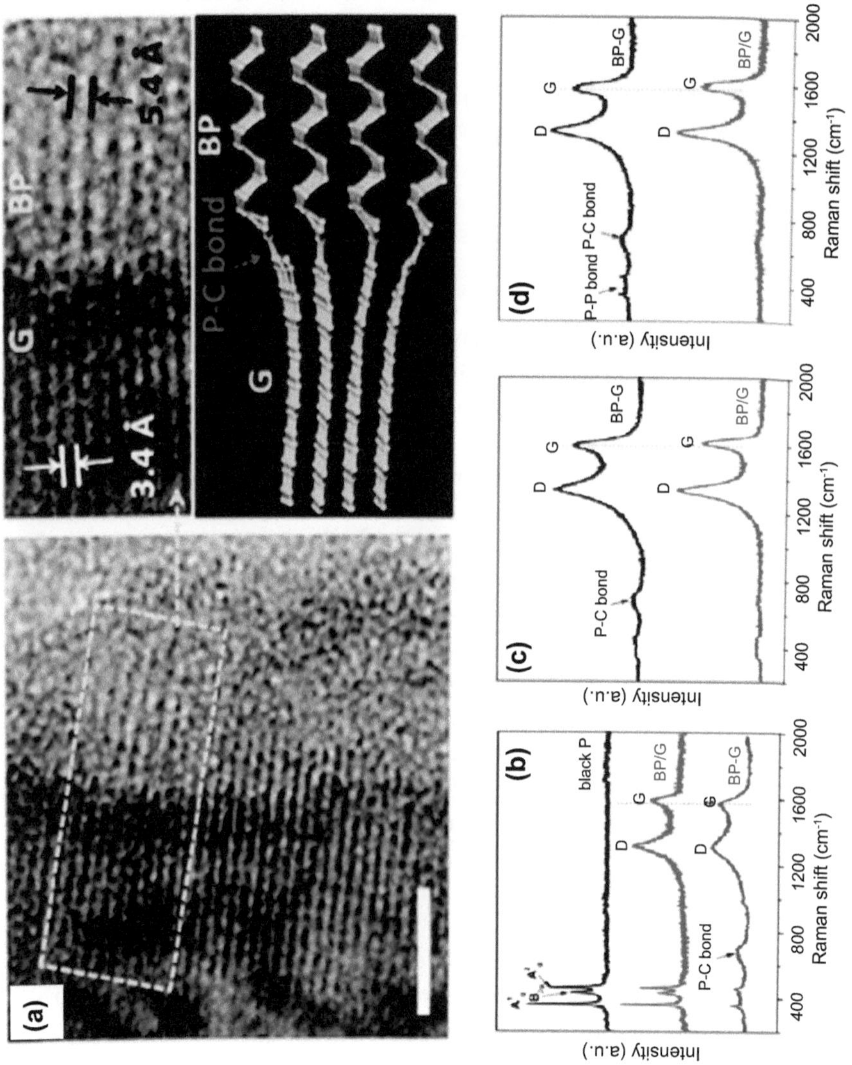
(a)
G
BP
3.4 Å
5.4 Å
P-C bond
(b)
black P
P-C bond
D
G
BP/G
BP-G
(c)
P-C bond
(d)
P-P bond P-C bond
Intensity (a.u.)
Raman shift (cm-1)
400
800
1200
1600
2000

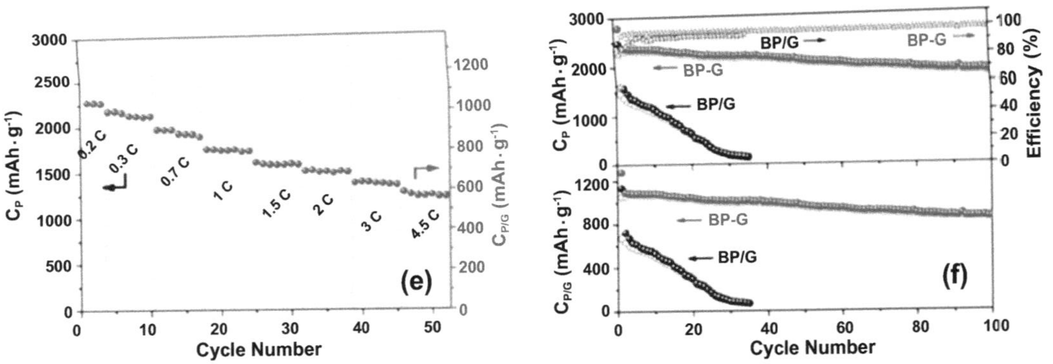

Figure 7.7 (a) The HRTEM photo and schematic atomic model of BP–G composite. Measured spacing between BP and G layers (top right). Scale bar: 2 nm. (b) Raman spectra of the fresh BP, BP/G, and BPG with a 514 nm wavelength laser. (c) Raman spectra of the BP/G and BP–G at the charged state for the first cycle. (d) Raman spectra of the BP/G and BP–G at the discharged state for the 30th cycle. (e) The cycling behavior of the capacity of the BP-G electrode at various current densities from 0.2 to 4.5 C. (f) The cycling performance and Coulombic efficiency of BP/G and BP-G electrodes at a current density of 0.2 C. Reprinted with permission from Sun et al.[30] Copyright (2014) by American Chemical Society.

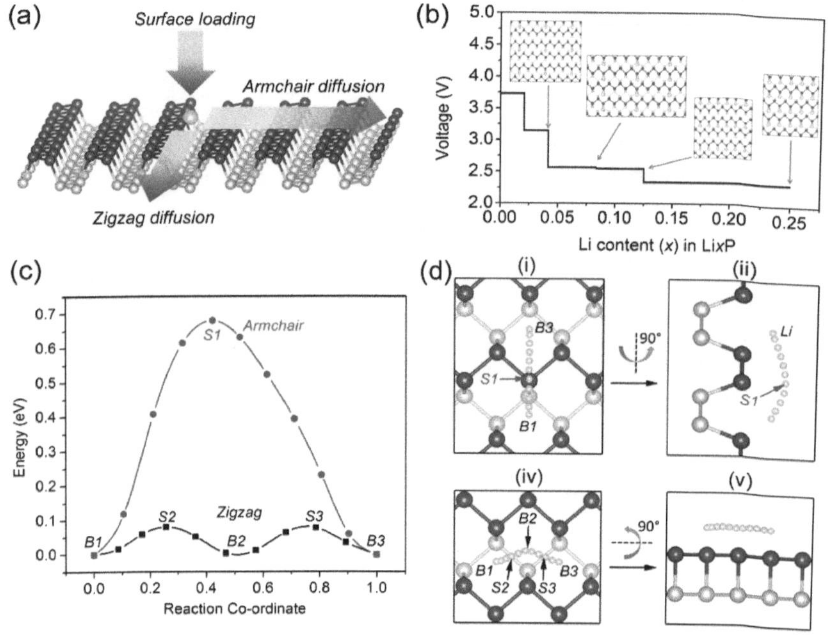

Figure 7.8 (a) Schematic atomic model of Li adsorption and diffusion above phosphorene. (b) Estimated voltage profile as a function of Li concentration from 0 to 0.25. (c) Energy barrier of Li diffusion along armchair and zigzag directions on phosphorene surface. (d) Top and side views of intermediate states of Li diffusion along the armchair (i, ii) and zigzag direction (iv, v). Reprinted with permission from Li et al.[53] Copyright (2015) by American Chemical Society.

metallic transition occurs resulting from the Li atoms intercalation and therefore giving rising to excellent electricity conductivity, which is beneficial for application as electrode materials.

The charge storage capability and structural stability of electrode materials are critically important to determine the performance of LIBs. Zhao et al. studied the potential application of phosphorene in LIBs using DFT calculations.[54] They show that the structure of phosphorene is largely preserved during the lithiation process with a volume change of only 0.2%. Moreover, they also demonstrated the theoretical specific capacity of phosphorene is 432.79 $mA\,h\,g^{-1}$, which is much higher than any values of commercial anode materials. These predicted high-capacity and

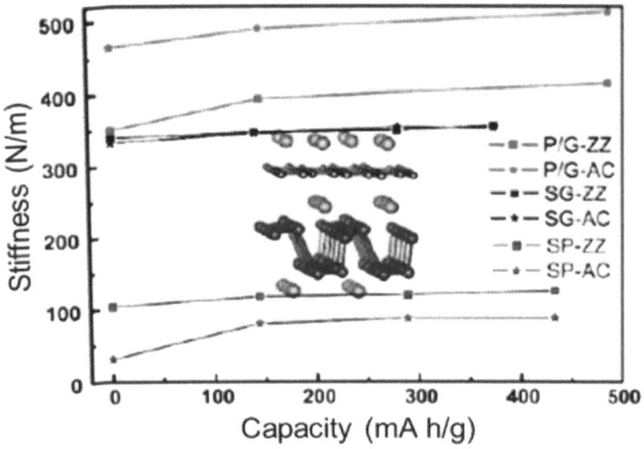

Figure 7.9 Stiffness of P/G, SG, and SP in the zigzag and armchair direction with different lithium intercalation concentrations. The inset is the optimized stable structure with 20 Li atoms inserted (P/G($Li_{20}C_{32}P_{24}$)). Reprinted with permission from Guo et al.[55] Copyright (2015) by American Chemical Society.

small volume changes suggest that phosphorene should be a good candidate for application in LIBs. Despite extensive investigations on them, phosphorene-based electrode materials with good electrical conductivity, high specific capacity, and excellent cyclic stability are still under development. Guo et al. have studied the Li adsorption and diffusion properties of the phosphorene/graphene composite (P/G) using DFT calculations.[55] The energetic stability of the P/G and the Li adsorption properties were examined. Their results show that the binding energy of Li on the pristine phosphorene is of 1.9 eV, while that of P/G heterostructure can be greatly increased to 2.6 eV without affecting the high mobility of Li within the layers. Moreover, the P/G composite was demonstrated to greatly improve the stiffness ($C_{ac} = 350$ N/m, $C_{zz} = 464$ N/m), which can potentially avoid the distortion of pristine phosphorene after the insertion of lithium (see Fig. 7.9). Therefore, the researchers suggested that the P/G composite can exhibit enhanced cycle life of the battery. Due to improved conductivity, high capacity, excellent Li mobility, and ultrahigh stiffness, P/G was shown to be a very promising anode material in LIBs.[55]

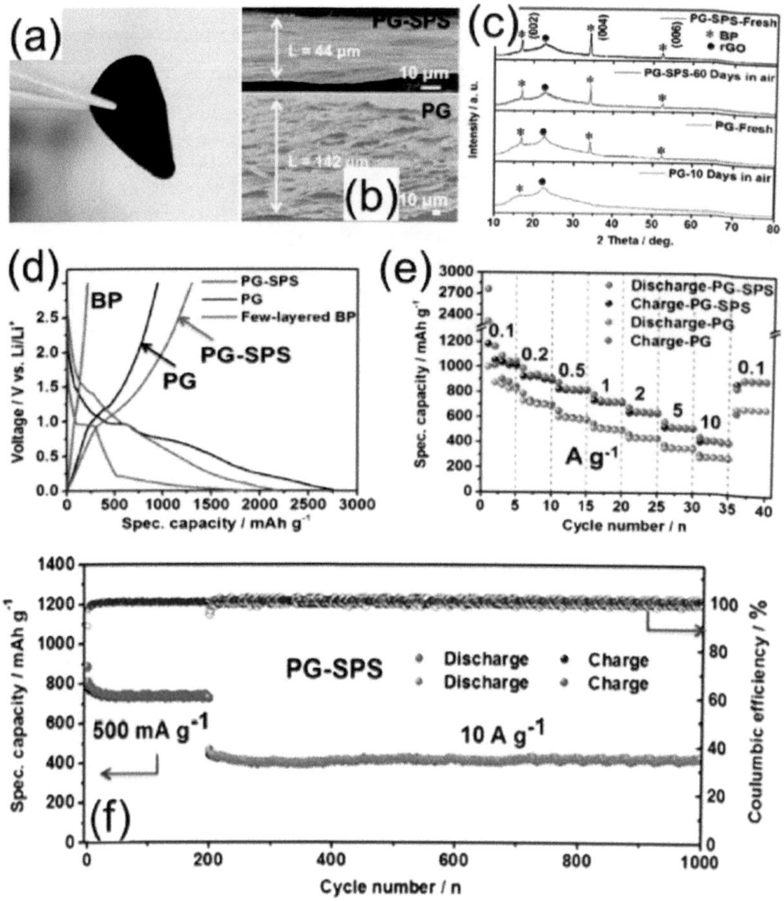

Figure 7.10 (a) Photograph of a PGO paper (diameter: 40 mm). (b) Cross-sectional SEM images of PG-SPS and PG samples. (c) XRD analysis of the air stability of PG-SPS versus PG. (d) The first galvanic charge/discharge curves of few layered BP, PG, and PG-SPS. (e) Rate capabilities of PG and PG-SPS electrodes based on volumetric capacities at a current density of 100 mA g^{-1}. (f) Long-term cycle life of PG-SPS electrode at different current densities with 500 mA g^{-1} (>200 cycles) and 10 A g^{-1} (<200 cycles). Spec. capacity stands for specific capacity; Vol. capacity stands for volumetric capacity at a current density of 100 mA g^{-1}. Reprinted with permission from Zhang et al.[56] Copyright (2016) by the John Wiley and Sons.

Although the theoretical predication suggests the advantages of phosphorene as a potential candidate for rechargeable LIBs, there are still some challenges toward real applications, such as large-scale fabrication and poor air stability. Zhang et al. reported phosphorene–graphene composites (PG-SPS) for LIBs applications. In this method, a high-yield synthesis of phosphorene with tunable size distributions together with good crystallinity was fabricated using liquid phase exfoliation of bulk BP in formamide.[56] In the procedure followed, a densely packed phosphorene–graphene composite (PG-SPS) was obtained via a facile and scalable spark plasma sintering (SPS) process. As shown in Fig. 7.10, the PG-SPS sample exhibits a much more densely packed structure with density of 0.6 g cm^{-3} as compared to the loosely stacked PG (0.16 g cm^{-3}). As we know that the as-exfoliated few-layer BP is likely to be degraded when exposed in ambient conditions due to the reactions with water and oxygen. The stability of the PG-SPS composites was demonstrated by XRD examination (see Fig. 7.10c). It shows that even after exposure to ambience for 60 days, the clear diffraction peaks from the well-crystalline few-layer BP can be collected. On the contrary, 95.2% of BP in PG sample was oxidized after only 10 days exposure to ambience. It therefore clearly clarifies that the chemical stability of modified PG-SPS composites is greatly improved compared to that of the PG samples. When the PG-SPS is working as anode material in LIBs, it shows much improved first-cycle Coloumbic efficiency of 60.2% compared to that of phosphorene (11.5%) and loosely stacked phosphorene–graphene composite (34.3%) (see Fig. 7.10d). A high specific capacity of 1306.7 mAh g^{-1} and volumetric capacity of 256.4 mAh cm^{-3} are achieved. Moreover, the PG-SPS also exhibits much good rate capabilities of 415.0 mAh g^{-1} at 10 A g^{-1} together with an outstanding long-term cycling life of 91.9% retention after 800 cycles at 10 A g^{-1} (see Fig. 7.10e,f).

7.2.2 Sodium-Ion Battery

Sodium-ion batteries have attracted a great attention due to its low-coat technology for large-scale applications, compared to that of lithium-ion batteries. Moreover, the performance of the recent developed sodium-ion battery is comparable to the counterpart of lithium-ion battery. Since the Na ions have a much larger radius (1.02 A) than that of Li ions (0.76 A) and preferably coordinate in octahedral or prismatic sites, it is hard to find crystalline host materials that perform both high electrochemical capacity and cyclability for Na ion insertion reaction.[57] Recent studies show that alloy-type anode materials, such as Si, Sn, Sb, etc., show high specific capacities but exhibit poor reversibility due to the large volume expansion and slow Na Kinetics,[58] while anode materials such as $Na_3V_2(PO_4)_3$ reveal prolonged cycle life, however with, limited capacity.[59] Therefore, it is highly desirable to find anode materials which co-exhibit high capacity, fast Na diffusion, and good structural stability.

Concerning the phosphorus-based materials for potential sodium-ion battery applications, white phosphorous (WP), composed of P_4 molecules, is quite reactive and toxic, and therefore it is not suitable for battery application.[60] Red phosphorous (RP) is shown to have a high Na storage capacity. In principle, each phosphorus atom can react with three Na atoms to form Na_3P compounds, giving a theoretical specific capacity of 2596 $mAhg^{-1}$,[61] which is much higher than those of Na alloys such as $Na_{15}Sn_4$ (847 $mA\,h\,g^{-1}$)[62,63] $Na_{15}Pb_4$ (485 $mA\,h\,g^{-1}$), Na_3Sb (660 $mA\,h\,g^{-1}$)[42,64] and almost two times higher than that of metallic sodium (1166 $mA\,h\,g^{-1}$). However, the amorphous RP undergoes large volume changes upon Na insertion (~491%), leading to a degraded capacity with cycling.[65,66] Moreover, amorphous RP is insulating with low electrical conductivity, which seriously limit the reversibility of Na insertion.[67,68] BP is the most thermodynamically stable allotrope with a layered crystal structure.[47,69] The flaky BP consisting of puckered layers of covalently bonded phosphorous atoms is a kind of good conductivity (~300 $S\,m^{-1}$).[47,69] Moreover, the large interlayer channel size of ~3.08 A enables both lithium (1.52 A) and sodium (2.04 A) ions to be stored between the phosphorene layers.

Kulish et al. studied the phosphorene as an anode material for Na-ion batteries and determined the corresponding Na adsorption energy, specific capacity and Na diffusion barriers on monolayer phosphorene using the first-principles calculations.[70] Four high-symmetry adsorption sites for single Na atom on phosphorene are

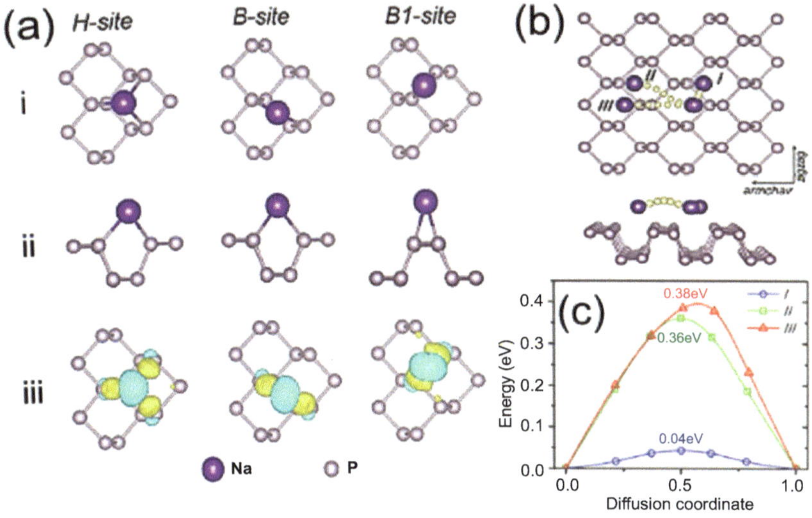

Figure 7.11 (a) Top (i) and side (ii) views of various adsorption sites of Na above phosphorene. (iii) corresponds to the differential charge density calculated by DFT. The blue and yellow colors represent loss and gain of electrons, respectively. (b) Top and side views of Na diffusion pathways: (I) H-B-H, (II) H-B1-H, and (III) H-T-H. (c) Energy barriers for Na diffusion on phosphorene. Reprinted with permission from Kulish et al.[70] Copyright (2016) by the Royal Society of Chemistry.

considered: (i) above the P hexagon center (H site), (ii) above the midpoint of the P–P bond, located in the P-sublayer far from the Na atom (B site), (iii) above the midpoint of the P–P bond, located in the P-sublayer near the Na atom (B1 site), and (iv) on the top of the P atom (T site) (see Fig. 7.11a). They calculated the Na mobility and diffusion pathways in phosphorene by evaluating the activation energies and found that H site is the most stable position for the Na atom on the phosphorene surface and the diffusion of Na on the phosphorene occurs through jumping between neighboring H sites using (B and B1) as saddle point. Therefore, the researchers considered the following three possible directions for Na diffusion: H-B-H, H-B1-H and H-T-H (Fig. 7.11b). The calculated diffusion pathways and estimated activation energies are shown in Fig. 7.11c. They found that H-B-H is the fastest Na diffusion pathway on the phosphorene surface with a calculated Perdew–Burke–Ernzerhof

(PBE) energy barrier of only 0.04 eV. They also evaluated the theoretical specific capacity of monolayer phosphorene, which demonstrated exceptionally high theoretical specific capacity, being capable of adsorbing Na atoms up to the compositions of NaP (865 $mA\,h\,g^{-1}$) and NaP_2 (433 $mA\,h\,g^{-1}$) as predicted by DFT-D_2 and DFT methods, respectively.

With respect to the experimental attempts, C. Zhang et al. found that for the conventional crystalline phosphorous, the P crystals are likely to be pulverized upon Na atom insertions as a result of ~200–500% volumetric expansion, thereby leading to phosphorene peeling off from current collector. To overcome this degradation, some breakthroughs have been made in stabilizing anode structure by designing amorphous P/carbon hybrids.[71,72] Qian et al. have compared the charge–discharge behaviors of amorphous phosphorus/carbon (a-P/C), RP, and BP in sodium cells using a 1.0 mol L^{-1} $NaPF_6$ solution as the electrolyte and sodium disks as counter electrode.[57] In the XRD patterns, the a-P/C composite shows a featureless diffraction pattern, completely different from those of crystalline RP and BP, as an indication of amorphous nature (Fig. 7.12a). The disordered structure is also confirmed by TEM image, in which discernible lattice fringe is not observe and its electron diffraction pattern shows a diffuse ring (Fig. 7.12b). As shown in Fig. 7.12c, the RP shows a quite large charge capacity in value of 897 $mA\,h\,g^{-1}$, but generates a limited discharge capacity of around 15 mA hg^{-1}, demonstrating inactivity for sodium insertion reaction resulting from its insulating electronic nature. In the case of BP, an improved charge capacity of 2035 $mA\,h\,g^{-1}$ is obtained, which is as high as nearly 80% of its theoretical three sodium ion insertion capacity, as result of the higher bulk conductivity.[69] Compared to the high charge capacity, the reversible discharge capacity is however 637 $mA\,h\,g^{-1}$. The degradation of the capacity is largely ascribed to the severe pulverization of the material resulted from the enormous volumetric change of 490% from phosphorous to Na_3P.[73] In the case of the a-P/C composite, it exhibits a greatly improved electrochemical performance with initial charge/discharge capacities of 2015 $mA\,h\,g^{-1}$ and 1764 $mA\,h\,g^{-1}$, as well as high initial coulombic efficiency of 87% (Fig. 7.12d). Similarly, Kim et al. demonstrated that a similar

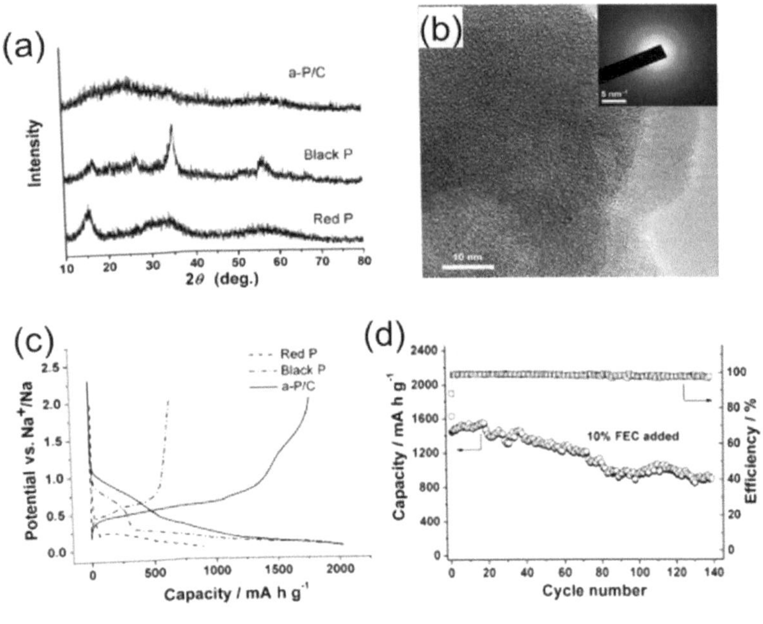

Figure 7.12 (a) Comparison of the XRD pattern of the as-prepared a-P/C composite with those of the commercial red phosphorus and black phosphorus, synthesized according to Park et al.[51] (b) HRTEM image of the a-P/C composite; inset shows the electron diffraction pattern. (c) Comparison of the initial charge/discharge curves of red phosphorus, black phosphorus, and a-P/C nanocomposite measured at a constant current density of 250 mAg^{-1} and in the voltage range of 0.01–2 V versus Na^+/Na. (d) Long-term cycle performance in 1 mol L^{-1} $NaPF^{6+}$EC-DEC electrolyte with 10% FEC. Reprinted with permission from Qian et al.[57] Copyright (2013) by the John Wiley and Sons.

amorphous phosphorus/carbon structure could deliver as high as 80% reversible capacity and less than 7% capacity fading after 30 cycles at a current density of 143 mA/g.[72] In addition, Chou et al. fabricated a composite anode via simple hand-grinding of commercial microsized red phosphorus and carbon nanotubes (CNTs); this showed a high capacity retention of 76.6% over 10 cycles.[74] The improvements in cycling stability in those reports were indeed remarkable, however, capacity retentions of less than 80% indicates that further efforts are still needed in order to meet the practical application requirements.[75]

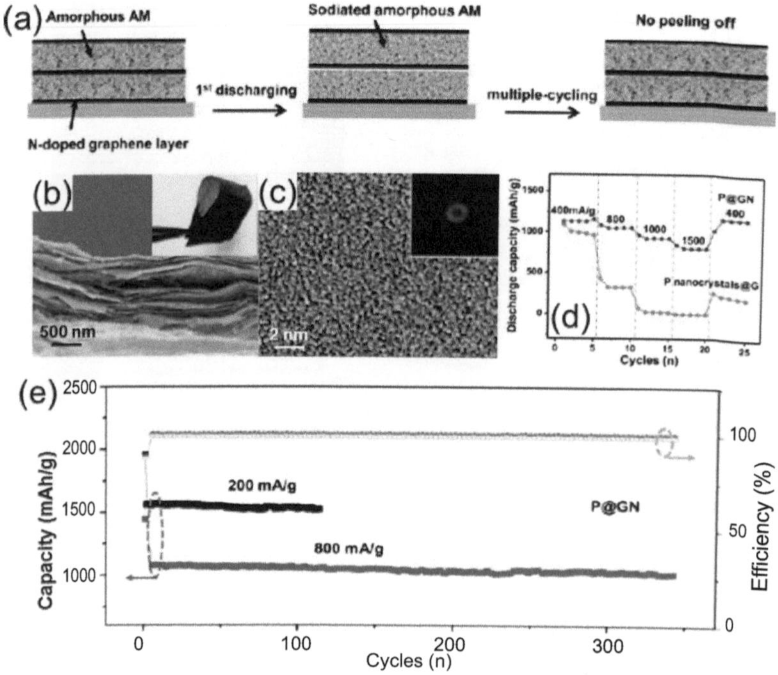

Figure 7.13 (a) Schematic of the procedures for fabricating a novel "butter-bread"-like anode structure consisting of amorphous P layer@N-doped graphene frameworks. (b) SEM image of the cross-section of a P@GN paper, the inset shows its paperlike appearance. (c) HRTEM image and the corresponding FFT pattern of the P@GN portion, showing an amorphous structure. (d) Comparison of the rate capability between P@GN and P nanocrystals@G (P NCs@G). (e) Cyclic performance and Coulombic efficiency of P@GN at 200 mA g^{-1} and 800 mA/g. Reprinted with permission from Zhang et al.[76] Copyright (2016) by the American Chemical Society.

In contrast to the previous work that assembled the P and C components via mechanical mixing (milling or grinding), C Zhao et al. designed a smart structure, that assembled the few-layered N-doped graphene (GN) as a substrate to support deposition of amorphous P.[76] In their work, the amorphous P@GN paper was fabricated through a "phase-transformation" route: P_4 vapors were formed through heating the bulk red P in a small sealed ampule. The P_4 vapors were then adsorbed and deposited onto the layer surface of GN sheets. After condensation process, amorphous red

P with "butter-bread"- like structure is formed on the interlayers of the GN surface (Fig. 7.13a and b).[77,78] The HRTEM image and the corresponding fast Fourier transform (FFT) patterns clarify the amorphous nature of P (Fig. 7.13c). In this novel hybrid structure, the input of GN is crucial to improve the battery capacity retention during electrochemical cycling. The GN is working as frameworks to not only confine the uniform growth of amorphous P, but also prevent volumetric expansion of P without mechanical damage or peeling off. Moreover, GN nanosheets are conductive, therefore can work as high conductivity electron transport networks to make P more electrochemically active. This smart designed structure was able to overcome limitation such as electrode chemical stability during the sodiation-desodiation and mechanical robustness after hybridization. Compared to the crystalized-P@GN, amorphous-P@GN shows an improved rate capability (see Fig. 7.13d). Moreover, the reversible capacity of 809 $mA\,h\,g^{-1}$ can be reached even at a high discharge current rate of 1500 $mA\,g^{-1}$, which is much better than counterpart of the crystallized P@GN with value of 10 $mA\,h\,g^{-1}$. In Fig. 7.13e, this amorphous P@GN structure enables the large volume-change anode with a superior capacity retention ($<85\%$ over 350 cycles), outstanding cycle stability (0.002% decay per cycle from 2nd to 350th cycle).

Y. Zhu reported a red P-single-walled carbon nanotube (SWCNT) (denoted as red P-SWCNT) composite, through a modified vaporization-condensation method (Fig. 7.14a).[45] In this composite, RP is uniformly distributed between tangled SWCNTs bundle, which serves as conductive network to enhance the conductivity of composite and stabilizes the solid electrolyte interphase (Fig. 7.14b,c). Figure 7.14d shows the variation of desodiation capacity for red P-SWCNT composite, P, SWCNTs, and red P/SWCNT mixture at current density of 50 $mA\,h\,g^{-1}$ for 50 cycles. Compared the cycling decay of pristine P and red P/SWCNT hand-mixture, the desodiation capacity of P-SWCNT composite gradually increases from 670 $mA\,h\,g^{-1}$ in the first cycle to 711 $mA\,h\,g^{-1}$ at the fifth cycle then maintains around 700 $mA\,h\,g^{-1}$ during the rest of 50 cycles. Figure 7.14e shows the rate capability of the red P-SWCNT composite at different current densities. It demonstrates that stable desodiation capacities of 700, 670, 640, 620, 590, 580, 500, and 300 $mA\,h\,g^{-1}$ are achieved at

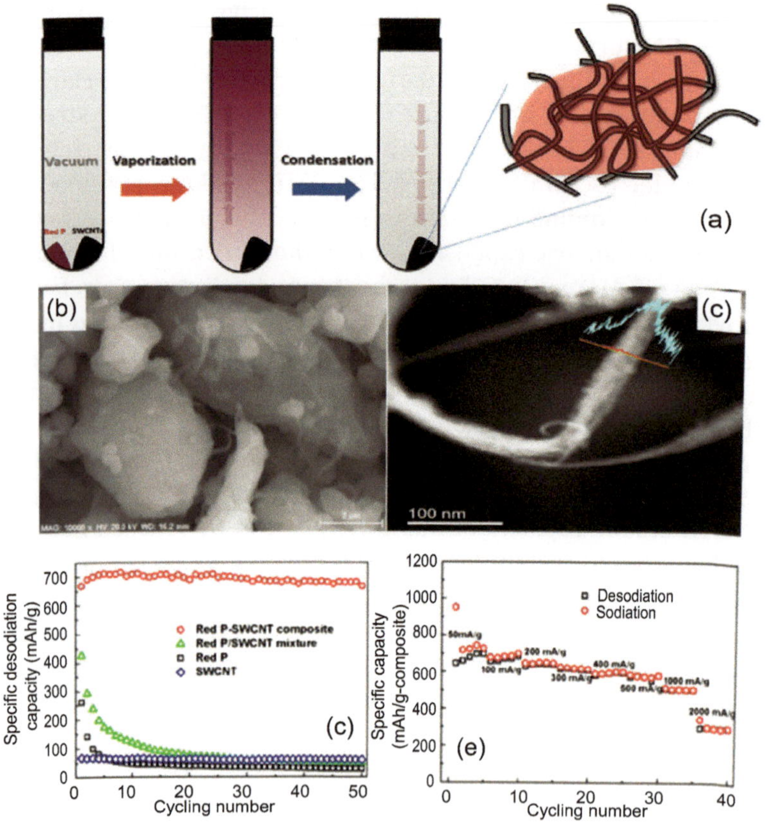

Figure 7.14 (a) Process for synthesizing red P-SWCNT composite. (b) SEM image of red P-SWCNT composite. (c) TEM image and EDS line scan of a bundle of SWCNTs in the composite (P signal is red and carbon signal is blue). (d) Cycling performance of red P-SWCNT composite, red P/SWCNT mixture hand-milled for 15 min, red P, and SWCNTs at a current density of 50 mA g^{-1}. (e) Sodiation/desodiation capacity of red P-SWCNT composite at different current densities. Reprinted with permission from Zhu et al.[45] Copyright (2015) by the American Chemical Society.

the current densities of 50, 100, 200, 300, 400, 500, 1000, and 2000 mA g^{-1}, respectively. The good rate capability was ascribed to the highly conductive network of SWCNTs, which provides fast and efficient electron transport during sodiation-desodiation reactions.

W. Li et al. recently have design and construct novel micro-structured/nanostructured P-based materials through confining nanosized amorphous red P into the mesoporous carbon matrix (P@CMK-3) using a vaporization–condensation–conversion process.[79] With this rational design, the corresponding lithium storage and sodium storage performance of RP can be greatly improved. In their fabrication procedure shown in Fig. 7.15a, P vapor generated at above sublimation temperature of RP tends to diffuse into the mesopores of CMK-3 by pressure differences and condense to be WP in the channels. After subsequent conversion of WP to RP, nanostructured amorphous RP would be confined in the channels of CMK-3. In their work, the employed CMK-3 presenting highly ordered porous structure with uniform pore diameter of ~4 nm (Fig. 7.15b,d). After impregnation of RP, most of RP has been encapsulated in CMK-3 without residing on the surface of CMK-3 (Fig. 7.15c and e). In this design, the porous CMK-3 matrix was aimed to accommodate the expansion of RP and provide large contact area for electrolyte access and good transport kinetics for both electrons/ions. It shows a high reversible specific capacity of ~2250 mA h g^{-1} based on the mass of red P (~971 mA h g^{-1} based on the composite) at 0.25 C and significantly enhanced cycle life of 1150 mA h g^{-1} based on the mass of RP (500 mA h g^{-1} based on the mass of the composite) at 5 C after 1000 cycles for LIBs (Fig. 7.15f and h). The behavior in terms of Na-ion storage performance of 1020 mA h g^{-1} based on the mass of RP (370 mA h g^{-1} based on the mass of the composite) maintained after 210 cycles at 5C is unprecedented (Fig. 7.15g,i). The improved electrochemical performance was attributed to the rationalized structure which enables easy access of electrolyte to the open channel structure, shorten the transport path of ions through carbon toward the red P, and increase the ionic and electronic conductivity.

Compared to the amorphous phosphorus, BP sheet are a good conductor with electricity of 300 S m^{-1}.[47,69] Moreover the interlayer channel size of 3.08 A, larger than both lithium (1.52 A) and sodium (2.04 A), enable BP to store lithium and sodium between the phosphorene layers, although the poor cycling performance needs to be improved.[30,51,80,81] J. Sun et al. recently reported a hybrid material made out of a few phosphorene layers sand-

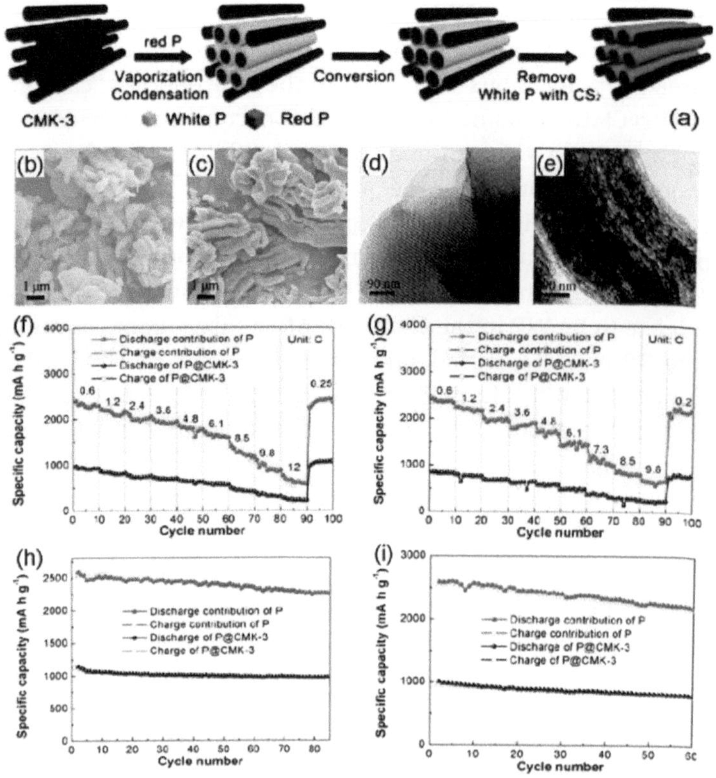

Figure 7.15 (a) Schematic illustration of the preparation process for the P@CMK-3 material. SEM micrographs of CMK-3 (b) and P@CMK-3 (c). TEM micrographs of CMK-3 (d) and P@CMK-3 (e). Capacity of P@CMK-3 composite as a function of discharge rate 0.6–12 C for (f) LIBs and 0.6–9.8 C for (g) NIBs. Cycle performance of P@CMK-3 electrodes for LIBs at 1.2 and 5 C (h) and NIBs at 1 and 5C (i) with activation first at low current density. Reprinted with permission from Li et al.[79] Copyright (2016) by the American Chemical Society.

wiched between graphene layers. The sodiation mechanism of black phosphorous is examined by in situ transmission electron microscopy (TEM) and ex situ diffraction (XRD) techniques. It shows that a large anisotropic volumetric expansion along the y and z axial directions occurs during the alloy reactions, which causes significant capacity degradation with cycling (Fig. 7.16). With the integration of graphene and BP, high capacity and stable cyclability can be simultaneously reached as: (i) graphene works as an elastic buffer layer that modulates the anisotropic volumetric expansion during the sodiation process, (ii) the high conductivity of graphene make hybrid material more electrochemically active, (iii) the layered structure of BP flake provide a short diffusion distance for sodium ions. Benefit from the improved electrical conductivity, the phosphorene-graphene sandwiched structure demonstrated an enhanced cycling capability (Fig. 7.16f). It shows that a reversible capacity of 2440 mA h g^{-1} was obtained at current density of 50 mA g^{-1}. With increasing the current density to 0.2 A g^{-1}, a reversible capacity of 2320 mA h g^{-1} is delivered. Even at the very high rates that is up to 26 A g^{-1}, a specific capacity of 645 mA h g^{-1} can be retained (26.4%). In addition, phosphorene-graphene sandwiched structure demonstrates enhanced capacity retention upon cycling (Fig. 7.16g). It shows that specific capacity of 2080 mA h g^{-1} can be retained, corresponding to an 85% capacity retention (0.16% decay per cycle). At relatively fast rates, such as 8 A g^{-1} and 26 A g^{-1}, the capacity retentions were 84 and 77%, respectively, after 100 cycles.

7.2.3 Lithium-Sulfur (Li-S) Batteries

To meet the demand for high capacity energy storage systems, a considerable effort has recently been made on the development of the lithium-sulfur (Li-S) battery, which has the advantages of a high theoretical energy density (2567 W h kg^{-1}) and the abundance, nontoxicity, and low cost of sulfur.[82–84] A typical Li-S battery cell is comprising of S as the positive electrode and Li as the negative electrode. A liquid electrolyte is working as both the charge transfer medium and the ionic conductor within the sulfur-containing cathode.[85–87] Despite the great promise, the application of the Li-S battery is still hindered by its limited cycle life and low power density. Some obstacles should be overcame before achieving its practical application, such as (i) low electrical conductivity of sulfur species, which leads to high over potentials and low utilization of active materials,[88–90] (ii) the diffusion of soluble intermediate polysulfide species into the electrolyte, which causes irreversible loss of active material, low coulombic efficiency, short lifecycle, and increases the device impedance over time,[91,92] (iii) the large volume

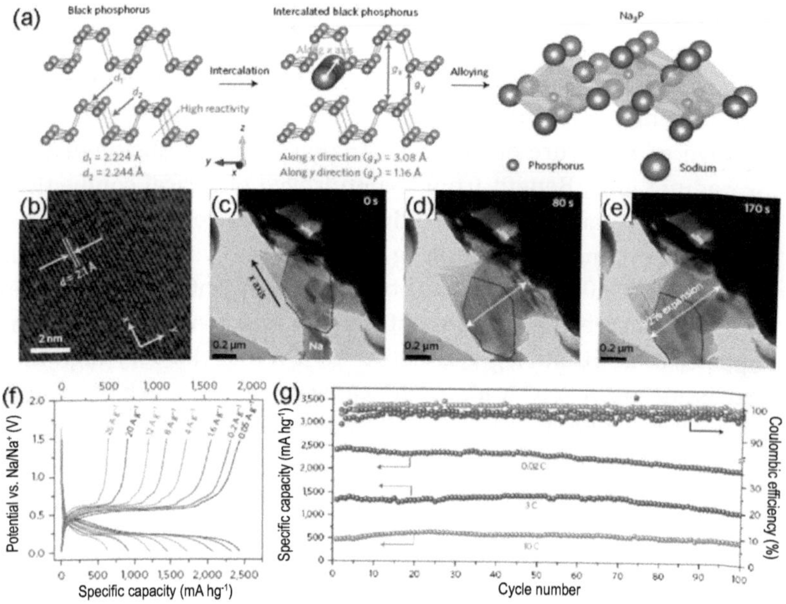

Figure 7.16 (a) Atomic model of black phosphorus before sodiation (left), with the first step of sodium-ion intercalation (middle), and the second step of alloy reaction to form Na_3P (right). (b) High-resolution and bright-field TEM image of black phosphorus before sodiation. (c–e) Time-lapse TEM images of sodiation in black phosphorus. Sodium ions transport along the *x*-axis channel, causing a volume expansion along the *y*-axis direction. (f) Volumetric and mass capacities at different current densities (from 0.05 to 26 A g^{-1}). (g) Reversible desodiation capacity and Coulombic efficiency for the first 100 galvanostatic cycles of the phosphorene/graphene (48.3 wt% P) anode tested under different currents. Reprinted with permission from Sun et al.[44] Copyright (2016) by Nature Publishing Group.

change of sulfur cathodes during the charge and discharge processes (up to 80%), which induces stress in the electrode and undermines its structural integrity, leading to the loss of electrical contact with the conductive additives and the detachment from the current collector.[93,94] To address these issues, the electrode structure and composition has been engineered in order to increase the electrical conductivity, as well as to prevent the dissolution of polysulfide through trapping the sulfur species within the electrodes. Some strategies have been proposed to solve these problems including designing new electrode structures, such as nanoporous carbon-sulfur composites,[32,95] conductive polymer-sulfur composites,[96] as well as metal oxide and sulfide coating have been explored for advanced electrode design.[97–99] Moreover, various routes for the physical trapping of soluble polysulfides beyond the electrode have been explored, including the insertion of a microporous carbon paper between cathode and separator[100,101] and the modification of separators with a carbon coating layer.[102,103]

Carbon materials were used as sulfur-entrapping host structures,[32,95] but their weak interaction with the polar Li_2S_x hinders their application as polysulfide traps.[95–97] L. Li et al. experimentally demonstrated the development of high-performance Li-S batteries incorporating with phosphorene nanosheets into a porous carbon nanofiber network (cathode matrix) which can significantly improve the cycle life of Li-S batteries.[104] In their work, the liquid exfoliation approach was employed to obtain few-layer phosphorene nanosheets (FLP) in thickness of 1–2 μm and the carbon nanofiber (CNF) network was used as cathode matrix (Fig. 7.17). The cyclic voltammetry (CV) was adopted to investigate the electrochemical reaction kinetics. As shown in Fig. 7.17d, two pairs of distinct and stable redox peaks were observed for the FLP-CNF electrode, in which the reduction peak (2.40–0.25 V) and the corresponding oxidation peak (2.38–2.50 V) belong to the transition between sulfur (S8) and high-order lithium polysulfides (Li_2S_x, $4 \leq x \leq 8$), while the cathodic peak (2.07–1.96 V) and the backward oxidation peak (2.27–2.38 V) correspond to the transformation between the high-order lithium polysulfides and Li_2S_2/Li_2S, respectively.[85] The electrochemical performance was evaluated in the coin cell format for FLP-CNF and a pristine CNF electrode was also used as referenced electrode. Their results showed that a specific capacity of 1262, 1092, 1027, 985, 865, and 785 $mA\,h\,g^{-1}$ was achieved at current densities of 0.2C, 0.5C, 0.8C, 1C, 2C, and 3C (1C $=$ 1675 $mA\,g^{-1}$), respectively. When the rate was restored to 1C after high current density testing, the specific capacity of the few-layer phosphorene and carbon nanofiber (FLP-CNF) electrode reverted to $\sim$967 $mA\,h\,g^{-1}$, which was close to the initial value (985 $mA\,h\,g^{-1}$) at 1C current density, indicating a good reversibility (see Fig. 7.17e). By contrast, the initial average

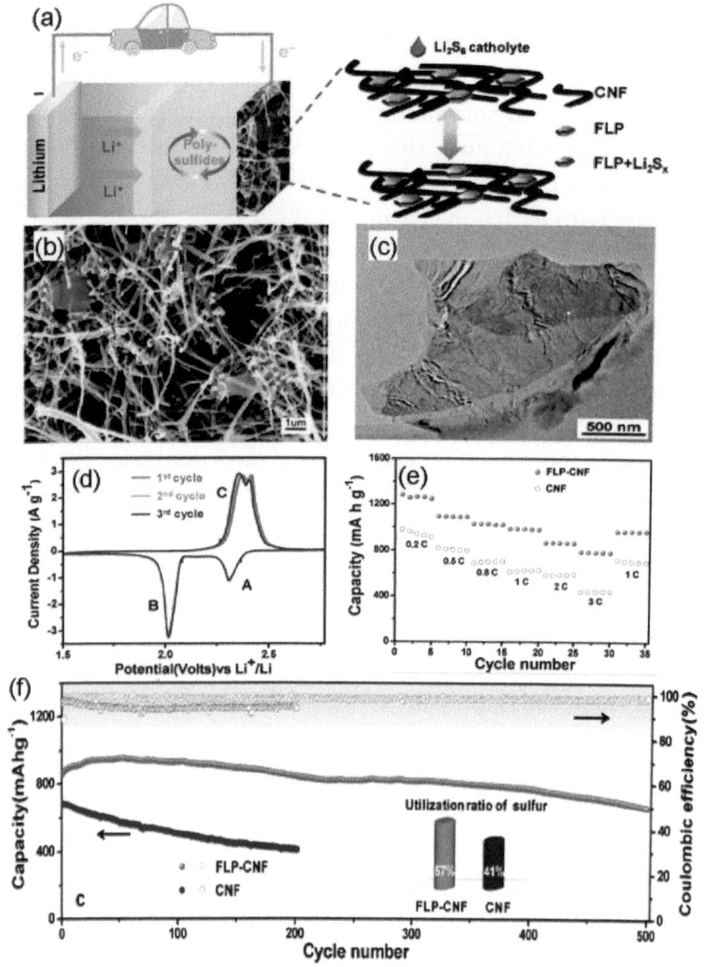

Figure 7.17 (a) Schematic of the FLP-CNF matrix used as the host for the lithium polysulphide catholyte. (b) Morphology and microstructure of FLP-CNF electrode (c) TEM image of FLP exfoliated in N-Methyl-2-pyrrolidone. (d) CV test of FLP-CNF. (e) Electrochemical performance of Li-S batteries with FLP-CNF electrode and pure CNF electrode. (f) Cycling stability and coulombic efficiency. Inset shows utilization of sulfur at a current density of 1C. Reprinted with permission from Li et al.[104] Copyright (2016) by the John Wiley and Sons.

discharge capacity of the pure CNF electrode (Fig. 7.4a) is only 944 $mA\,h\,g^{-1}$, which is ~300 $mA\,h\,g^{-1}$ lower than the FLP-CNF. Fig. 7.17f compares the long-term cycling performance between FLP-CNF and pure CNF electrodes at a current density of 1C. It shows that the specific capacity of the FLP-CNF electrode is retained above 660 $mA\,h\,g^{-1}$ with only 0.053% average capacity decay per cycle, while that of the pure CNF electrode decays into 0.25% per cycle after 200 cycles, clearly indicating that the presence of FLP improves the electrochemical stability of the Li-S battery. The promoted electrochemical performance is ascribed to that presence of phosphorene lowers the polarization accelerates the redox reaction and improves sulfur utilization in the battery.

Sun et al. recently assembled a new Li-S battery, in which the functional separator is prepared through depositing BP nanoflakes onto a Celgard, acting as sulfur-capturing layers.[105] Figure 7.18a shows a scheme of a coin cell with a conventional and BP-coated separator. The coated BP layer binds and intercepts the polysulfides in order to prevent their diffusion through the polypropylene separator (Fig. 7.18b,c). The first galvanostatic discharge–charge curves of LS cell is shown in Fig. 7.18d, and the corresponding Raman spectra were collected at five charge states: at open circuit voltage (OCV) 2.42 V (A), 2.10 V (B), 1.70 V (C), 2.39 V (D), and 2.60 V (E), respectively, during the first discharge/charge cycle, in order to monitor the evolution of the chemical composition on the BP-modified separator surface (Fig. 7.18e). Their results unravel that the BP coating layer can effectively anchor the various sulfur species and therefore prevents the diffusion of the polysulfide anions. Moreover, the sulfur species were shown to be effectively trapped by the BP layer and are electrochemically reactive. The electrochemical performance of the BP-modified separator on of a Li-S battery is shown in Fig. 7.18f, which is also compared to the counterpart of cell using a graphene-coated separator and conventional separator. Their results shows that the initial discharge capacity of the BP-modified cell can reach up to 930 $mA\,h\,g^{-1}$, which is 80 $mA\,h\,g^{-1}$ higher than the graphene-modified cell, resulting from the stronger interaction of BP for trapping the sulfur species. The cycling stability of the cells using the different separators is shown in Fig. 7.18g. The cell using the conventional separator exhibits a rapid capacity decay

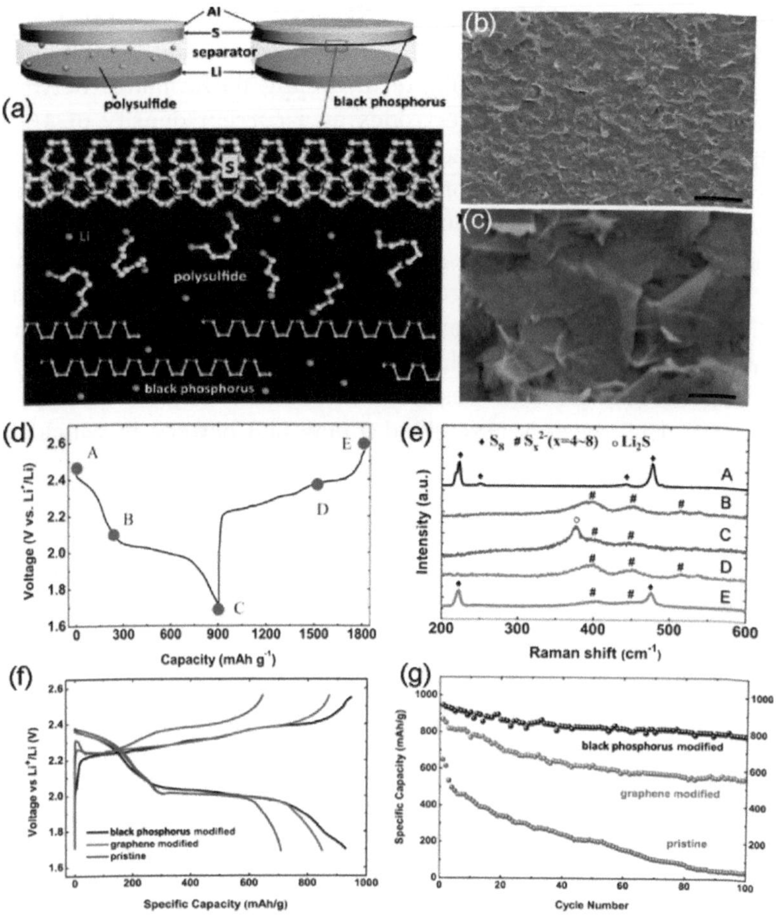

Figure 7.18 (a) Schematic cell configuration of the Li-S cell employing a commercial separator and the BP-coated separator as well as the operating principle of the BP coating layer in Li-S battery. (b) SEM image of the as-prepared BP-coated separator. (c) High-magnification SEM image of the as-prepared BP-coated separator. The scale bars in (b) and (c) are 10 and 1 μm, respectively. (d) First galvanostatic discharge–charge curves of lithium-sulfur cell. (e) Corresponding typical Raman spectra of the cathode at five different potentials in (d). (f) First galvanostatic discharge-charge curves of the Li-S battery with BP-modified, graphene-modified and conventional separators, respectively, at a current density of 0.4 A g^{-1}. (g) Reversible lithiation capacity and Coulombic efficiency for the first 100 galvanostatic cycles of the three Li-S cells, between 2.6 and 1.7 V at a current density of 0.4 A g^{-1}. Reprinted with permission from Sun et al.[105] Copyright (2016) by the John Wiley and Sons.

to ~300 mA h g^{-1} within 40 cycles, while exhibits a capacity of 800 mA h g^{-1} after 100 cycles. The retention rate of 86% is much higher than that of graphene-modified cell of 66%. The much improved electrochemical performances of the BP-modified cell were ascribed to high electron conductivity and ultrahigh Li diffusivity of BP coating layer, which trap and activate the soluble polysulfides in Li-S batteries.

7.3 Application in Supercapacitors

Supercapacitors or ultracapacitors have recently attracted considerable attention owing to their high power density, high charge-discharge rates, and long cycle life performance, which perfectly fill the gap between rechargeable batteries and conventional electrostatic capacitors.[106–109] Based on the types of energy storage, supercapacitors are classified into electric double-layer capacitors (EDLCs) and pseudocapacitors.[110] Generally, large specific surface areas of electrode materials are needed for achieving a high capacitance in electrical EDLCs due to the nature of electrosorption of ions on electrodes. Layered materials were promising electrode materials in electrochemical energy storage devices in which efficient intercalation of ions is required.[111–113] Recent investigations[114–117] show that flexible films or papers fabricated by restacking 2D exfoliated nanosheets of layered materials are beneficial for outstanding efficient intercalation of various ions between the restacking layers, and therefore able to enhance intercalation capacitances.[113–115]

Recently, liquid phase exfoliation of 2D materials[57] has been well established, allowing the preparation of BP nanoflakes or nanosheets in large quantity. Compared to the clean surfaces derived from the mechanically exfoliated BP nanoflakes, liquid-exfoliated BP nanosheets involves unwanted surface contamination with solvent molecule, which, however, are possible to protect the exfoliated BP nanoflakes from degradation upon ambient exposure, and enable to prepare flexible additive-free BP films with liquid-exfoliated dispersion of BP nanoflakes.[54,118] Moreover, the large spacing enabling the intercalation of ions between the adjacent

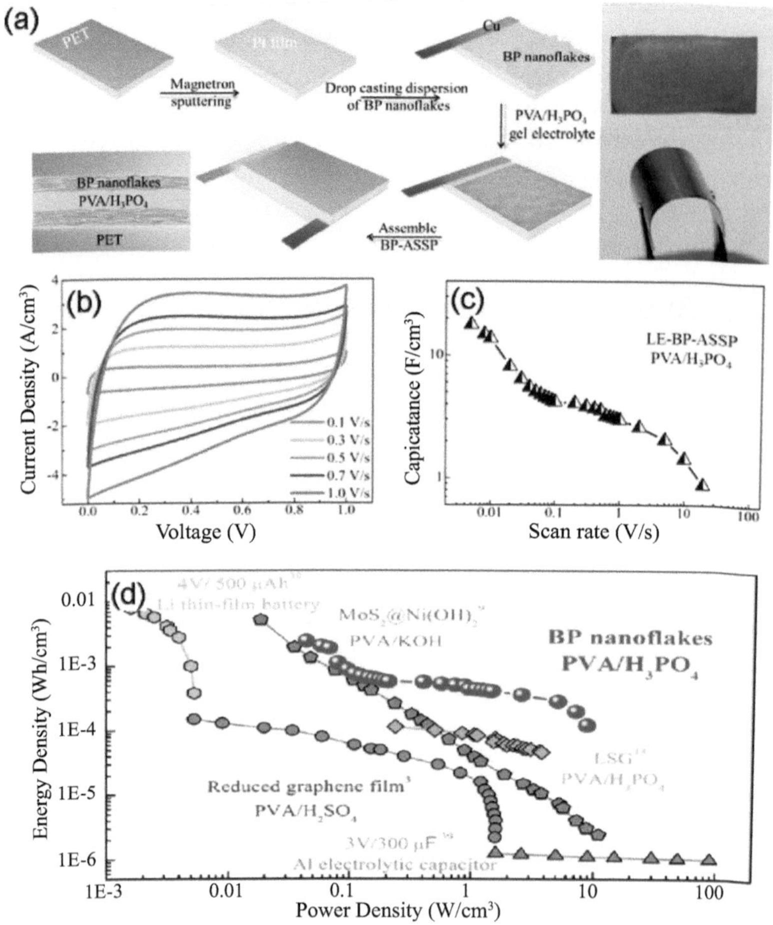

Figure 7.19 (a) Schematic illustration for fabrication process of the BP-ASSP devices through a drop-casting method. In the right side, the photographs of a flexible BP film in the flat and bent configurations are shown. (b) Cyclic voltammograms at various scan rates for the LE-BP-ASSP device with ~1.5 μm thick BP film electrodes. (c) Stack capacitances calculated from the CV curves at different scan rates. (d) Ragone plot of an LE-BP-ASSP device (the BP film electrodes have a thickness of ~1.5 μm), compared with several commercial energy storage systems and reported devices. Reprinted with permission from Kim et al.[72] Copyright (2013) by the John Wiley and Sons. Reprinted with permission from Li et al.[79] and Chowdhury et al.[121] Copyright (2016) by the John Wiley and Sons.

puckered layers has been confirmed by experimental and theoretical studies on BP as electrode material.[70,119] Similar to restacked nanosheets of graphene[56,72] and TMDs,[51,120] C. Hao restacked the liquid-exfoliated BP nanoflakes as flexible electrodes, and enhanced high-performance electrochemical energy storage has been clarified due to the efficient intercalation of ions between restacked BP nanoflakes. Figure 7.19a displays the fabrication procedure of flexible all-solid-state supercapacitors (ASSPs) with BP films and poly(vinyl alcohol)/H_3PO_4 (PVA/H_3PO_4) as gel electrolyte. The PVA/H_3PO_4 is sandwiched between two as-prepared BP films on Pt-coated PET substrate and is aimed to avoid harmful leakage current generated from liquid electrolyte or any other additional binding materials.[30,33,43] The liquid-exfoliated BP nanoflakes were transparent under the electron beam, as indicative of a thin thickness nature. The sizes of BP nanoflakes are not uniformly distributed ranging from several tens to several hundreds of nanometers. Using a facile drop-casting method, a flexible additive free BP films are prepared on a Pt-coated PET substrate with a film thickness in 1.5 μm and can be freely deformed in flat and bent ways (see Fig. 7.19a). The CV curves at different scan rates for an LE-BP-ASSP device are shown in Fig. 7.19b. It shows that the CV curves tend to perform a rectangular shape at the increased scan rate from 0.1 to 1 V s^{-1}, demonstrating a pretty good capacitive behavior. The corresponding calculated stack volumetric capacitance in Fig. 7.19c shows a slow drop with the scan rate from 0.1 to 1 V s^{-1}. A high stack capacitance up to 13.75 F cm^{-3} (45.8 F g^{-1}) was achieved at the scan rate of 0.01 Vs^{-1}. With a rapid scan rate above 1.0 V s^{-1}, the calculated capacitances rapidly decrease again. Significantly, a high capacitance of 1.43 F cm^{-3} (4.8 F g^{-1}) is still retained at a large scan rate of 10 V s^{-1}, which is much higher than that of graphene-based supercapacitor[72] and multiwalled carbon nanotubes electrodes ASSP. Ragone plot in Fig. 7.19d demonstrates the performance of LE-BP-ASSP devices. It shows that the energy density of LE-BP-ASSP device is in ranging from 0.123 to 2.47 mW h cm^{-3} as well as a maximum power density of 8.83 W cm^{-3}. These values are comparable to the conventional lithium-ion thin batteries and superior to the graphene-based sandwich solid-state supercapacitor.[79] In contrast to the expected degradation of BP in presence of water or O_2, it was noted that an outstanding mechanical flexibility and cycling stability over 30,000 cycles are achieved in this work (15.5% capacitance decay after 10,000 cycles, and 28.2% after 30,000 cycles). The improved stability performance of LE-BP-ASSP device is probably contributed from the presence of oxygen functional groups and H_3PO_4 on the surfaces of BP nanoflakes, which efficiently limits the direct contact between the diffused water/oxygen molecules and BP nanoflakes, therefore eliminating the reaction-induced damages on BP nanoflakes. These results may suggest a feasible application of liquid-exfoliated BP nanoflakes as flexible electrodes in high-performance electrochemical energy-storage devices.

References

1. Yau S-L, Moffat TP, Bard AJ, Zhang Z, Lerner MM, (1992). STM of the (010) surface of orthorhombic phosphorus. *Chem. Phys. Lett.*, **198**:383–388.
2. Hanlon D, Backes C, Doherty E, Cucinotta CS, Berner NC, Boland C, Lee K, Harvey A, Lynch P, Gholamvand Z, Zhang S, Wang K, Moynihan G, Pokle A, Ramasse QM, McEvoy N, Blau WJ, Wang J, Abellan G, Hauke F, Hirsch A, Sanvito S, O'Regan DD, Duesberg GS, Nicolosi V, Coleman JN, (2015). Liquid exfoliation of solvent-stabilized few-layer black phosphorus for applications beyond electronics. *Nat. Commun.*, **6**:8563.
3. Ziletti A, Carvalho A, Campbell DK, Coker DF, Castro Neto AH, (2015). Oxygen defects in phosphorene. *Phys. Rev. Lett.*, **114**:04680.
4. Farmer DB, Lin YM, Avouris P, (2010). Graphene field-effect transistors with self-aligned gates. *Appl. Phys. Lett.*, **97**:013103.
5. Wang G, Slough WJ, Pandey R, Karna SP, (2016). Degradation of phosphorene in air: understanding at atomic level. *2D Mater.*, **3**:025011.
6. Castellanos-Gomez A, Vicarelli L, Prada E, Island JO, Narasimha-Acharya KL, Blanter SI, Groenendijk DJ, Buscema M, Steele GA, Alvarez JV, Zandbergen HW, Palacios JJ, Zant HSJ, (2014). Isolation and characterization of few-layer black phosphorus. *2D Mater.*, **1**:025001.
7. Du Y, Ouyang C, Shi S, Lei M, (2010). Ab initio studies on atomic and electronic structures of black phosphorus. *J. Appl. Phys.*, **107**:093718.

8. Island JO, Steele GA, Zant HSJ, Castellanos-Gomez A, (2015). Environmental instability of few-layer black phosphorus. *2D Mater.*, **2**:011002.

9. Yasaei P, Behranginia A, Foroozan T, Asadi M, Kim K, Khalili-Araghi F, Salehi-Khojin A, (2015). Stable and selective humidity sensing using stacked black phosphorus flakes. *ACS Nano*, **9**:9898–9905.

10. Kang J, Wells SA, Wood JD, Lee JH, Liu X, Ryder CR, Zhu J, Guest JR, Husko CA, Hersam MC, (2016). Stable aqueous dispersions of optically and electronically active phosphorene. *Proc. Natl. Acad. Sci. U.S.A.*, **113**:11688–11693.

11. Huang Y, Qiao J, He K, Bliznakov S, Sutter E, Chen X, Luo D, Meng F, Su D, Decker J, Ji W, Ruoff RS, Sutter P, (2016). Interaction of black phosphorus with oxygen and water. *Chem. Mater.*, **28**:8330–833.

12. Song YL, Zhang Y, Zhang JM, Lu DB, Xu KW, (2011). First principles study of the structural and electronic properties of armchair silicene nanoribbons with vacancies. *J. Mol. Struct.*, **990**:75–78.

13. Yazyev OV, Helm L, (2007). Defect-induced magnetism in graphene. *Phys. Rev. B*, **75**:125408.

14. Ataca C, Sahin H, Akturk E, Ciraci S, (2011). Mechanical and electronic properties of MoS_2 nanoribbons and their defects. *J. Phys. Chem. C*, **115**:3934–3941.

15. Cai Y, Ke Q, Zhang G, Yakobson BI, Zhang Y-W, (2016). Highly itinerant atomic vacancies in phosphorene. *J. Am. Chem. Soc.*, **138**:10199–10206.

16. Kistanov AA, Cai Y, Zhou K, Dmitriev SV, Zhang Y-W, (2017). The role of H_2O and O_2 molecules and phosphorus vacancies in the structure instability of phosphorene. *2D Mater.*, **4**:015010.

17. Yoshihara A, Fujimura T, Oka Y, Fujisaki H, Shirotani I, (1986). Surface Brillouin scattering in black phosphorus. *Phys. Rev. B*, **34**:7467–7470.

18. Favron A, Gaufrès E, Fossard F, Phaneuf-L'Heureux A-L, Tang NY-W, Lévesque PL, Loiseau A, Leonelli R, Francoeur S, Martel R, (2015). Photooxidation and quantum confinement effects in exfoliated black phosphorus. *Nat. Mater.*, **14**:826–832.

19. Cai Y, Zhang G, Zhang Y-W, (2015). Electronic properties of phosphorene/graphene and phosphorene/hexagonal boron nitride heterostructures. *J. Phys. Chem. C*, **119**:13929–13936.

20. Cao Y, Mishchenko A, Yu GL, Khestanova E, et al. (2015). Quality heterostructures from two-dimensional crystals unstable in air by their assembly in inert atmosphere. *Nano Lett.*, **15**:4914–4921.

21. Na J, Lee YT, Lim JA, Hwang DK, Kim G-T, Choi WK, Song Y-W, (2014). Few-layer black phosphorus field-effect transistors with reduced current fluctuation. *ACS Nano*, **8**:P11753–11762.

22. Wood JD, Wells SA, Jariwala D, Chen K-S, Cho E, Sangwan VK, Liu X, Lauhon LJ, Marks TJ, Hersam MC, (2014). Effective passivation of exfoliated black phosphorus transistors against ambient degradation. *Nano Lett.*, **14**:6964–6970.

23. Pei J, Gai X, Yang J, Wang X, Yu Z, Choi D-Y, Luther-Davies B, Lu Y, (2016). Producing air-stable monolayers of phosphorene and their defect engineering. *Nat. Commun.*, **7**:10450.

24. Whittingham MS, (1976). Electrical energy storage and intercalation chemistry. *Science*, **192**:1126–1127.

25. Besenhard JO, Eichinger G, (1976). High energy density lithium cells: Part I. Electrolytes and anodes. *J. Electroanal. Chem. Interfacial Electrochem.*, **68**:1–18.

26. Eichinger G, Besenhard JO, (1976). High energy density lithium cells: Part II. Cathodes and complete cells. *J. Electroanal. Chem. Interfacial Electrochem.*, **72**:1–31.

27. Tang Y, Zhang Y, Li W, Ma B, Chen X, (2015). Rational material design for ultrafast rechargeable lithium-ion batteries. *Chem. Soc. Rev.*, **44**:5926–5940.

28. Goodenough JB, Kim Y, (2010). Challenges for rechargeable Li batteries. *Chem. Mater.* **22**:587.

29. Corbridge DEC, (2013). *Phosphorus: Chemistry, Biochemistry and Technology*, CRC Press.

30. Sun J, Zheng G, Lee H, Liu N, Wang H, Yao H, Yang W, Cui Y, (2014). Formation of stable phosphorus–carbon bond for enhanced performance in black phosphorus nanoparticle–graphite composite battery anodes. *Nano Lett.*, **14**:4573–4580.

31. Karttunen AJ, Linnolahti M, Pakkanen TA, (2008). Structural principles of polyhedral allotropes of phosphorus. *ChemPhysChem*, **9**:2550–2558.

32. Ji X, Lee KT, Nazar LF, (2009). A highly ordered nanostructured carbon-sulphur cathode for lithium-sulphur batteries. *Nat. Mater.*, **8**:500–506.

33. Marino C, Debenedetti A, Fraisse B, Favier F, Monconduit L, (2011). Activated-phosphorus as new electrode material for Li-ion batteries. *Electrochem. Commun.*, **13**:346–349.

34. Marino C, Boulet L, Gaveau P, Fraisse B, Monconduit LJ, (2012). Nanoconfined phosphorus in mesoporous carbon as an electrode

for Li-ion batteries: performance and mechanism. *Mater. Chem.*, **22**:22713–22720.

35. Wang L, He X, Li J, Sun W, Gao J, Guo J, Jiang C, (2012). Nano-structured phosphorus composite as high-capacity anode materials for lithium batteries. *Angew. Chem. Int. Ed.*, **51**:9034–9037.
36. Li W, Yang Z, Jiang Y, Yu Z, Gu L, Yu Y, (2014). Crystalline red phosphorus incorporated with porous carbon nanofibers as flexible electrode for high performance lithium-ion batteries. *Carbon*, **78**:455–462.
37. Fan Q, Chupas PJ, Whittingham MS, (2007). Characterization of amorphous and crystalline tin–cobalt anodes. *Electrochem. Solid-State Lett.*, **10**:A274–A278.
38. Hatchard TD, Dahn JR, (2004). Study of the electrochemical performance of sputtered $Si_{1-x}Sn_x$ films. *J. Electrochem. Soc.*, **151**:A1628–A1635.
39. Idota Y, Kubota T, Matsufuji A, Maekawa Y, Miyasaka T, (1997). Tin-based amorphous oxide: a high-capacity lithium-ion-storage material. *Science*, **276**:1395–1397.
40. Chen Z, Cao Y, Qian J, Ai X Yang H, (2010). Facile synthesis and stable lithium storage performances of Sn- sandwiched nanoparticles as a high capacity anode material for rechargeable Li batteries. *J. Mater. Chem.*, **20**:7266–7271.
41. Noh M, Kwon Y, Lee H, Cho J, Kim Y, Kim MG, (2005). Amorphous carbon-coated tin anode material for lithium secondary battery. *Chem. Mater.*, **17**:1926–1929.
42. Qian J, Chen Y, Wu L, Cao Y, Ai X, Yang H, (2012). High capacity Na-storage and superior cyclability of nanocomposite Sb/C anode for Na-ion batteries. *Chem. Commun.*, **48**:7070–7072.
43. Qian J, Qiao D, Ai X, Cao Y, Yang H, (2012). Reversible 3-Li storage reactions of amorphous phosphorus as high capacity and cycling-stable anodes for Li-ion batteries. *Chem. Commun.*, **48**:8931–8933.
44. Sun J, Lee H-W, Pasta M, Yuan H, Zheng G, Sun Y, Li Y, Cui Y, (2015). A phosphorene–graphene hybrid material as a high-capacity anode for sodium-ion batteries. *Nat. Nanotechnol.*, **10**:980–985.
45. Zhu Y, Wen Y, Fan X, Gao T, Han F, Luo C, Liou S-C, Wang C, (2015). Red phosphorus–single-walled carbon nanotube composite as a superior anode for sodium ion batteries. *ACS Nano*, **9**:3254–3264.
46. Yu Z, Song J, Gordin ML, Yi R, Tang D, Wang D, (2015). Phosphorus-graphene nanosheet hybrids as lithium-ion anode with exceptional high-temperature cycling stability. *Adv. Sci.*, **2**:1400020.

47. Hultgren R, Gingrich NS, Warren BE, (1935). The atomic distribution in red and black phosphorus and the crystal structure of black phosphorus. *J. Chem. Phys.*, **3**:351.
48. Brown A, Rundqvist S, (1965). Refinement of the crystal structure of black phosphorus. *Acta Crystallogr.*, **19**:684.
49. Jamieson JC, (1963). Crystal structures adopted by black phosphorus at high pressures. *Science*, **139**:1291.
50. Kikegawa T, Iwasaki H, (1983). An X-ray diffraction study of lattice compression and phase transition of crystalline phosphorus. *Acta Crystallogr.*, **B39**:158–164.
51. Park C, Sohn H, (2007). Black phosphorus and its composite for lithium rechargeable batteries. *Adv. Mater.*, **19**:2465–2468.
52. Wei Q, Peng X, (2014). Superior mechanical flexibility of phosphorene and few-layer black phosphorus. *Appl. Phys. Lett.*, **104**:251915.
53. Li W, Yang Y, Zhang G, Zhang Y-W, (2015). Ultrafast and directional diffusion of lithium in phosphorene for high-performance lithium-ion battery. *Nano Lett.*, **15**:1691–1697.
54. Zhao S, Kang W, Xue J, (2014). The potential application of phosphorene as an anode material in Li-ion batteries. *J. Mater. Chem. A*, **2**:19046–19052.
55. Guo G-C, Wang D, Wei X-L, Zhang Q, Liu H, Lau W-M, Liu L-M, (2015). First-principles study of phosphorene and graphene heterostructure as anode materials for rechargeable Li batteries. *J. Phys. Chem. Lett.* **6**:5002–5008.
56. Zhang Y, Wang H, Luo Z, Tan HT, Li B, Sun S, Li Z, Zong Y, Xu ZJ, Yang Y, Khor KA, Yan Q, (2016). An air-stable densely packed phosphorene–graphene composite toward advanced lithium storage properties. *Adv. Energy Mater.*, **6**:1600453.
57. Qian J, Wu X, Cao Y, Ai X, Yang H, (2013). High capacity and rate capability of amorphous phosphorus for sodium ion batteries. *Angew. Chem.*, **125**:4731–4734.
58. Carvalho A, Rodin A, Neto AC, (2014). Phosphorene nanoribbons. *Europhys. Lett.*, **108**:47005.
59. Ong SP, Chevrier VL, Hautier G, Jain A, Moore C, Kim S, Ma X, Ceder G, (2011). Voltage, stability and diffusion barrier differences between sodium-ion and lithium-ion intercalation materials. *Energy Environ. Sci.*, **4**:3680–3688.

60. Cotton A, Wilkinson G, Murillo CA, Bochmann M, (1999). *Advanced Inorganic Chemistry*, Wiley.
61. Sangster J, (2010). Na-P (sodium-phosphorus) system. *J. Phase Equilib. Diffus.*, **31**:62-67.
62. Datta MK, Epur R, Saha P, Kadakia K, Park SK, Kumta PN, (2013). Tin and graphite based nanocomposites: Potential anode for sodium ion batteries. *J. Power Sources*, **225**:316 –322.
63. Wang JW, Liu XH, Mao SX, Huang JY, (2012). Microstructural evolution of tin nanoparticles during in situ sodium insertion and extraction. *Nano Lett.*, **12**:5897–5902.
64. Xiao LF, Cao YL, Xiao J, Wang W, Kovarik L, Nie ZM, Liu J, (2012). High capacity, reversible alloying reactions in SnSb/C nanocomposites for Na-ion battery applications. *Chem. Commun.*, **48**:3321–3323.
65. Palomares V, Casas-Cabanas M, Castillo-Martinez E, Han MH, Rojo T, (2013). Update on Na-based battery materials. A growing research path. *Energy Environ. Sci.*, **6**:2312–2337.
66. Slater MD, Kim D, Lee E, Johnson CS, (2013). Sodium-ion batteries. *Adv. Funct. Mater.*, **23**:947–958.
67. Kim SW, Seo DH, Ma XH, Ceder G, Kang K, (2012). Electrode materials for rechargeable sodium-ion batteries: potential alternatives to current lithium-ion batteries. *Adv. Energy Mater.*, **2**:710–721.
68. Ellis BL, Nazar LF, (2012). Sodium and sodium-ion energy storage batteries. *Curr. Opin. Solid State Mater. Sci.*, **16**:168–177.
69. Morita A, (1986). Semiconducting black phosphorus. *Appl. Phys. A*, **39**:227–242.
70. Kulish VV, Malyi OI, Persson C, Wu P, (2015). Phosphorene as an anode material for Na-ion batteries: a first-principles study. *Phys. Chem. Chem. Phys.*, **17**:13921–13928.
71. Qian JF, Wu XY, Cao YL, Ai XP, Yang HX, (2013). High capacity and rate capability of amorphous phosphorus for sodium ion batteries. *Angew. Chem., Int. Ed.*, **52**:4633–4636.
72. Kim Y, Park Y, Choi A, Choi NS, Kim J, Lee J, Ryu JH, Oh SM, Lee KT, (2013). An amorphous red phosphorus/carbon composite as a promising anode material for sodium ion batteries. *Adv. Mater.*, **25**:3045–3049.
73. Park C-M, Kim J-H, Kim H, Sohn H-J, (2010). Li-alloy based anode materials for Li secondary batteries. *Chem. Soc. Rev.*, **39**:3115–3141.

74. Li W-J, Chou S-L, Wang J-Z, Liu H-K, Dou S-X, (2013). Simply mixed commercial red phosphorus and carbon nanotube composite with exceptionally reversible sodium-ion storage. *Nano Lett.*, **13**:5480–5484.
75. Luo F, Liu B, Zheng J, Chu G, Zhong K, Li H, Huang X, Chen LJ, (2015). Review-nano-silicon/carbon composite anode materials towards practical application for next generation Li-ion batteries. *Electrochem. Soc.*, **162**:A2509–A2528.
76. Zhang C, Wang X, Liang Q, Liu X, Weng Q, Liu J, Yang Y, Dai Z, Ding K, Bando Y, Tang J, Golberg D, (2016). Amorphous phosphorus/nitrogen-doped graphene paper for ultrastable sodium-ion batteries. *Nano Lett.*, **16**:2054–2060.
77. Pauling L, (1988). *General Chemistry*, Pan American: Los Angeles, CA, pp. 211–215.
78. Roth WL, DeWitt TW, Smith AJ, (1947). Polymorphism of red phosphorus. *J. Am. Chem. Soc.*, **69**:2881–2885.
79. Li W, Yang Z, Li M, Jiang Y, Wei X, Zhong X, Gu L, Yu Y, (2016). Amorphous red phosphorus embedded in highly ordered mesoporous carbon with superior lithium and sodium storage capacity. *Nano Lett.*, **16**:1546–1553.
80. Nagao M, Hayashi A, Tatsumisago M, (2011). All-solid-state lithium secondary batteries with high capacity using black phosphorus negative electrode. *J. Power Sources*, **196**:6902–6905.
81. Sun L-Q, Li M-J, Sun K, Yu S-H, Wang R-S, Xie H-M, (2012). Electrochemical activity of black phosphorus as an anode material for lithium-ion batteries. *J. Phys. Chem. C*, **116**:14772–14779.
82. Li N, Weng Z, Wang Y, Li F, Cheng H-M, Zhou H, (2014). An aqueous dissolved polysulfide cathode for lithium–sulfur batteries. *Energy Environ. Sci.*, **7**:3307.
83. Bruce PG, Freunberger SA, Hardwick LJ, Tarascon J-M, (2011). Li-O_2 and Li-S batteries with high energy storage. *Nat. Mater.*, **11**:172.
84. Bresser D, Passerini S, Scrosati B, (2013). Recent progress and remaining challenges in sulfur-based lithium secondary batteries: a review. *Chem. Commun.*, **49**:10545.
85. Guo JC, Xu YH, Wang CS, (2011). Sulfur-impregnated disordered carbon nanotubes cathode for lithium-sulfur batteries. *Nano Lett.*, **11**:4288–4294.
86. Xin S, Gu L, Zhao NH, Yin YX, Zhou LJ, Guo YG, Wan LJ, (2012). Smaller sulfur molecules promise better lithium-sulfur batteries. *J. Am. Chem. Soc.*, **134**:18510–18513.

87. Zhang SS, Read JA, (2012). A new direction for the performance improvement of rechargeable lithium/sulfur batteries. *J. Power Sources*, **200**:77–82.

88. Shim J, Striebel KA, Cairns EJ, (2002). The lithium/sulfur rechargeable cell effects of electrode composition and solvent on cell performance. *J. Electrochem. Soc.*, **149**:A1321.

89. Mikhaylik YV, Akridge JR, (2004). Polysulfide shuttle study in the Li/S battery system. *J. Electrochem. Soc.*, **151**:A1969.

90. Yao H, Zheng G, Hsu P-C, Kong D, Cha JJ, Li W, She ZW, McDowell MT, Yan K, Liang Z, Narasimhan VK, Cui Y, (2014). Improving lithium–sulphur batteries through spatial control of sulphur species deposition on a hybrid electrode surface. *Nat. Commun.*, **5**:3943.

91. Zhou G, Pei S, Li L, Wang D-W, Wang S, Huang K, Yin L-C, Li F, Cheng H-M, (2014). A graphene–pure-sulfur sandwich structure for ultrafast, long-Life lithium–sulfur batteries. *Adv. Mater.*, **26**:625.

92. Xiao L, Cao Y, Xiao J, Schwenzer B, Engelhard MH, Saraf LV, Nie Z, Exarhos GJ, Liu J, (2012). A soft approach to encapsulate sulfur: polyaniline nanotubes for lithium-sulfur batteries with long cycle life. *Adv. Mater.*, **24**:1176.

93. Yang Y, Zheng G, Cui Y, (2013). Nanostructured sulfur cathodes. *Chem. Soc. Rev.*, **42**:3018.

94. Seh ZW, Li W, Cha JJ, Zheng G, Yang Y, McDowell MT, Hsu P-C, Cui Y, (2013). Sulphur–TiO_2 yolk–shell nanoarchitecture with internal void space for long-cycle lithium–sulphur batteries. *Nat. Commun.*, **4**: 1331.

95. Seh ZW, Wang H, Liu N, Zheng G, Li W, Yao H, Cui Y, (2014). High-capacity Li_2S–graphene oxide composite cathodes with stable cycling performance. *Chem. Sci.*, **5**:1396.

96. Zheng G, Zhang Q, Cha JJ, Yang Y, Li W, She ZW, Cui Y, (2013). Amphiphilic surface modification of hollow carbon nanofibers for improved cycle life of lithium sulfur batteries. *Nano Lett.*, **13**:1265.

97. Zhang Q, Wang Y, She ZW, Fu Z, Zhang R, Cui Y, (2015). Understanding the anchoring effect of two-dimensional layered materials for lithium–sulfur batteries. *Nano Lett.*, **15**:3780.

98. Tao X, Wang J, Ying Z, Cai Q, Zheng G, Gan Y, Huang H, Xia Y, Liang C, Zhang W, Cui Y, (2014). Strong sulfur binding with conducting magneli-phase Ti_nO_{2n-1} nanomaterials for improving lithium–sulfur batteries. *Nano Lett.*, **14**:5288.

99. She ZW, Yu JH, Li W, Hsu P-C, Wang H, Sun Y, Yao H, Zhang Q, Cui Y, (2014). Two-dimensional layered transition metal disulphides for

effective encapsulation of high-capacity lithium sulphide cathodes. *Nat. Commun.*, **5**:5017.

100. Su Y-S, Manthiram A, (2012). Lithium–sulphur batteries with a microporous carbon paper as a bifunctional interlayer. *Nat. Commun.*, **3**:1166.
101. Su Y-S, Manthiram A, (2012). A new approach to improve cycle performance of rechargeable lithium–sulfur batteries by inserting a free-standing MWCNT interlayer. *Chem. Commun.*, **48**:8817.
102. Chung S, Manthiram A, (2014). High-performance Li–S batteries with an ultra-lightweight MWCNT-coated separator. *J. Phys. Chem. Lett.*, **5**:1978.
103. Wang X, Wang Z, Chen L, (2013). Reduced graphene oxide film as a shuttle-inhibiting interlayer in a lithium–sulfur battery. *J. Power Sources*, **242**:65.
104. Li L, Chen L, Mukherjee S, Gao J, Sun H, Liu Z, Ma X, Gupta T, Singh CV, Ren W, Cheng H-M, Koratkar N, (2016). Phosphorene as a polysulfide immobilizer and catalyst in high-performance lithium–sulfur batteries. *Adv. Mater.*, **29**:1602734.
105. Sun J, Sun Y, Pasta M, Zhou G, Li Y, Liu W, Xiong F, Cui Y, (2016). Entrapment of polysulfides by a black-phosphorus modified separator for lithium–sulfur batteries. *Adv. Mater.*, **28**:9797–9803.
106. Miller JR, Simon P, (2008). Electrochemical capacitors for energy management. *Science*, **321**:651–652.
107. Simon P, Gogotsi Y, (2008). Materials for electrochemical capacitors. *Nat. Mater.*, **7**:845–854.
108. Liu C, Li F, Ma LP, Cheng HM, (2010). Advanced materials for energy storage. *Adv. Mater.*, **22**:E28–E62.
109. Zhao X, Zhang L, Murali S, Stoller MD, Zhang Q, Zhu Y, Ruoff RS, (2012). Incorporation of manganese dioxide within ultraporous activated graphene for high-performance electrochemical capacitors. *ACS Nano*, **6**:5404–5412.
110. Winter M, Brodd R, (2004). What are batteries, fuel cells, and supercapacitors? *J. Chem. Rev.*, **104**:4245–4269.
111. Burke A, (2000). Ultracapacitors: why, how, and where is the technology? *J. Power Sources*, **91**:37–50.
112. Huang L, Chen D, Ding Y, Feng S, Wang ZL, Liu M, (2013). Nickel–cobalt hydroxide nanosheets coated on $NiCo_2O_4$ nanowires grown on carbon fiber paper for high-performance pseudocapacitors. *Nano Lett.*, **13**:3135–3139.

113. Shi W, Zhu J, Sim DH, Tay YY, Lu Z, Zhang X, Sharma Y, Srinivasan M, Zhang H, Hng HH, Yan Q, (2011). Achieving high specific charge capacitances in Fe_3O_4/reduced graphene oxide nanocomposites. *J. Mater. Chem.*, **21**:3422–3427.

114. Li YZ, Zhao X, Xu Q, Zhang QH, Chen D, (2011). Facile preparation and enhanced capacitance of the polyaniline/sodium alginate nanofiber network for supercapacitors. *Langmuir*, **27**:6458–6463.

115. Fan LZ, Maier J, (2006). High-performance polypyrrole electrode materials for redox supercapacitors. *Electrochem. Commun.*, **8**:937–940.

116. Stenger-Smith JD, Webber CK, Anderson N, Chafin AP, Zong KK, Reynolds JR, (2002). Poly (3, 4-alkylenedioxythiophene)-based supercapacitors using ionic liquids as supporting electrolytes. *J. Electrochem. Soc.*, **149**:A973–A977.

117. Frackowiak E, Khomenko V, Jurewicz K, Lota K, Beguin F, (2006). Supercapacitors based on conducting polymers/nanotubes composites. *J. Power Sources*, **153**:413–418.

118. Kou L, Chen C, Smith CS, (2015). Phosphorene: fabrication, properties, and applications. *J. Phys. Chem. Lett.*, **6**:2794–2805.

119. Zhang X, Hou L, Ciesielski A, Samorì P, (2016). 2D materials beyond graphene for high-performance energy storage applications. *Adv. Energy Mater.*, **6**:1600671.

120. Stan MC, Zamory JV, Passerini S, Nilges T, Winter M, (2013). Puzzling out the origin of the electrochemical activity of black P as a negative electrode material for lithium-ion batteries. *J. Mater. Chem. A*, **1**:5293–5300.

121. Chowdhury C, Karmakar S, Datta A, (2016). Capping black phosphorene by h-BN enhances performances in anodes for Li and Na ion batteries. *ACS Energy Lett.*, **1**:253–259.

Index